# Lecture Notes in Computational Science and Engineering

# 92




For further volumes:
http://www.springer.com/series/3527

Hester Bijl • Didier Lucor • Siddhartha Mishra
Christoph Schwab
Editors

# Uncertainty Quantification in Computational Fluid Dynamics

Springer

*Editors*
Hester Bijl
Faculty of Aerospace Engineering
Delft University of Technology
Delft, The Netherlands

Didier Lucor
d'Alembert Institute
Université Pierre et Marie
Curie-Paris VI - CNRS
Paris, France

Siddhartha Mishra
Christoph Schwab
Seminar für Angewandte Mathematik
ETH Zürich
Zürich, Switzerland

ISSN 1439-7358
ISBN 978-3-319-00884-4 ISBN 978-3-319-00885-1 (eBook)
DOI 10.1007/978-3-319-00885-1
Springer Heidelberg New York Dordrecht London

Library of Congress Control Number: 2013947366

Math. Subj. Class. (2010): 65M08, 65M75 , 65M60, 76G25, 76J20, 76K05, 35L65, 35L70



Printed on acid-free paper

Springer is part of Springer Science+Business Media (www.springer.com)

# Preface

The present volume addresses methods and computational aspects of efficient Uncertainty Quantification (UQ) in Computational Fluid Dynamics (CFD). While the general area of computational uncertainty quantification in engineering simulations has experienced a massive development in recent years, and is under strong expansion currently, by now key computational issues have been identified and analysis and implementation have progressed to the point where, for several broad classes of PDEs with uncertainty, computational methodologies are available which are also backed by numerical and mathematical analysis. Against this background, and consistent with the scientific focus of the 2011 Von Karmann Institute workshop which initiated the development of the chapters in the present volume, the present volume combines several contributions on efficient methods for UQ in CFD which address specific computational issues which arise in the use of the general computational methodologies in CFD problems; some (but not all) of these are: highly nonlinear, unsteady nature of the governing equations, singularity (shock) formation in finite time in pathwise solutions, and the impact of discontinuities on the accuracy and the regularity of statistical quantities even when all data in the problem are smooth. Other issues are the corresponding low regularity in the space of uncertain parameters, massive parallelism in forward simulations, necessity for multiscaling and multi-modelling in forward simulations (in particular in the presence of turbulence), uncertain topography and geometry of the flow domain.

The low solution regularity and the propagation of singularities in solutions of the governing equations prompt the development of the numerical techniques which are specifically adapted to deal with these pheonomena; among them are Finite Volume methods in the stochastic parameter domain, WENO reconstruction and limiting methods for positivity enforcement in computation of probability density functions of random solution quantities, to name but a few. Most of these techniques are *nonintrusive* since, unlike the situation encountered in computational UQ in solids and wave propagation, the strong nonlinearity of the governing equation narrows applicable UQ methods to essentially those of collocation type. Due to the low, pathwise regularity of solutions of nonlinear hyperbolic conservation laws, however, the (in general high) regularity properties of parametric stochastic solutions required for example by spectral collocation methods must be carefully verified in practice. A logical conclusion of these remarks is the prominent role which will be played by *stochastic collocation* methods and by *Monte-Carlo* sampling approaches; in particular, *Multilevel Monte-Carlo* sampling approaches have proved quite efficient and powerful strategies when solving UQ problems in CFD. We are confident that the methods which we found to be viable and robust for the CFD problems considered here will also prove to be applicable to other, "hard" and fully nonlinear computational models in engineering and in the sciences.

The notes address computational technicalities of specific issues arising in UQ in current CFD applications, in particular UQ in output functionals such as lift-, drag- and other, integral quantities of the primitive uncertain variables, estimation of statistical moments, in particular of variance, and the probability of computation of extremal events, and the assessment of the accuracy of statistical quantities in the presence of discretization and other, numerical errors.

While these notes focus on computational and implementation aspects of discretization, stability, parallelization of computational UQ for problems in CFD. Naturally, they impinge on a number of related issues in numerical analysis and high performance computing; we only mention load balancing issues in massively parallel UQ simulations and the mathematical regularity of statistical quantities of output functionals; here, the most prominent example is that of statistics of shock locations and profiles where additional regularity of outputs is generated by ensemble averaging of random entropy solutions, so that for example the statistics of shock locations can become Lipschitz continuous or more regular, even for hyperbolic equations without any viscosity.

As can be expected in a field which is currently undergoing rapid development, the present notes represent only a snapshot in time of this evolving field of computational science and engineering. The present notes are intended to present the key ideas, the description of UQ algorithms, and prototypical implementations on a high technical level, which should be accessible, nevertheless by graduate students and researchers in computational science as well as in CFD-related areas of engineering.

The background knowledge of the intended readership of this volume is knowledge of elementary probability and statistics and solid knowledge of computational fluid dynamics.

We very much hope that these notes stimulate further algorithmic and theoretical developments in UQ for CFD and, due to the interdisciplinarity nature of UQ, also in the adjacent areas of statistics, high-performance computing, and the analysis of partial differential equations with random input data.

Delft, The Netherlands — Hester Bijl
Paris, France — Didier Lucor
Zürich, Switzerland — Siddhartha Mishra
Christoph Schwab

# Contents

**Non-intrusive Uncertainty Propagation with Error Bounds for Conservation Laws Containing Discontinuities** ........ 1
Timothy Barth
1 Introduction ........ 2
2 A Framework for Non-intrusive Uncertainty Propagation with Error Bounds for Statistics ........ 3
2.1 A Deterministic Model ........ 3
2.2 A Stochastic Model ........ 3
2.3 Output Quantities of Interest ........ 4
2.4 Error Bounds for the Expectation and Variance of Outputs of Interest ........ 6
2.5 A Software Framework for Non-intrusive Uncertainty Propagation with Computable Error Bounds ........ 8
3 Discontinuous Solutions of Nonlinear Conservation Laws with Uncertainty ........ 9
3.1 Burgers' Equation with Uncertain Initial Data ........ 10
4 The Numerical Calculation of Statistics ........ 12
4.1 Stochastic and Probabilistic Collocation Methods ........ 13
4.2 Node-Nested Quadratures ........ 18
4.3 Uncertainty Propagation Using Adaptive Piecewise Polynomial Approximation and Subscale Recovery ........ 25
4.4 The HYGAP Algorithm ........ 35
5 Further Selected Applications ........ 38
5.1 The Propagation of Functional Uncertainty with Estimated Error Bounds for Subsonic Compressible Euler Flow ........ 39
5.2 NACA0012 Airfoil Transonic Euler Flow ........ 44
5.3 The Reynolds-Averaged Navier-Stokes Applications ........ 44
References ........ 55

**Uncertainty Quantification in Aeroelasticity** ... 59
Philip Beran and Bret Stanford
1 Introduction ... 59
1.1 Assessing Flutter Safety ... 61
1.2 Designing for Flutter Safety ... 61
1.3 Big Picture ... 62
1.4 Chapter Organization and Scope ... 63
2 Fundamental Mathematical Concepts in Aeroelasticity ... 64
2.1 Equations of Motion ... 64
2.2 Condition of Flutter ... 66
2.3 Description of Panel Problem ... 67
2.4 Simulation of LCO ... 69
2.5 Time-Linearized Behavior ... 70
2.6 Eigen-Analysis ... 70
3 Computation of Flutter Points and Their Sensitivities ... 75
3.1 Direct Evaluation of System Damping ... 75
3.2 Sensitivities of Flutter Speed to Parameters via Perturbation Analysis ... 76
3.3 Computation of Flutter Points Through a Bifurcation Approach ... 79
4 Computation of LCOs and Their Sensitivities ... 80
4.1 LCO Formulation ... 80
4.2 LCO Results ... 83
4.3 LCO Sensitivities ... 83
5 Uncertainty Quantification of Aeroelastic Responses ... 86
5.1 Assessment of Flutter for Panels of Random Thickness ... 87
5.2 Impact of Other Structural Nonlinearities on Flutter ... 90
5.3 Computing Flutter Probability of Failure using Monte-Carlo Simulation ... 93
5.4 Computing Flutter Probability of Failure Using the First-Order Reliability Method ... 94
5.5 Uncertainty Quantification of LCOs ... 97
6 Summary ... 100
References ... 101

**Robust Uncertainty Propagation in Systems of Conservation Laws with the Entropy Closure Method** ... 105
Bruno Després, Gaël Poëtte, and Didier Lucor
1 Introduction ... 106
2 The Moment Method for Uncertain Systems of Conservation Laws ... 107
3 Proof of Spectral Accuracy for a Non-linear Scalar Hyperbolic Case ... 110
3.1 Numerical Application ... 116
4 Loss of Hyperbolicity of the Discretized Problem ... 118
4.1 Example 1: Shallow Water Equations ... 118
4.2 Example 2: Euler Equations ... 120

5 Ensuring Hyperbolicity via Entropy Closure ..... 124
5.1 Reformulation ..... 126
5.2 Wave Velocities ..... 129
5.3 Entropy Choice ..... 132
5.4 Numerical Applications ..... 133
6 Parametric Uncertainty of the Model ..... 137
6.1 First Case: The Model Parameter Is a Random Variable ..... 140
6.2 Second Case: The Model Parameter Is a Random Process ..... 144
6.3 Modeling Parameter Uncertainties in Eulerian Systems ..... 145
7 Conclusion and Open Problems ..... 145
References ..... 148

**Adaptive Uncertainty Quantification for Computational Fluid Dynamics** ..... 151
Richard P. Dwight, Jeroen A.S. Witteveen, and Hester Bijl
1 Introduction ..... 152
2 Method 1: Adaptive Stochastic Finite Elements with Newton-Cotes Quadrature ..... 152
2.1 Background ..... 153
2.2 Adaptive Stochastic Finite Elements ..... 154
2.3 Numerical Results ..... 161
3 Method 2: Gradient-Enhanced Kriging with Adaptivity for Uncertainty Quantification ..... 168
3.1 Uncertainty Quantification Problem ..... 170
3.2 Gradient Evaluation via the Adjoint Method ..... 171
3.3 Gradient-Enhanced Kriging for Uncertainty Quantification ..... 172
3.4 Limitations of Gradients for Response Surfaces in CFD ..... 177
3.5 Numerical Results ..... 181
References ..... 191

**Implementation of Intrusive Polynomial Chaos in CFD Codes and Application to 3D Navier-Stokes** ..... 193
Chris Lacor, Cristian Dinescu, Charles Hirsch, and Sergey Smirnov
1 Introduction ..... 193
2 Polynomial Chaos Methodology ..... 194
3 Mathematical Formulation ..... 197
4 Application to 3D Compressible Navier-Stokes ..... 198
5 Simplifications of the Non-deterministic Navier-Stokes Equations ..... 201
5.1 Pseudo Spectral Approach ..... 201
5.2 Steady State Solutions of the Navier-Stokes Equations ..... 203
5.3 Truncation of Higher Order Terms ..... 204
6 Applications ..... 207
6.1 Supersonic 1D Nozzle ..... 207
6.2 Lid Driven Cavity Flow ..... 211
6.3 NASA Rotor 37 ..... 213
References ..... 222

**Multi-level Monte Carlo Finite Volume Methods for Uncertainty Quantification in Nonlinear Systems of Balance Laws** .... 225
Siddhartha Mishra, Christoph Schwab, and Jonas Šukys
1 Introduction ... 226
1.1 Weak Solutions of Systems of Balance Laws ... 226
1.2 Numerical Methods ... 227
1.3 Uncertainty Quantification (UQ) ... 228
1.4 Objectives of These Notes ... 231
2 Random Entropy Solutions ... 232
2.1 Random Fields ... 232
2.2 $k$-th Moments ... 233
2.3 Random Initial Data ... 234
2.4 Random Flux Functions for SCL ... 235
2.5 Random Entropy Solutions of Scalar Conservation Laws ... 237
3 Monte Carlo Finite Volume Method ... 240
4 Multi-level Monte Carlo Finite Volume Method ... 242
4.1 MLMC-FVM Error Analysis ... 242
4.2 Sparse Tensor Approximations of $k$-Point Correlations ... 246
5 Efficient Implementation of MLMC-FVM ... 250
5.1 Step 1: Hierarchy of Nested Grids ... 250
5.2 Step 2: Sample ... 250
5.3 Step 3: Solve ... 251
5.4 Stable Computation of Sample Statistics ... 252
6 Performance Studies of the MLMC-FVM for Conservation Laws ... 254
6.1 Euler Equations with Uncertain Initial Data ... 255
6.2 MHD Equations of Plasma Physics ... 257
6.3 Shallow-Water Equations with Uncertain Bottom Topography ... 263
6.4 Burgers' Equation with Random Flux ... 268
6.5 Two Phase Flows in a Porous Medium with Uncertain Permeabilities ... 270
6.6 Euler Equations with Uncertain Equation of State ... 274
6.7 Verification of the Derived Constants in the Asymptotic Error Bounds ... 276
6.8 Sparse Tensor MLMC Estimation of Two-Point Correlations ... 277
7 MLMC Approximation of Probabilities ... 285
7.1 MLMC Estimation of Probabilities ... 285
7.2 Shallow Water Equation in 2d: Perturbation of a Steady-State ... 287
8 Conclusion ... 289
References ... 292

**Essentially Non-oscillatory Stencil Selection and Subcell Resolution in Uncertainty Quantification** ... 295
Jeroen A.S. Witteveen and Gianluca Iaccarino
1 Introduction ... 296
2 Simplex Stochastic Collocation ... 300

3 Essentially Non-oscillatory Stencil Selection .... 303
3.1 Interpolation Stencil Selection .... 303
3.2 Efficient Implementation .... 304
4 Subcell Resolution .... 306
4.1 Discontinuous Representation .... 306
4.2 Discontinuous Derivatives .... 310
4.3 Sampling Strategy .... 311
5 Discontinuous Test Function .... 313
6 Linear Advection Equation .... 316
7 Shock Tube Problem .... 320
8 Transonic Flow Over the RAE 2822 Airfoil .... 325
9 Conclusions .... 330
References .... 331

# Non-intrusive Uncertainty Propagation with Error Bounds for Conservation Laws Containing Discontinuities

**Timothy Barth**

**Abstract** The propagation of statistical model parameter uncertainty in the numerical approximation of nonlinear conservation laws is considered. Of particular interest are nonlinear conservation laws containing uncertain parameters resulting in stochastic solutions with discontinuities in both physical and random variable dimensions. Using a finite number of deterministic numerical realizations, our objective is the accurate estimation of output uncertainty statistics (e.g. expectation and variance) for quantities of interest such as functionals, graphs, and fields. Given the error in numerical realizations, error bounds for output statistics are derived that may be numerically estimated and included in the calculation of output statistics. Unfortunately, the calculation of output statistics using classical techniques such as polynomial chaos, stochastic collocation, and sparse grid quadrature can be severely compromised by the presence of discontinuities in random variable dimensions. An alternative technique utilizing localized piecewise approximation combined with localized subscale recovery is shown to significantly improve the quality of calculated statistics when discontinuities are present. The success of this localized technique motivates the development of the HYbrid Global and Adaptive Polynomial (HYGAP) method described in Sect. 4.4. HYGAP employs a high accuracy global approximation when the solution data varies smoothly in a random variable dimension and local adaptive polynomial approximation with local postprocessing when the solution is non-smooth. To illustrate strengths and weaknesses of classical and newly proposed uncertainty propagation methods, a number of computational fluid dynamics (CFD) model problems containing various sources of parameter uncertainty are calculated including 1-D Burgers' equation, subsonic and transonic flow over 2-D single-element and multi-element airfoils, transonic Navier-Stokes flow over a 3-D ONERA M6 wing, and supersonic Navier-Stokes flow over a greatly simplified Saturn-V rocket.

T. Barth (✉)
NASA Ames Research Center, Moffett Field, CA 94035, USA
e-mail: Timothy.J.Barth@nasa.gov

H. Bijl et al. (eds.), *Uncertainty Quantification in Computational Fluid Dynamics*, Lecture Notes in Computational Science and Engineering 92,
DOI 10.1007/978-3-319-00885-1_1, 

# 1 Introduction

A mathematical model is often an approximate representation of a more complex system. Models of complex systems often utilize a large number of model parameters. The value of these parameters may be approximately determined through the fitting of model predictions with calibration data obtained from laboratory experiments, first principle arguments, ab initio calculations, more refined models, etc. Unfortunately, repeating a given experiment multiple times may yield different results that are suitably described by a statistical distribution. Ab initio chemistry calculations often utilize a statistical microscale description rather than attempting to deterministically enumerate all possible states, configurations, and/or interactions. Consequently, model parameters obtained from calibration data sources are often described statistically. A major task at hand is to propagate this model parameter uncertainty throughout subsequent calculations to quantify the statistical behavior of output quantities of interest.

Techniques for the propagation of uncertainty such as polynomial chaos [12, 40, 42, 43], stochastic collocation [1, 22, 23, 39], and sparse grid quadrature [26, 38] have proven to be powerful approaches that are now routinely used in computations. Nevertheless, these methods implicitly require that outputs of interest vary smoothly with respect to uncertain input parameters. When this is not the case, these methods may exhibit a significant deterioration in accuracy. Figure 1 (left) provides such an example of deterioration in the stochastic collocation method for transonic flow over an airfoil with uncertainty in the inflow Mach number. In this example, the inflow Mach number uncertainty is characterized by a Gaussian probability density truncated at four standard deviations. The stair-stepped oscillations in surface pressure coefficient statistics shown in Fig. 1 (left) are spurious numerical artifacts linked to the discontinuous dependence of the local surface pressure coefficient with respect to the uncertain inflow Mach number parameter. These oscillations in approximated statistics are a well-known pathology that is often observed in global polynomial chaos, stochastic collocation, and sparse grid quadrature when discontinuities are present. Using the HYGAP method developed in Sect. 4.3, nonoscillatory statistics can be approximated as graphed in Fig. 1 (right) *using the same number of evaluations as the stochastic collocation method while retaining the high order accuracy of stochastic collocation when the output of interest varies smoothly with respect to uncertain parameters.*

In the next section, a general framework is developed for non-intrusive uncertainty propagation including error bounds that are amenable to numerical estimation. For the methods considered herein, the estimated output statistics for quantities of interest contain errors originating from finite-dimensional approximation error in the numerical solution of realizations and quadrature error in the calculation of statistics. The uncertainty propagation framework provides estimated error bounds for output statistics when given an estimate of the realization error. The remainder of this article is then devoted to the development and testing of specialized uncertainty propagation techniques for conservation laws with uncertain parameters that admit discontinuities in the associated random variable dimensions.

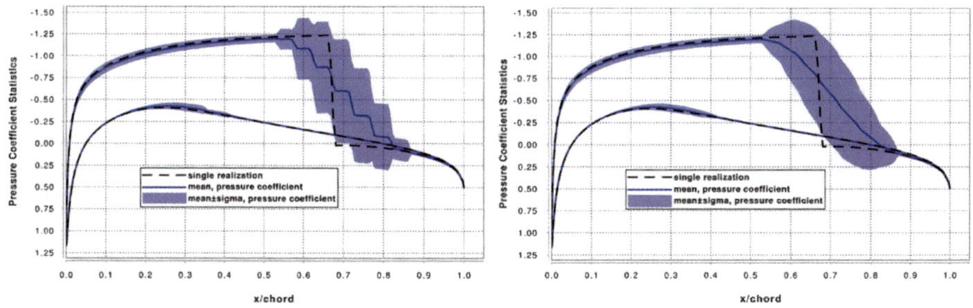


**Fig. 1** Transonic Euler equation flow over a NACA 0012 airfoil with $M_\infty = \text{Gaussian}_{4\sigma}$ $(m = 0.8, \sigma = 0.01)$ inflow Mach number uncertainty and a flow angle of attack of 2.26°. Mean and standard deviation envelopes for the surface pressure coefficient are graphed for calculations using stochastic collocation (*left*) and the HYGAP method of Sect. 4.4 (*right*)

## 2 A Framework for Non-intrusive Uncertainty Propagation with Error Bounds for Statistics

### 2.1 A Deterministic Model

Our starting point is a well-posed deterministic system of $m$ conservation laws in space dimension $d$ that depends on $M$ parameters, $\boldsymbol{\xi} \in \mathbb{R}^M$

$$\partial_t \boldsymbol{u}(\boldsymbol{x}, t) + \sum_{i=1}^{d} \partial_{x_i} \boldsymbol{f}_i(\boldsymbol{u}(\boldsymbol{x}, t); \boldsymbol{\xi}) = 0$$

$$\boldsymbol{u}(\boldsymbol{x}, 0) = \boldsymbol{u}_0(\boldsymbol{x}; \boldsymbol{\xi}) \tag{1}$$

with $\boldsymbol{x} \in \Omega \subset \mathbb{R}^d$ and $\boldsymbol{u}, \boldsymbol{f}_i \in \mathbb{R}^m$. This system together with suitable spatial boundary conditions (that may also depend on $\boldsymbol{\xi}$) is representative of many conservation law systems arising in computational science such as the equations of compressible flow considered herein.

### 2.2 A Stochastic Model

Let $(\Theta, \Sigma, P)$ denote the probability space of event outcomes, $\sigma$-algebra, and probability measure, respectively. Suppose the parameters $\boldsymbol{\xi}$ are now random variables depending on random events $\theta \in \Theta$. A stochastic form of the conservation law system is now given by

$$\partial_t \boldsymbol{u}(\boldsymbol{x},t,\theta) + \sum_{i=1}^{d} \partial_{x_i} \boldsymbol{f}_i(\boldsymbol{u}(\boldsymbol{x},t,\theta), \boldsymbol{\xi}(\theta)) = 0$$

$$\boldsymbol{u}(\boldsymbol{x},0,\theta) = \boldsymbol{u}_0(\boldsymbol{x}, \boldsymbol{\xi}(\theta)) \ . \tag{2}$$

with the support of $\boldsymbol{\xi}(\theta)$ denoted by $\Xi$. The statistical behavior of $\boldsymbol{\xi}(\theta)$ is characterized herein by a probability density $p_{\boldsymbol{\xi}}(\boldsymbol{\xi})$ such that $dP(\theta) = p_{\boldsymbol{\xi}}(\boldsymbol{\xi}) d\boldsymbol{\xi}(\theta)$. For simplicity, it is assumed in later examples that the probability density is of product form

$$p_{\boldsymbol{\xi}}(\boldsymbol{\xi}) = \prod_{i=1}^{M} p_{\xi_i}(\xi_i) \ . \tag{3}$$

## 2.3 Output Quantities of Interest

A primary objective is the estimation of uncertainty for outputs of interest

$$J(\boldsymbol{u}(\boldsymbol{x},t,\theta), \boldsymbol{\xi}(\theta)) \tag{4}$$

in terms of low order statistics such as expectation

$$E[J(\boldsymbol{u})](\boldsymbol{x},t) = \int_{\Theta} J(\boldsymbol{u}(\boldsymbol{x},t,\theta), \boldsymbol{\xi}(\theta)) \, dP(\theta) \tag{5}$$

and variance

$$V[J(\boldsymbol{u})](\boldsymbol{x},t) = E[J^2(\boldsymbol{u})](x,t) - (E[J(\boldsymbol{u})](\boldsymbol{x},t))^2 \ . \tag{6}$$

Outputs of interest may include stochastic functionals, graphs, and fields. Unfortunately, the exact stochastic solution $\boldsymbol{u}(\boldsymbol{x},t,\theta)$ is generally not known and the required statistics integrals may not be integrated in closed form. Conceptually, we introduce the notion of a finite-dimensional numerical approximation $\boldsymbol{u}_h(\boldsymbol{x},t,\theta)$ depending on a discretization parameter $h$. From the numerical approximation $\boldsymbol{u}_h(\boldsymbol{x},t,\theta)$, numerically approximated outputs of interest

$$J(\boldsymbol{u}_h(\boldsymbol{x},t,\theta), \boldsymbol{\xi}(\theta)) \tag{7}$$

as well as the finite-dimensional error in outputs of interest

$$\epsilon_h(\boldsymbol{x},t,\theta) \equiv J(\boldsymbol{u}(\boldsymbol{x},t,\theta), \boldsymbol{\xi}(\theta)) - J(\boldsymbol{u}_h(\boldsymbol{x},t,\theta), \boldsymbol{\xi}(\theta)) \tag{8}$$

are defined. Rather than directly constructing these stochastic functions, non-intrusive uncertainty propagation methods calculate a finite set of $N$ decoupled deterministic numerical realizations for distinct parameter values

$$\{\boldsymbol{\xi}^{(1)}, \ldots, \boldsymbol{\xi}^{(N)}\} \tag{9}$$

with $\boldsymbol{\xi}^{(i)}$ chosen in a way that facilitates the evaluation of output statistics. This yields $N$ realizations of the output quantity of interest

$$\{J(\boldsymbol{u}_h(\boldsymbol{x}, t; \boldsymbol{\xi}^{(1)}); \boldsymbol{\xi}^{(1)}), \ldots, J(\boldsymbol{u}_h(\boldsymbol{x}, t; \boldsymbol{\xi}^{(N)}); \boldsymbol{\xi}^{(N)})\} \tag{10}$$

and (optionally) estimates of the error magnitude $|\boldsymbol{\epsilon}_h|$

$$\{|\boldsymbol{\epsilon}_h(\boldsymbol{x}, t; \boldsymbol{\xi}^{(1)})|, \ldots, |\boldsymbol{\epsilon}_h(\boldsymbol{x}, t; \boldsymbol{\xi}^{(N)})|\} \ . \tag{11}$$

In practice, this error magnitude $|\boldsymbol{\epsilon}_h|$ may be estimated using a number of techniques including

- Richardson extrapolation [31] or other extrapolation techniques [6] from mesh or basis hierarchies $\{\boldsymbol{u}_h, \boldsymbol{u}_{2h}, \boldsymbol{u}_{4h}, \ldots\}$,
- A posteriori error estimation using dual (adjoint) problems [4, 11, 29] discussed further in Sect. 5.1,
- A posteriori error estimates obtained using superconvergent patch recovery [10, 44].

Output statistics are then approximated from these $N$ realizations

- Directly using $N$-point numerical quadrature denoted by $Q_N[\cdot]$ with weight $w_i$

$$E[J(\boldsymbol{u}_h)](\boldsymbol{x}, t) \approx Q_N[E[J(\boldsymbol{u}_h)]](\boldsymbol{x}, t) = \sum_{i=1}^{N} w_i \ J(\boldsymbol{u}_h(\boldsymbol{x}, t; \boldsymbol{\xi}^{(i)}); \boldsymbol{\xi}^{(i)}) \tag{12}$$

and

$$V[J(\boldsymbol{u}_h)](\boldsymbol{x}, t) \approx Q_N[E[J^2(\boldsymbol{u}_h)]](\boldsymbol{x}, t) - (Q_N[E[J(\boldsymbol{u}_h)]](\boldsymbol{x}, t))^2 \ , \tag{13}$$

- Indirectly by first constructing a finite-dimensional response surface using the $N$ realizations

$$\hat{J}(\boldsymbol{x}, t, \hat{\boldsymbol{\xi}}) = \sum_{i=1}^{N} J_i(\boldsymbol{x}, t) \, \Psi_i(\hat{\boldsymbol{\xi}}) \tag{14}$$

using either global or piecewise basis representations $\Psi_i(\hat{\boldsymbol{\xi}}) : \mathbb{R}^m \mapsto \mathbb{R}$ together with a product factorization of physical dimensions and parameter dimensions. A particularly convenient choice are nodal basis functions such that $\Psi_i(\boldsymbol{\xi}^{(j)}) = \delta_{ij}$ so that $J_i(\boldsymbol{x},t) = J(\boldsymbol{u}_h(\boldsymbol{x},t;\boldsymbol{\xi}^{(i)});\boldsymbol{\xi}^{(i)})$ are simply the $N$ computed realizations. Once the response surface is constructed, statistics such as expectation and variance can be numerically approximated using various forms of numerical quadrature ranging from dense and sparse grid quadrature to random (e.g. Monte Carlo) sampling.

Optional error bounds for expectation and variance statistics are then estimated using formulas derived in Sect. 2.4.

A low arithmetic complexity calculation of statistics for both a small or a large number of uncertain parameters is a challenging task. The difficulty is further compounded when the statistics integrand is non-smooth so that specialized techniques must be employed. A detailed discussion of numerical quadrature for smooth and non-smooth data is given in Sect. 4.

### *2.4 Error Bounds for the Expectation and Variance of Outputs of Interest*

Let $I[f]$ denote the weighted definite integral

$$I[f] = \int_{\Xi} f(\hat{\boldsymbol{\xi}})\, p(\hat{\boldsymbol{\xi}})\, d\hat{\boldsymbol{\xi}} \tag{15}$$

for a non-negative weighting function $p(\hat{\boldsymbol{\xi}})$. Let $Q_N[f]$ denote an $N$-point weighted numerical quadrature approximation to $I[f]$ with weights $w_i$ and evaluation points $\boldsymbol{\xi}^{(i)}$

$$Q_N[f] = \sum_{i=1}^{N} w_i\, f(\boldsymbol{\xi}^{(i)}) \tag{16}$$

with numerical quadrature error denoted by $R_N[f]$, i.e.

$$R_N[f] = I[f] - Q_N[f] \ . \tag{17}$$

Using this notation, the following lemma provides the basis for error bounds in expectation and variance statistics when a finite-dimensional approximation $\boldsymbol{u}_h(\boldsymbol{x},t,\theta)$ and numerical quadrature $Q_N[\cdot]$ are utilized.

**Lemma 1.** *Let $X$ and $X_h$ denote two random variables with bounded first and second moments with respect to the probability measure $p(\boldsymbol{\xi})d\boldsymbol{\xi}$ over the set $\Xi$ for all values of the discretization parameter $h$. Further define the error difference*

*$\epsilon_h \equiv X - X_h$. Let the expectation and variance of $X_h$ be approximated by $N$-point quadrature $Q_N[E[X_h]]$ and $Q_N[V[X_h]]$. Suppose $X_h$ and the error difference magnitude $|\epsilon_h|$ are given, the following expectation error bound is satisfied*

$$|E[X] - Q_N[E[X_h]]| \leq |Q_N[E[|\epsilon_h|]]| + |R_N[E[|\epsilon_h|]]| + |R_N[E[X_h]]| \quad (18)$$

*and the variance error bound is satisfied*

$$\begin{aligned}|V[X] - Q_N[V[X_h]]| \leq\ & 2\left((|Q_N[E[|\epsilon_h|^2]]| + |R_N[E[|\epsilon_h|^2]]|)(|Q_N[V[X_h]]| + |R_N[V[X_h]]|)\right)^{\frac{1}{2}} \\ & + |Q_N[E[|\epsilon_h|^2]]| + |R_N[E[|\epsilon_h|^2]]| + |R_N[V[X_h]]| \ . \end{aligned} \quad (19)$$

*Proof.* The expectation error bound follows directly

$$\begin{aligned}|E[X] - Q_N[E[X_h]]| &= |E[\epsilon_h] + R_N[E[X_h]]| \\ &\leq E[|\epsilon_h|] + |R_N[E[X_h]]| \\ &\leq |Q_N[E[|\epsilon_h|]]| + |R_N[E[|\epsilon_h|]]| + |R_N[E[X_h]]| \ . \end{aligned} \quad (20)$$

Note that the outer absolute value appearing in $|Q_N[E[|\epsilon_h|]]|$ arises because quadratures with negative weights may result in negative estimates of the positive expectation $E[|\epsilon_h|]$. In constructing a variance difference bound, we use the well-known covariance identity for arbitrary random variables $X$ and $Y$

$$V[X+Y] = V[X] + V[Y] + 2\,COV[X,Y] \quad (21)$$

and the Cauchy-Schwarz inequality for random variables $X$ and $Y$ with finite variance

$$|COV[X,Y]|^2 \leq V[X]\ V[Y] \ . \quad (22)$$

The variance error bound then follows from the following steps

$$\begin{aligned}|V[X] - Q_N[V[X_h]]| &= |2\,COV[\epsilon_h, X_h] + V[\epsilon_h] + R_N[V[X_h]]| \\ &\leq 2\,(V[\epsilon_h]\ V[X_h])^{1/2} + V[\epsilon_h] + |R_N[V[X_h]]| \\ &\leq 2\,(E[|\epsilon_h|^2]\ V[X_h])^{1/2} + E[|\epsilon_h|^2] + |R_N[V[X_h]]| \\ &\leq 2\left((|Q_N[E[|\epsilon_h|^2]]| + |R_N[E[|\epsilon_h|^2]]|)(|Q_N[V[X_h]]| + |R_N[V[X_h]]|)\right)^{\frac{1}{2}} \\ &\quad + |Q_N[E[|\epsilon_h|^2]]| + |R_N[E[|\epsilon_h|^2]]| + |R_N[V[X_h]]| \end{aligned} \quad (23)$$

so that the Lemma 1 bounds are obtained. □

Setting $X = J(\boldsymbol{u}) \equiv J(\boldsymbol{u}(\boldsymbol{x}, t, \theta), \boldsymbol{\xi}(\theta))$ and $X_h = J(\boldsymbol{u}_h) \equiv J(\boldsymbol{u}_h(\boldsymbol{x}, t, \theta), \boldsymbol{\xi}(\theta))$, the following expectation error bound follows from Lemma 1
*Expectation error bound:*

$$\begin{aligned} |E[J(\boldsymbol{u})] - Q_N[E[J(\boldsymbol{u}_h)]]| &= |E[\epsilon_h] + R_N[E[J(\boldsymbol{u}_h)]]| \\ &\leq |Q_N[E[|\epsilon_h|]]| + |R_N[E[|\epsilon_h|]]| + |R_N[E[J(\boldsymbol{u}_h)]]| \end{aligned} \tag{24}$$

as well as the following variance error bound
*Variance error bound:*

$$\begin{aligned} |V[J(\boldsymbol{u})] - Q_N[V[J(\boldsymbol{u}_h)]]| &\leq 2\,(|Q_N[E[|\epsilon_h|^2]]| + |R_N[E[|\epsilon_h|^2]]|)^{1/2} \\ &\quad \times (|Q_N[V[J(\boldsymbol{u}_h)]]| + |R_N[V[J(\boldsymbol{u}_h)]]|)^{1/2} \\ &\quad + |Q_N[E[|\epsilon_h|^2]]| + |R_N[E[|\epsilon_h|^2]]| + |R_N[V[J(\boldsymbol{u}_h)]]| \ . \end{aligned} \tag{25}$$

*Remark 1.* The bounds (24) and (25) are not computable because they require the evaluation of quadrature error $R_N[\cdot]$ which is not generally available. Section 4 examines node-nested quadratures such as Gauss-Kronrod and Clenshaw-Curtis quadrature that permit computationally efficient estimates for the quadrature error $R_N[\cdot]$. Unfortunately, these estimates are not reliable unless the underlying integrand exhibits sufficient smoothness. Section 4.4 introduces the HYGAP method that replaces global quadrature with local piecewise polynomial approximation and quadrature whenever the data is non-smooth. The HYGAP method also permits an estimate of quadrature error but a formal proof of reliability is still lacking for non-smooth data.

## 2.5 A Software Framework for Non-intrusive Uncertainty Propagation with Computable Error Bounds

A non-intrusive uncertainty propagation framework with optional error bounds is summarized in the Fig. 2 flowchart. A user specifies sources of uncertainty, provides realizations for outputs of interest, $J(u_h(\boldsymbol{x}, t; \boldsymbol{\xi}^{(i)}); \boldsymbol{\xi}^{(i)})$, and optionally an estimate of the error, $|J(u(\boldsymbol{x}, t; \boldsymbol{\xi}^{(i)}); \boldsymbol{\xi}^{(i)}) - J(u_h(\boldsymbol{x}, t; \boldsymbol{\xi}^{(i)}); \boldsymbol{\xi}^{(i)})|, i = 1, \ldots, N$. This information is sufficient to estimate statistics for outputs of interest and optional estimated error bounds.

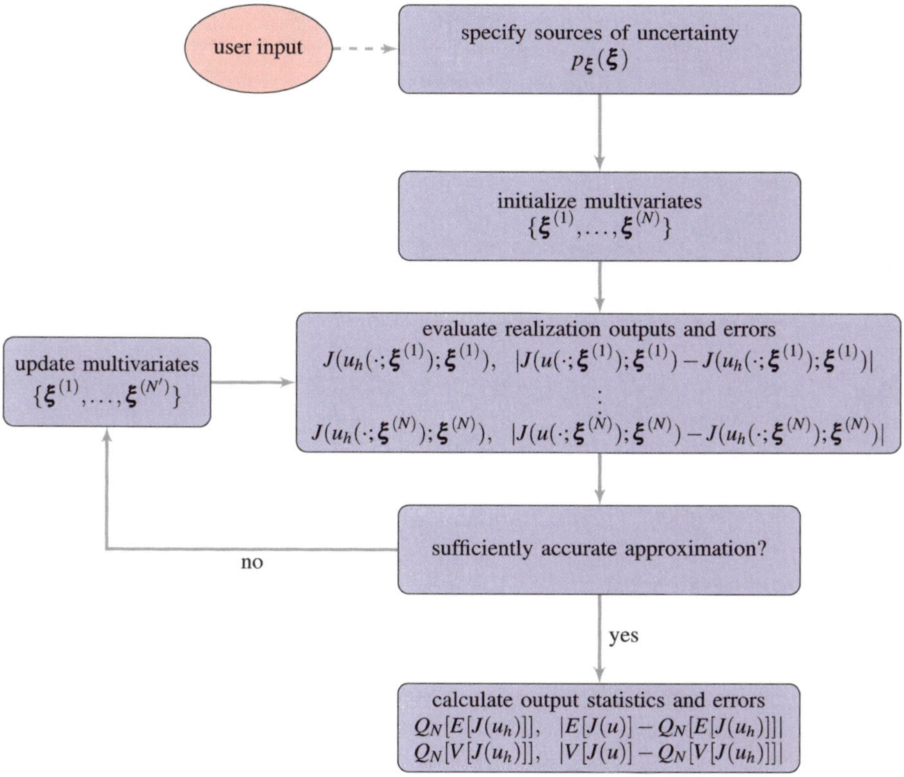


**Fig. 2** Flowchart of the generalized non-intrusive uncertainty propagation framework

## 3 Discontinuous Solutions of Nonlinear Conservation Laws with Uncertainty

An important feature of nonlinear conservation laws depending on uncertain parameters is the formation of solution discontinuities in *both* physical and random variable dimensions from smooth initial data. This makes the numerical approximation of output uncertainty particularly difficult. To illustrate this feature, consider the non-viscous Burgers' equation

$$\frac{\partial}{\partial t}u(x,t) + \frac{\partial}{\partial x}(u^2(x,t)/2) = 0 \tag{26}$$

with sinusoidal initial data for a given amplitude $A$

$$u(x,0) = A\,\sin(2\pi x) \tag{27}$$

in the domain $(x, t) \in [0, 1] \times [0, T]$ with periodicity assumed in space. This smooth initial data steepens as time progresses and eventually forms a discontinuity at $x = 1/2$. An exact solution to this problem for a given amplitude $A$ can be constructed

$$u(x, t) = u_{Burgers}(x, t; A) \tag{28}$$

in a piecewise sense using the method of characteristics.[1]

## 3.1 Burgers' Equation with Uncertain Initial Data

It is instructive to then modify this model problem by introducing statistically uncertain amplitudes and phase shifts into the initial data. Let $(\Theta, \Sigma, P)$ denote a probability space with event outcomes in $\Theta$, a $\sigma$-algebra $\Sigma$, and probability measure $P$. Introducing a random event $\theta$ from $\Theta$ and random variables $\boldsymbol{\xi}(\theta) \in \mathbb{R}^2$ corresponding to uncertain amplitude and phase, the initial data with uncertainty becomes

$$u(x, 0, \theta) = \xi_1(\theta) \sin(2\pi(x + g(\xi_2(\theta)))) \tag{29}$$

for an arbitrary function $g(\cdot) : \mathbb{R} \mapsto \mathbb{R}$ chosen so that the solution is not exactly representable by finite-dimensional polynomials in $\xi_2$-$x$ planes. The initial data with uncertainty is then propagated forward in time via the stochastic Burgers' equation

$$\frac{\partial}{\partial t} u(x, t, \theta) + \frac{\partial}{\partial x}(u^2(x, t, \theta)/2) = 0 \ . \tag{30}$$

An exact solution of (30) with initial data (29) is readily obtained in terms of the solutions of (26) with initial data (27), i.e.

$$u(x, t, \theta) = u_{Burgers}(x + g(\xi_2(\theta)), t; \xi_1(\theta)) \ . \tag{31}$$

The solution of this Burgers' equation problem with amplitude and phase uncertainty is initially smooth but as time progresses eventually develops a discontinuity that traverses obliquely through both physical and random variable dimensions as shown in Figs. 3 and 4. This discontinuity is a genuinely nonlinear discontinuity in the physical dimension and a degenerate discontinuity in the random variable dimensions.

Suppose the probability measure is characterized by the probability density

$$p_{\boldsymbol{\xi}}(\boldsymbol{\xi}) = p_{\xi_1}(\xi_1) p_{\xi_2}(\xi_2)$$

---

[1] This does require the numerical solution of a scalar implicit function relation for each characteristic that is easily solved to any desired accuracy.

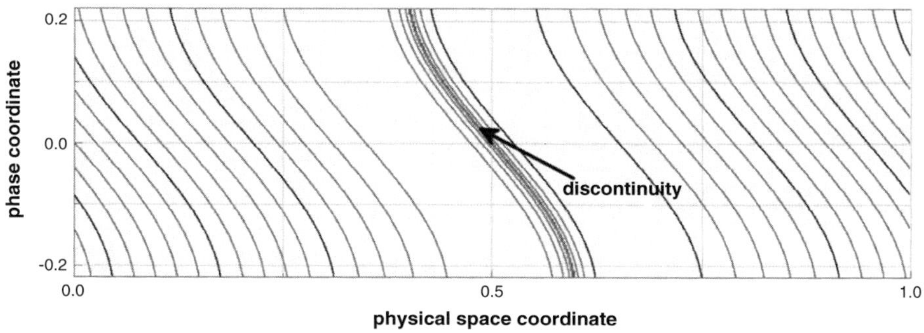


**Fig. 3** Solution contours in the $\xi_2$-$x$ plane at $\xi_1 = 0.5$ for the Burgers' equation problem (30) with combined amplitude and phase uncertain sinusoidal initial data (29) at time $t = 0.35$ with $g(\xi_2) = \frac{1}{10}\sin(2\pi\xi_2)$

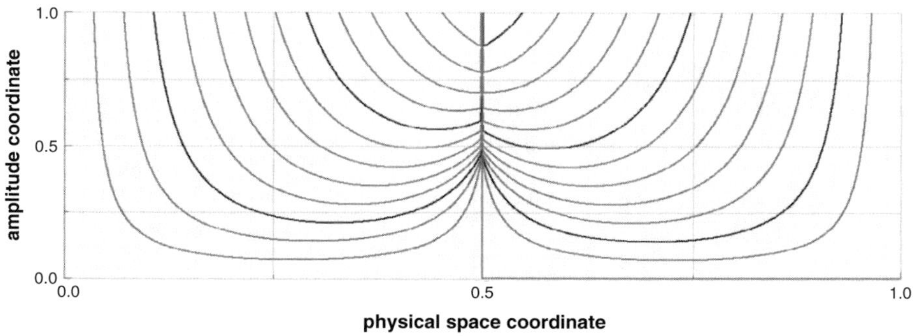


**Fig. 4** Solution contours in $\xi_1$-$x$ plane at $\xi_2 = 0$ for the Burgers' equation problem (30) with combined amplitude and phase uncertain sinusoidal initial data (29) at time $t = 0.35$ with $g(\xi_2) = \frac{1}{10}\sin(2\pi\xi_2)$

with $p_{\xi_1}$ a probability density for amplitude uncertainty and $p_{\xi_2}$ a probability density for phase uncertainty. The expectation and variance are then calculated from

$$E[u](x,t) = \int_{\Theta} u(x,t,\theta)\, dP(\theta) = \int_{\Xi} u(x,t,\boldsymbol{\xi}(\theta))\, p_{\boldsymbol{\xi}}(\boldsymbol{\xi})\, d\boldsymbol{\xi}(\theta) \tag{32}$$

and

$$V[u](x,t) = \sigma^2[u](x,t) = E[u^2](x,t) - (E[u](x,t))^2 \ . \tag{33}$$

Using the formulas (32) and (33) and the exact solution (31), the expectation and variance can be evaluated to any desired accuracy using adaptive quadrature such as found in QUADPACK [28] with care taken to avoid performing quadratures

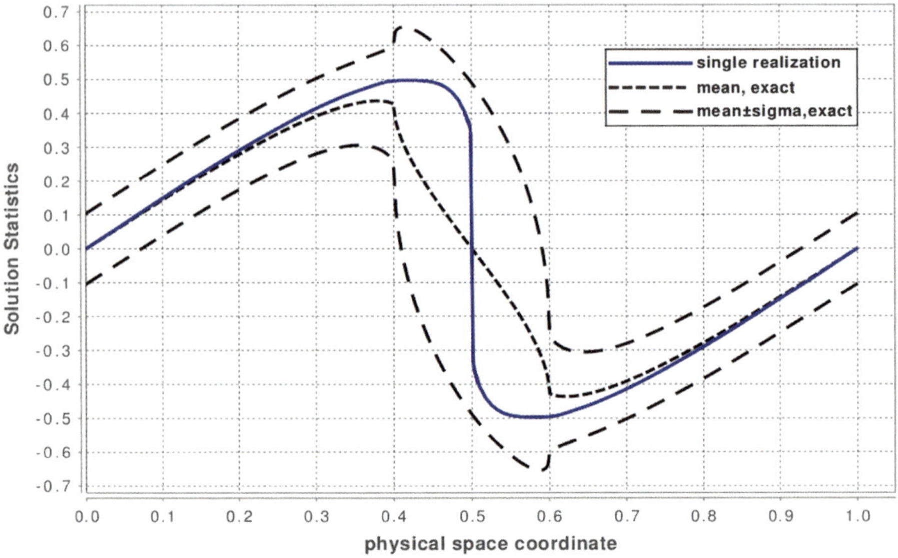


**Fig. 5** Graphs of expectation (mean) $E[u](x,t)$ and standard deviation envelopes $E[u](x,t) \pm \sigma[u](x,t)$ for the Burgers' equation problem (30) with sinusoidal initial data (29) at time $t = 0.35$ with $g(\xi_2) = \frac{1}{10}\sin(2\pi\xi_2)$ assuming uniform probability density $\mathscr{U}(0.2, 0.8)$ amplitude uncertainty and uniform probability density $\mathscr{U}(-0.25, 0.25)$ phase uncertainty. A single realization corresponding to $(\xi_1 = 0.5, \xi_2 = 0.0)$ is also graphed

across analytically known discontinuity locations in physical and random variable dimensions. For illustration, a uniform probability density $\mathscr{U}(0.2, 0.8)$ has been chosen for amplitude uncertainty and a uniform probability density $\mathscr{U}(-0.25, 0.25)$ has been chosen for phase uncertainty. The mean (expectation) and standard deviation envelope at time $t = 0.35$ are graphed in Fig. 5.

The presence of discontinuities in random variable dimensions, such as depicted in Fig. 3, has enormous consequences in the performance of many classical techniques in uncertainty quantification using dense and sparse grid quadrature. Section 4 is devoted to this topic and offers some alternative algorithmic approaches when discontinuities are present in random variable dimensions.

## 4 The Numerical Calculation of Statistics

In this section, several techniques for the calculation of output statistics are evaluated for smooth and non-smooth data such as described in Sect. 3. In a non-intrusive setting, these approaches can be interpreted as producing linear and nonlinear (data dependent) quadrature formulas for the evaluation of output statistics.

## 4.1 Stochastic and Probabilistic Collocation Methods

A popular class of non-intrusive uncertainty propagation methods for PDEs with smooth solutions are the stochastic and probabilistic collocation methods [1, 22, 23, 25, 39]. Assume that the solution in $d$ space dimensions and $M$ random variables dimensions is of product form

$$\boldsymbol{u}_h(\boldsymbol{x}, t, \theta) = \sum_{i_1=1}^{N_1} \cdots \sum_{i_M=1}^{N_M} C_{i_1 \dots i_M}(\boldsymbol{x}, t)\, \Psi_{i_1}(\xi_1(\theta)) \dots \Psi_{i_M}(\xi_M(\theta)) \ , \quad N = \prod_{i=1}^{M} N_i \tag{34}$$

with $\Psi_i(\xi)$ a 1-D nodal Lagrange basis

$$\Psi_i(\xi) = \prod_{l=1, l \neq i}^{N_i} \frac{\xi - \xi^{(l)}}{\xi^{(i)} - \xi^{(l)}} \ . \tag{35}$$

Evaluating (34) at $N$ collocation points $\{\boldsymbol{\xi}^{(1)}, \dots, \boldsymbol{\xi}^{(N)}\}$ then uniquely determines the coefficients $C_{i_1 \dots i_M}(\boldsymbol{x}, t)$ in terms of deterministic solutions, i.e.

$$C_{i_1 \dots i_M}(\boldsymbol{x}, t) = \boldsymbol{u}_h(\boldsymbol{x}, t; \xi_1^{(i_1)}, \dots, \xi_M^{(i_M)}) \ . \tag{36}$$

There is still freedom in the choice of collocation point locations. A particularly convenient choice are the locations for optimal Gauss quadratures of moment statistics integrals given the specific probability densities. Some example probability densities and optimal quadratures in a single random variable dimension include

- Normal probability density, $p(\xi) = \frac{1}{\sqrt{2\pi}\sigma} e^{-\frac{(\xi-\mu)^2}{\sigma^2}}$. $m = 1, 2$ moment statistics are calculated from

$$E[f^m] = \int_{-\infty}^{\infty} \frac{f^m(\xi)}{\sqrt{2\pi}\sigma} e^{-\frac{(\xi-\mu)^2}{\sigma^2}} \, d\xi \ .$$

  Let $\xi(y) \equiv \mu + \sigma y$, a change of variable yields the following canonical form which is efficiently approximated by Gauss-Hermite quadrature with weights $w_1, \dots, w_N$ and quadrature locations $y_1, \dots, y_N$

$$E[f^m] = \int_{-\infty}^{\infty} \frac{f^m(\xi(y))}{\sqrt{2\pi}} e^{\frac{-y^2}{2}} \, dy \approx \sum_{n=1}^{N} w_n f^m(\xi(y_n)) \ . \quad \text{(Gauss-Hermite Quadrature)}$$

- Log-normal probability density, let $\xi > 0$ and $p(\xi) = \frac{1}{\xi\sqrt{2\pi}\sigma} e^{-\frac{(\ln \xi - \mu)^2}{2\sigma^2}}$. $m = 1, 2$ moment statistics are calculated from

$$E[f^m] = \int_{0}^{\infty} \frac{f^m(\xi)}{\xi\sqrt{2\pi}\sigma} e^{-\frac{(\ln \xi - \mu)^2}{2\sigma^2}} \, d\xi \ .$$

Let $\xi(y) \equiv e^{\mu+\sigma y}$, a change of variable yields the canonical form which is efficiently approximated by Gauss-Hermite quadrature with weights $w_1, \ldots, w_N$ and quadrature locations $y_1, \ldots, y_N$

$$E[f^m] = \int_{-\infty}^{\infty} \frac{f^m(\xi(y))}{\sqrt{2\pi}} e^{\frac{-y^2}{2}}\, dy \approx \sum_{n=1}^{N} w_n f^m(\xi(y_n)) \ . \text{ (Gauss-Hermite Quadrature)}$$

- Uniform probability densities. $m = 1, 2$ moment statistics are calculated from

$$E[f^m] = \frac{1}{\xi_{max} - \xi_{min}} \int_{\xi_{min}}^{\xi_{max}} f^m(\xi)\, d\xi$$

Let $\xi(y) \equiv \xi_{min} + (\xi_{max} - \xi_{min})\, y$, a change of variable yields the canonical form which is efficiently approximated by Gauss-Legendre quadrature with weights $w_1, \ldots, w_N$ and quadrature locations $y_1, \ldots, y_N$

$$F[f^m] = \int_0^1 f^m(\xi(y))\, dy \approx \sum_{n=1}^{N} w_n f^m(\xi(y_n)) \ . \text{ (Gauss-Legendre Quadrature)}$$

- Non-classical probability densities and/or truncated random variable domains. A stable procedure for computing orthogonal polynomials and optimal Gaussian quadratures given nonclassical weights and/or domains is presented in Sack and Donovan [33] and Wheeler [41]. Truncated domains are often used so than physically unrealizable random variable states (e.g. negative viscosity) do not arise in calculations. One such example are Gaussian probability densities truncated at $n$ standard deviations and renormalized to have unit total probability, Gaussian$_{n\sigma}(m, \sigma)$, such as employed in the examples of Sect. 5.

Since these moment statistics can be approximated via quadrature without ever explicitly constructing the Lagrange interpolant, one can dispense with the notion of stochastic solutions altogether and reinterpret the stochastic collocation method as a non-intrusive uncertainty propagation method that calculates a set of $N$ decoupled deterministic numerical realizations for parameter values

$$\{\boldsymbol{\xi}^{(1)}, \ldots, \boldsymbol{\xi}^{(N)}\} \tag{37}$$

with $\boldsymbol{\xi}$ chosen in a way that facilitates the evaluation of output statistics. This yields $N$ realizations of the solution data and outputs of interest

$$\{J(\boldsymbol{u}_h(\boldsymbol{x}, t; \boldsymbol{\xi}^{(1)}); \boldsymbol{\xi}^{(1)}), \ldots, J(\boldsymbol{u}_h(\boldsymbol{x}, t; \boldsymbol{\xi}^{(N)}); \boldsymbol{\xi}^{(N)})\} \tag{38}$$

and optionally estimates of the error magnitude $|\boldsymbol{\epsilon}_h|$

$$\{|\boldsymbol{\epsilon}_h(\boldsymbol{x}, t; \boldsymbol{\xi}^{(1)})|, \ldots, |\boldsymbol{\epsilon}_h(\boldsymbol{x}, t; \boldsymbol{\xi}^{(N)})|\} \ . \tag{39}$$

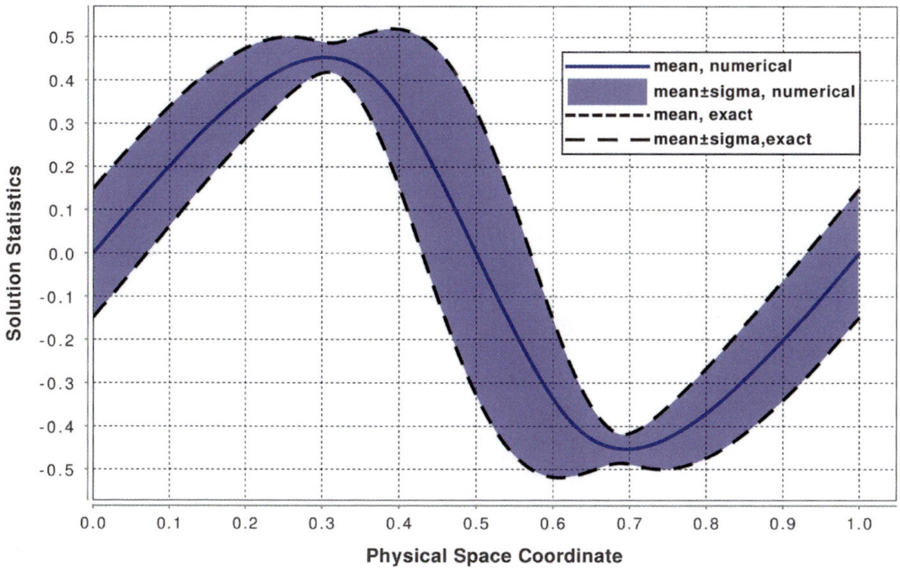


**Fig. 6** Expectation and standard deviation envelopes approximated using stochastic collocation ($N = 9$) for the Burgers' equation problem (30) with phase uncertain initial data (29) at the time $t = 0.15$

Output statistics with estimated error bounds are then numerically approximated using the formulas (24) and (25) developed in Sect. 2.4.

### 4.1.1 Stochastic Collocation for Nonlinear Conservation Laws

The stochastic collocation method has been used to approximate statistics for the Burgers' equation problem (30) assuming uniform probability density $\mathscr{U}(-0.25, 0.25)$ phase uncertain initial data (29) with deterministic amplitude $A = 1/2$. Realizations are approximated on a mesh containing 128 intervals in space using a WENO [17] finite-volume method with $P_2$ quadratic and later $P_4$ quartic polynomials in space together with fourth-order accurate time integration. When the solution remains globally smooth, the stochastic collocation method gives excellent results at small cost. Figure 6 shows a graph of statistics for the Burgers' equation problem at the time $t = 0.15$ before discontinuities have formed in realizations. Statistics have been approximated using $N = 9$ Gauss-Legendre quadrature points in stochastic collocation. The agreement with the exact statistics is excellent. Figure 7 graphs measured errors using $N = 4, \ldots, 10$ quadrature points and finite-volume discretization in space using both $P_2$ quadratic and $P_4$ quartic

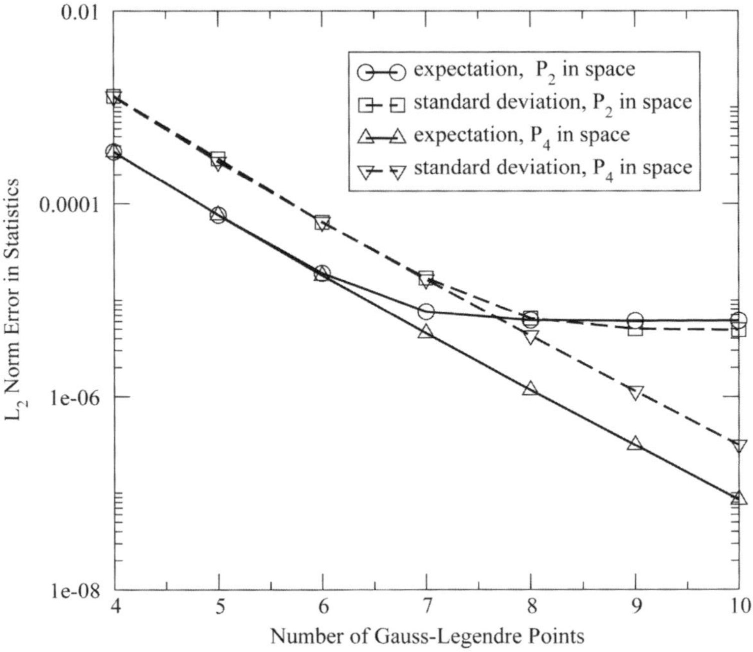


**Fig. 7** Measured expectation and variance errors at the time $t = 0.15$ before a discontinuity forms in the domain for the Burgers' equation problem (30) with phase uncertain initial data (29) using stochastic collocation and $P_2$ quadratic and $P_4$ quartic polynomial finite-volume realizations in space

polynomials. The measured errors in statistics initially decrease very rapidly with an increasing number of stochastic collocation points consistent with Gauss-Legendre quadrature which exhibits a well-known [14] quadrature error

$$R_N[f] = \frac{2^{2N}(N!)^4}{(2N+1)[(2N)!]^3} f^{(2N)}(\eta), \quad \eta \in [0,1] \ . \tag{40}$$

However, the total error in statistics contains components arising from spatial discretization error in the finite-volume method and quadrature error in the calculation of statistics. Consequently, the measured errors in statistics using the $P_2$ finite-volume method eventually cease decreasing at approximately $N = 7$ because the error in statistics is now dominated by the spatial discretization error so decreasing the quadrature error further is ineffective.

A pitfall of stochastic collocation is the resolution of discontinuities in random variable dimensions. The previous calculations have been repeated, but now at a later time $t = 0.35$ after discontinuities have formed. Although one does not usually explicitly construct the Lagrange interpolant arising in the stochastic collocation

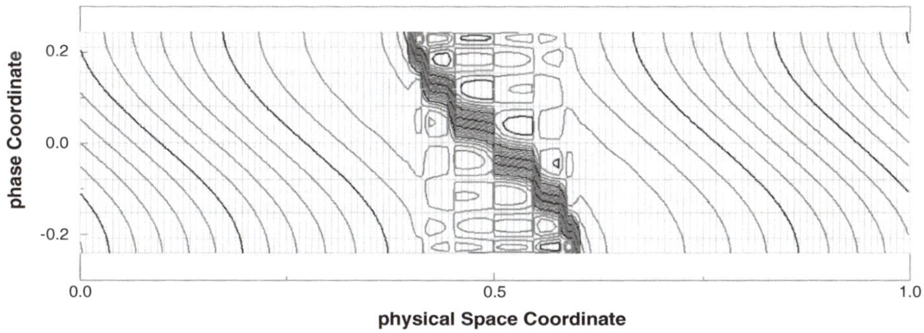


**Fig. 8** Contours of the stochastic collocation Lagrange interpolant ($N = 9$) for the Burgers' equation solution (30) with phase uncertain initial data (29) at time $t = 0.35$

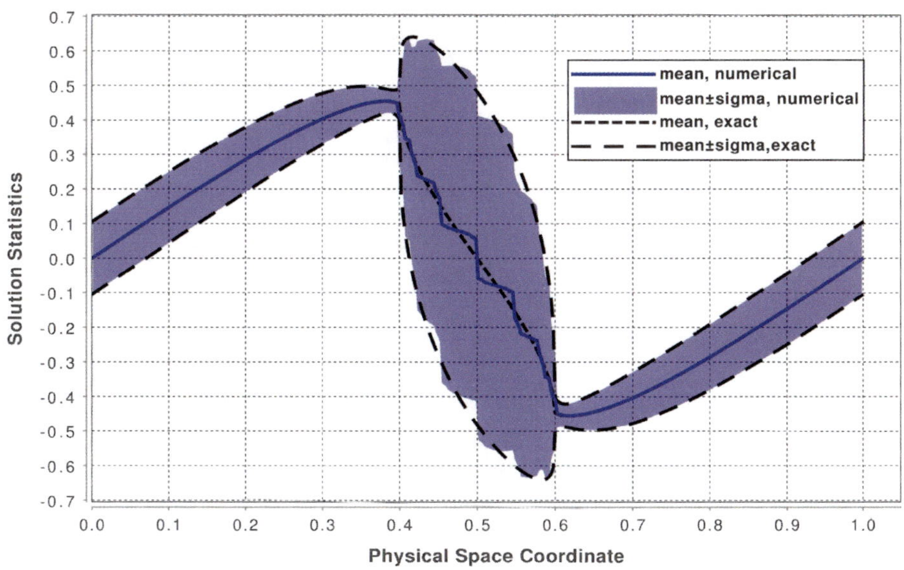


**Fig. 9** Expectation and standard deviation envelopes approximated using stochastic collocation ($N = 9$) for the Burgers' equation problem (30) with phase uncertain initial data (29) at time $t = 0.35$

method, it is useful to do so for purposes of understanding the behavior of the stochastic collocation method. Contours of the collocation interpolant are shown in Fig. 8. Severe oscillations in the interpolant are clearly visible whenever the Lagrange interpolant spans the discontinuity in the phase uncertainty dimension. Figure 9 graphs the expectation and standard deviation statistics obtained using this Gauss-Legendre quadrature. Spurious stair-stepped oscillations in the approximated statistics are clearly observed whenever the underlying Lagrange interpolant is oscillatory. This shortcoming of stochastic collocation explains the stair-stepped oscillatory behavior observed earlier in Fig. 1.

## 4.2 *Node-Nested Quadratures*

When only output statistics are sought, the stochastic collocation method reduces to the calculation of $N$ decoupled deterministic solutions followed by numerical quadrature. Gauss quadrature is a rather natural candidate given the optimal performance of these quadratures for specific probability densities. In the following section, alternative dense and sparse quadratures are considered. Particular attention is given to node-nested quadratures because these quadratures permit very convenient estimates of quadrature error that can be used in the error bound formulas (24) and (25). Unfortunately, these dense and sparse quadratures also suffer from oscillations when discontinuities are present. The HYGAP method developed in Sect. 4.4 addresses this oscillation problem for dense quadratures.

### 4.2.1 Dense Product Global Quadratures

In this section, quadrature formulas are considered that have particularly convenient estimates of quadrature error for use in the error bound formulas (24) and (25). The task of estimating quadrature error is greatly simplified and efficiently implemented through the use of nested quadratures. Two nested quadratures often used are

- Gauss-Kronrod quadrature. $N$-point Gauss quadratures exhibit the well-known property that $2N - 1$ degree polynomials are integrated exactly as seen from (40). Gauss-Kronrod quadratures are a variant of Gauss quadrature such that by adding $N + 1$ new points to an existing $N$-point Gauss quadrature (see Fig. 10) the result is a quadrature that integrates $3N + 1$ degree polynomials exactly. The Gauss quadrature error $R_N^G[f]$ is often estimated by the forward evaluation formula that uses the $2N + 1$ Gauss-Kronrod points

$$|R_N^G[f]| \le C_N^{GK} |Q_{2N+1}^{GK}[f] - Q_N^G[f]| \qquad \text{(forward estimate)} \tag{41}$$

  with even more accurate specialized nonlinear formulas such as

$$|R_7^G[f]| \le (200 |Q_{15}^{GK}[f] - Q_7^G[f]|)^{3/2} \tag{42}$$

  often used in adaptive quadrature libraries such as QUADPACK [28]. To avoid additional computation, we can instead start from the $2N + 1$ Gauss-Kronrod quadrature approximation and estimate the quadrature error by a backward formula

$$|R_{2N+1}^{GK}[f]| \le c_{2N+1}^{GK}|Q_{2N+1}^{GK}[f] - Q_N^G[f]| \qquad \text{(backward estimate)} \tag{43}$$

where the constant $c_{2N+1}^{GK}$ depends strongly on $N$ and should be estimated. Let

$$C_s(R_N) = \sup_{\|f^{(s)}\|_\infty \le 1} |R_N[f]| \tag{44}$$

and suppose the ratio of $Q_{2N+1}^{GK}[f]$ and $Q_N^G[f]$ quadrature errors is adequately approximated by the ratio of $C_{2N}(R_{2N+1}^{GK})$ and $C_{2N}(R_N^G)$. Let this ratio is bounded by a constant depending on $N$, i.e.

$$\left|\frac{R_{2N+1}^{GK}[f]}{R_N^G[f]}\right| \approx \frac{C_{2N}(R_{2N+1}^{GK})}{C_{2N}(R_N^G)} \le \kappa_{2N+1}^{GK} \ . \tag{45}$$

It then follows that

$$\begin{aligned} |I[f] - Q_{2N+1}^{GK}[f]| &\le \kappa_{2N+1}^{GK}|I[f] - Q_N^G[f]| \\ &= \kappa_{2N+1}^{GK}|I[f] - Q_N^{GK}[f] + Q_N^{GK}[f] - Q_{\lfloor N/2\rfloor+1}^{GK}[f]| \\ &\le \kappa_{2N+1}^{GK}(|I[f] - Q_N^{GK}[f]| + |Q_N^{GK}[f] - Q_{\lfloor N/2\rfloor+1}^{GK}[f]|) \end{aligned} \tag{46}$$

yielding

$$|R_{2N+1}^{GL}[f]| \le \frac{\kappa_{2N+1}^{GK}}{1-\kappa_{2N+1}^{GK}}|Q_{2N+1}^{GK}[f] - Q_N^G[f]| \tag{47}$$

thus permitting $c_{2N+1}^{GK}$ to be written in terms of $\kappa_{2N+1}^{GK}$

$$c_{2N+1}^{GK} = \frac{\kappa_{2N+1}^{GK}}{1-\kappa_{2N+1}^{GK}} \ . \tag{48}$$

Brass and Förster [5] provide the following estimate for $\kappa_{2N+1}^{GK}$

$$\kappa_{2N+1}^{GK} \le C_{BF}N^{1/4}\left(\frac{1}{3.493\ldots}\right)^N \tag{49}$$

so that an explicit estimate for $c_{2N+1}^{GK}$ can be obtained once the constant $C_{BF}$ is chosen.

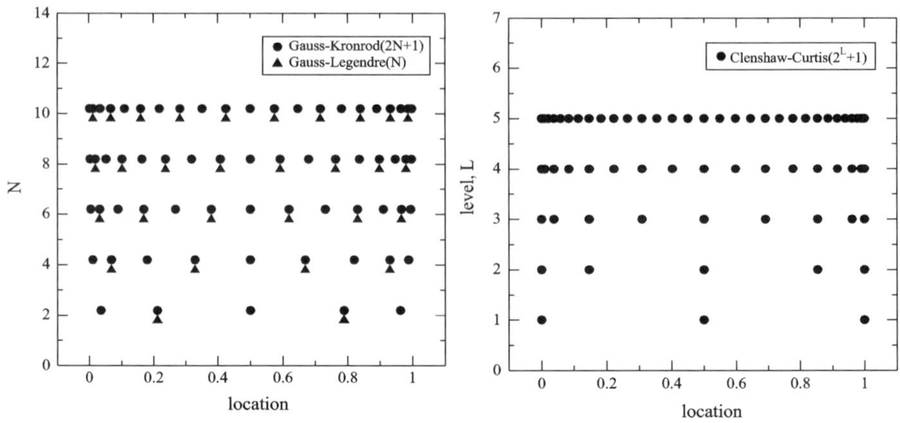


**Fig. 10** Node nested quadratures. The *left figure* shows Gauss quadrature locations using $N$ points as well as the $2N+1$ Gauss-Kronrod quadrature point locations. The *right figure* shows Clenshaw-Curtis quadrature point locations for various values of level $L$ with $2^L+1$ points

- Clenshaw-Curtis quadrature [8]. The quadrature point locations are the extreme points of Tchebysheff polynomials of the first kind

$$\xi^{(i)} = \frac{1}{2}\left(1 - \cos\left(\frac{\pi i}{N+1}\right)\right) \quad , \quad i = 1, \ldots, N \ . \tag{50}$$

These locations are nested (see Fig. 10) and relatively straightforward to compute. The weights are determined by interpolation conditions. $N$-point univariant Clenshaw-Curtis quadrature [15]

  - Integrates $N-1$ degree polynomials exactly when $N$ is an even number,
  - Integrates $N$ degree polynomials exactly when $N$ is an odd number,
  - Exhibits a quadrature error

$$|R_N^{CC}[f]| = \mathcal{O}(N^{-r}) \quad \text{for } f \in C^r([0,1]) \ . \tag{51}$$

In practice, the number of points is chosen by level $L$ such that $N = 2^L + 1$ is an odd number. The quadrature error $R_N^{CC}[f]$ can be accurately estimated by the forward evaluation formula that reuses all previous evaluations but requires that new evaluations be calculated

$$|R_N^{CC}[f]| \le C_N^{CC} |Q_{2N-1}^{CC}[f] - Q_N^{CC}[f]| \qquad \text{(forward estimate)} \tag{52}$$

with $C_N^{CC}$ often chosen equal to unity. Another less accurate estimate that only uses previous evaluation information is given by the backward estimate

$$|R_N^{CC}[f]| \le c_N^{CC} |Q_N^{CC}[f] - Q_{\lfloor N/2 \rfloor + 1}^{CC}[f]| \qquad \text{(backward estimate)} \tag{53}$$

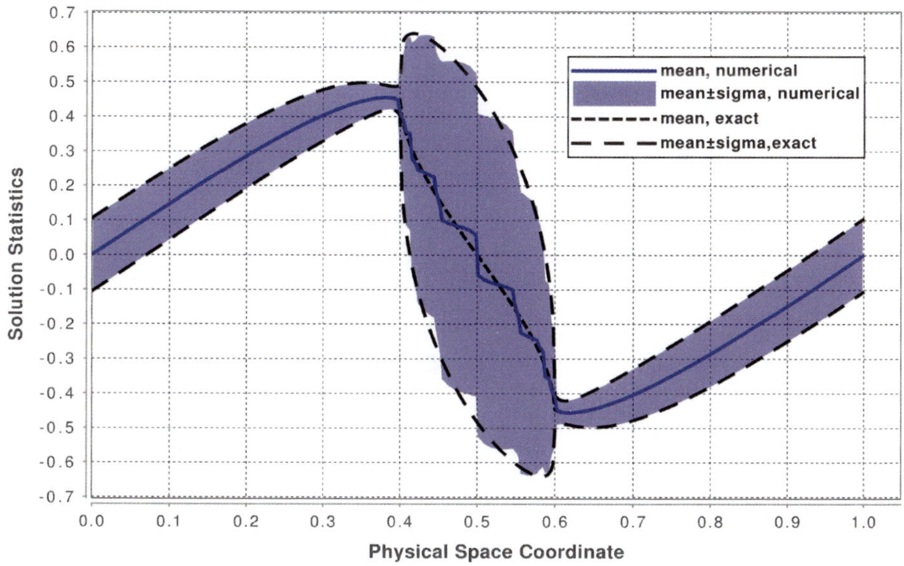


**Fig. 11** Expectation and standard deviation envelopes approximated using $N = 9$ Gauss-Kronrod quadrature points in stochastic collocation for the Burgers' equation problem (30) with phase uncertain initial data (29) at time $t = 0.35$

where now $c_N^{CC}$ is generally small and should be estimated. Assuming the ratio of $Q_N^{CC}[f]$ and $Q_{\lfloor N/2 \rfloor+1}^{CC}[f]$ Clenshaw-Curtis quadrature errors is bounded by constant depending on $N$

$$\left| \frac{R_N^{CC}[f]}{R_{\lfloor N/2 \rfloor+1}^{G}[f]} \right| \leq \kappa_N^{CC} \;, \tag{54}$$

we then follow the same path as in (46) to conclude that

$$|R_N^{CC}[f]| \leq \frac{\kappa_N^{CC}}{1-\kappa_N^{CC}} |Q_N^{CC}[f] - Q_{\lfloor N/2 \rfloor+1}^{CC}[f]| \tag{55}$$

thus permitting $c_N^{CC}$ to be written in terms of $\kappa_N^{CC}$. This constant must be estimated for specific problems.

Revisiting the Burgers' equation problem (30) assuming uniform probability density $\mathscr{U}(-0.25, 0.25)$ phase uncertain initial data (29) considered previously in Sect. 4.1.1, statistics have been approximated using $N = 9$ Gauss-Kronrod quadrature points in the stochastic collocation method as graphed in Fig. 11 and $N = 9$ Clenshaw-Curtis quadrature points in the stochastic collocation method as graphed in Fig. 12. As expected, both Gauss-Kronrod and Clenshaw-Curtis

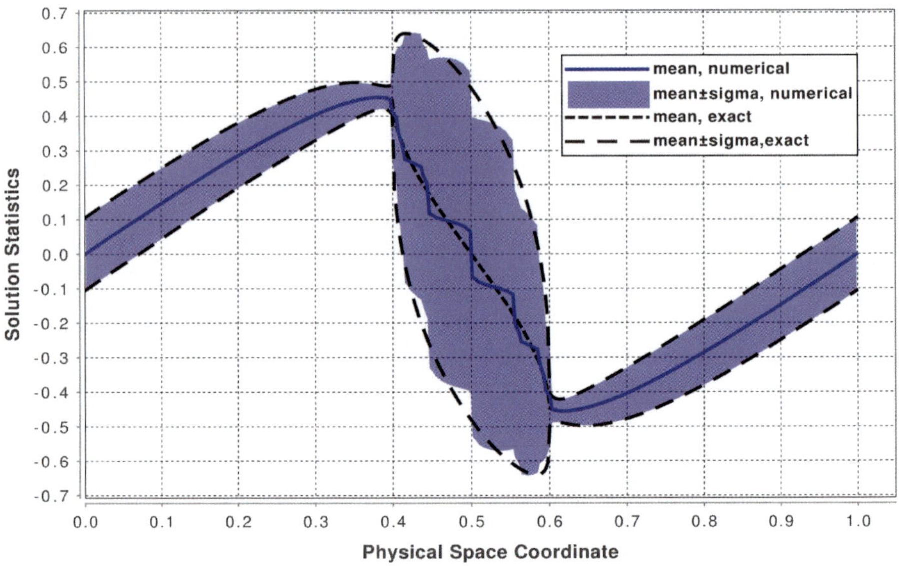


**Fig. 12** Expectation and standard deviation envelopes approximated using $N = 9$ Clenshaw-Curtis quadrature points in stochastic collocation for the Burgers' equation problem (30) with phase uncertain initial data (29) at time $t = 0.35$

quadratures yield oscillatory estimates of output statistics for the Burgers' equation problem of Sect. 4.1.1. Even so, these quadratures are extremely advantageous when error bounds are sought. The HYGAP method developed in Sect. 4.4 addresses the oscillation problem for both Gauss-Kronrod and Clenshaw-Curtis quadratures.

### 4.2.2 Sparse Grid Quadrature

Unfortunately, dense product quadratures exhibit exponential complexity growth with respect to an increasing number of dimensions. A quadrature of just 2 points in $M$ dimensions requires

$$N^{dense} = \mathcal{O}(2^M) \ , \qquad \text{(dense product quadratures)} \qquad (56)$$

evaluations. In contrast, complete polynomials of degree $P$ in $M$ dimensions contain only

$$N^{poly} = \binom{P+M}{M} \approx \frac{M^P}{P!} \ , \qquad \text{(complete polynomials)} \qquad (57)$$

degrees of freedom. This strongly indicates that dense product quadratures contain many unneeded evaluations for modest order $P$ and large dimension $M$. The sparse

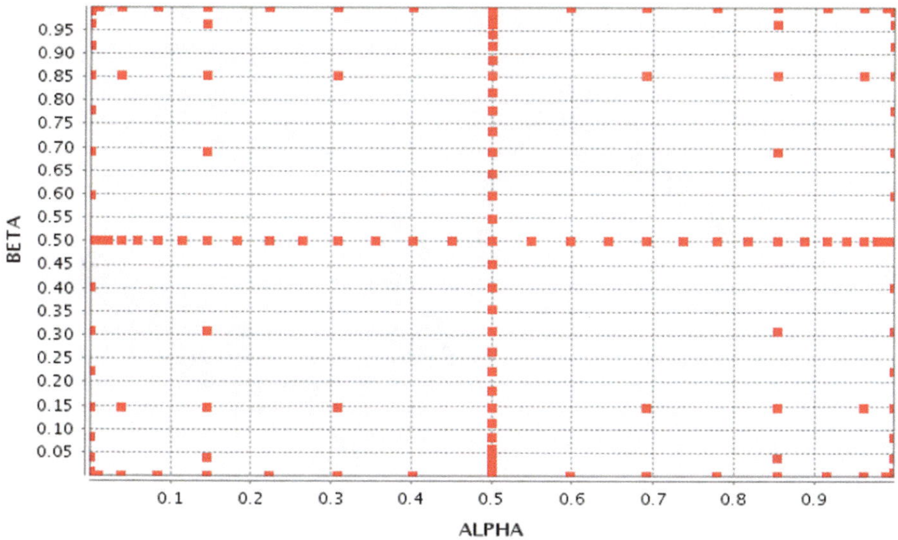


**Fig. 13** Clenshaw-Curtis Sparse Grid, $M = 2$, $L = 5$, $N = 145$

grid quadrature of Smolyak [38] addresses this inefficiency and offers a dramatic reduction in the number of evaluations required for a given precision $P$ and dimension $M$ when compared to dense product quadrature.

Let $U_i$ denote an indexed family of univariant quadrature formulas where $i$ denotes the 1-D fill level. Product formulas are compactly written in terms of a multi-index, $\boldsymbol{i} \in \mathbb{N}^M$, so that a given product rule may be written as $U_{i_1} \otimes \cdots \otimes U_{i_M}$ with product level $|\boldsymbol{i}| = \sum_{j=1}^{M} i_j$. Using this compact notation, Smolyak sparse grid quadratures with maximum level $L$ in $M$ dimensions have the form

$$SG_{L,M} = \sum_{L-M+1 \leq |\boldsymbol{i}| \leq L} (-1)^{L-|\boldsymbol{i}|} \binom{M-1}{L-|\boldsymbol{i}|} (U_{i_1} \otimes \cdots \otimes U_{i_M}) \tag{58}$$

Choosing 1-D Clenshaw-Curtis quadrature ($U_i = Q^{CC}_{2^i+1}$), sparse grid Clenshaw-Curtis (SGCC) quadrature is given by

$$Q^{SGCC}_{L,M} = \sum_{L-M+1 \leq |\boldsymbol{i}| \leq L} (-1)^{L-|\boldsymbol{i}|} \binom{M-1}{L-|\boldsymbol{i}|} (Q^{CC}_{2^{i_1}+1} \otimes \cdots \otimes Q^{CC}_{2^{i_M}+1}) \; . \tag{59}$$

Figure 13 shows an sample sparse grid for $M = 2$ and $L = 5$ requiring $N = 145$ evaluation points in sharp contrast to the dense product quadrature form which would require $N = 1{,}089$ evaluation points. Sparse grid Clenshaw-Curtis quadrature [26] attains a polynomial precision $P$ equal to $2L + 1$ and requires

$$N^{SGCC} \approx \frac{(2M)^P}{P!} \tag{60}$$

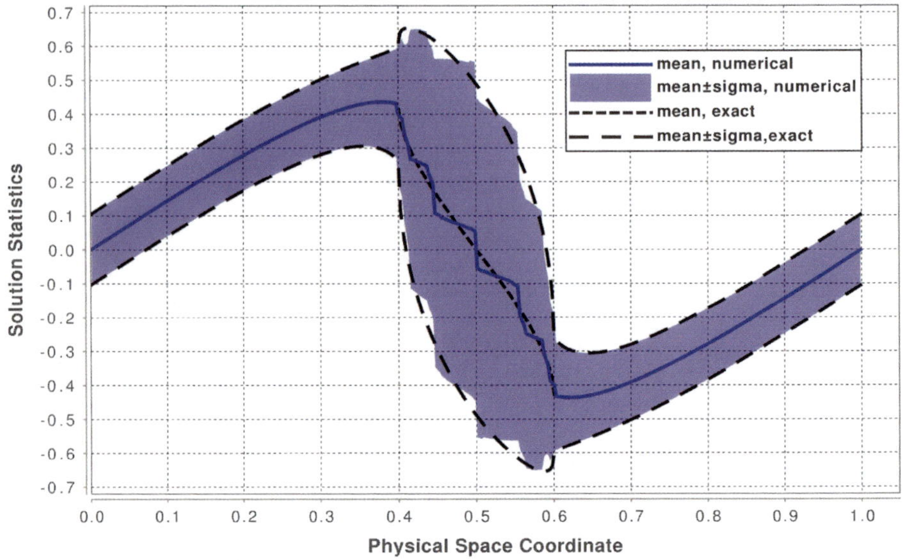


**Fig. 14** Expectation and standard deviation envelopes approximated using a 3-level 29-point Clenshaw-Curtis sparse grid quadrature for the Burgers' equation problem (30) with amplitude and phase uncertain initial data (29) at time $t = 0.35$

evaluations. This is a vast improvement over dense products (56) and differs from the use of complete polynomials (57) by a factor $2^P$. The error in sparse grid Clenshaw-Curtis quadrature, $R^{SGCC}_{L,M}[f] = I[f] - Q^{SGCC}_{L,M}[f]$, can be estimated using a forward formula using successive levels $L$ and $L+1$

$$|R^{SGCC}_{L,M}[f]| \le C^{SGCC}_{L,M} |Q^{SGCC}_{L+1,M}[f] - Q^{SGCC}_{L,M}[f]| \qquad \text{(forward estimate)} \qquad (61)$$

requiring additional evaluations of $f$ or the less accurate backward formula using levels $L$ and $L-1$

$$|R^{SGCC}_{L,M}[f]| \le c^{SGCC}_{L,M} |Q^{SGCC}_{L,M}[f] - Q^{SGCC}_{L-1,M}[f]| \qquad \text{(backward estimate)} \qquad (62)$$

for constants $C_{L,M}$ and $c_{L,M}$ that must be estimated. When discontinuities are present in random variable dimensions, sparse grid quadratures using the Smolyak formula (58) have the combined negative attributes of global approximation and negative quadrature weights. This may result in oscillations and poor accuracy. To evaluate the behavior of Clenshaw-Curtis sparse grid quadrature when discontinuities are present, statistics have been estimated for the Burgers' equation problem of Sect. 3 containing both amplitude and phase uncertainty. Specifically, a uniform probability density $\mathscr{U}(0.2, 0.8)$ has been chosen for amplitude uncertainty and a uniform probability density $\mathscr{U}(-0.25, 0.25)$ has been chosen for phase uncertainty. Exact solution contours were given previously in Figs. 3 and 4. In Fig. 14, statistics approximated from $P_2$ polynomial finite-volume method realizations using 3-level 29-point Clenshaw-Curtis sparse quadrature are graphed. The spurious oscillations are quite similar to the dense quadrature formula of Sect. 4.2.1 but this may be a consequence of the discontinuity being aligned with a coordinate dimension as depicted in Fig. 4.

## 4.3 Uncertainty Propagation Using Adaptive Piecewise Polynomial Approximation and Subscale Recovery

In this section, an alternative non-intrusive uncertainty propagation approach is described that yields a non-oscillatory approximation of output statistics when discontinuities are present in random variable dimensions. The approach utilizes a piecewise polynomial approximation but the key to success is combining this local polynomial approximation with a local subscale recovery technique.

Begin by defining a parameter response surface that is a product factorization of physical dimensions and $M$ parameter dimensions, $\boldsymbol{\xi} \in \mathbb{R}^M$

$$u_h(\boldsymbol{x},t,\boldsymbol{\xi}) = \sum_{i_1=1}^{N_1} \cdots \sum_{i_m=1}^{N_m} C_{i_1 \ldots i_M}(\boldsymbol{x},t)\, h_{i_1}(\xi_1) \ldots h_{i_M}(\xi_M) \;, \qquad N = \prod_{i=1}^{M} N_i \quad (63)$$

with the 1-D interpolants $h_i(\xi)$ satisfying the nodal interpolation property $h_i(\xi^{(j)}) = \delta_{ij}$. Evaluating (63) at $N$ interpolation points $\{\boldsymbol{\xi}^{(1)},\ldots,\boldsymbol{\xi}^{(N)}\}$ uniquely determines the coefficients $C_{i_1 \ldots i_M}(\boldsymbol{x},t)$ in terms of deterministic solutions, i.e.

$$C_{i_1 \ldots i_M}(\boldsymbol{x},t) = u_h(\boldsymbol{x},t;\xi_1^{(i_1)},\ldots,\xi_M^{(i_M)}) \;. \quad (64)$$

There is freedom in the choice of interpolation basis $h_i(\xi)$ and the interpolation points $\{\boldsymbol{\xi}^{(1)},\ldots,\boldsymbol{\xi}^{(N)}\}$. In general, we require

- High order accuracy for smooth solutions,
- Non-oscillatory approximation of discontinuities,
- Convenient calculation of statistics.

To achieve these requirements, global polynomial approximations are replaced by nonoscillatory piecewise polynomial approximations.

### 4.3.1 Piecewise Polynomial Approximation

Conceptually, each parameter dimension $\xi$ is partitioned into non-overlapping variably-spaced intervals, $\Delta\xi_{j+1/2} \equiv \xi_{j+1} - \xi_j$, $j = 1,\ldots,N-1$. In each interval, nonoscillatory piecewise polynomial approximations are constructed. In this implementation, the differential 1-D mapping $p(\xi)\,d\xi = d\mu$ depicted in Fig. 15

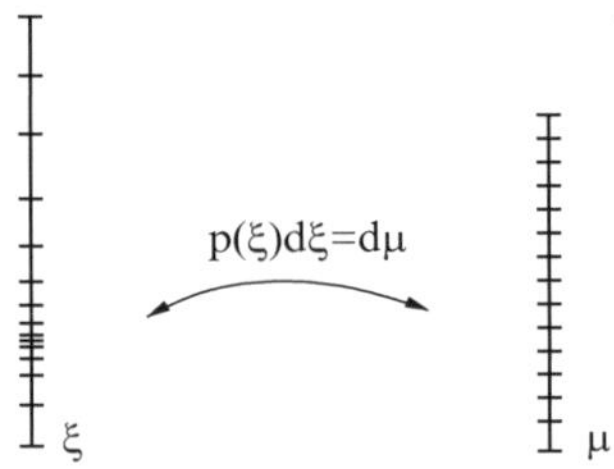


**Fig. 15** The differential mapping $p(\xi)\,d\xi = d\mu$

for a given probability density $p(\xi)$ will be used to simplify the calculation of statistics. The mapping $\mu(\xi)$ is calculated directly from the cumulative density function, $\mu(\xi) = \int_{-\infty}^{\xi} p(\eta)\,d\eta$.

### 4.3.2 Construction of Piecewise Polynomials from Pointwise Data

The present strategy is to construct an adaptive piecewise polynomial approximation from pointwise data using a variant of the weighted non-oscillatory (WENO) reconstruction algorithm developed by Jiang and Shu [17]. An excellent overview of WENO reconstruction with implementational details can be found in Shu [37]. Piecewise polynomial approximations of maximal degree $q$ in the $\xi$ coordinate have the form

$$h_{j+1/2}(\xi) = \sum_{0 \le s \le q} \tilde{h}^{(s)}_{j+1/2} \left( \frac{\xi - \overline{\xi}_j}{\Delta \xi_j} \right)^s \quad , \qquad \xi \in [\xi_j, \xi_{j+1}] \tag{65}$$

with $\overline{\xi}_j \equiv (\xi_j + \xi_{j+1})/2$.

These piecewise polynomials will eventually be used in $N$-point quadratures $Q_N[\cdot]$ of statistics for each interval $[\xi_j, \xi_{j+1}]$. Thus, the task at hand is to evaluate $h_{j+1/2}(\xi)$ at a quadrature point $QP$ in the interval $[\xi_j, \xi_{j+1}]$ using $q'$ shifted $q'$-order polynomial approximations. For simplicity, let $q$ be an odd number and $q' = (q+1)/2$. The reconstruction process is outlined below for $q = 5, q' = 3$.

Consider $q'$ shifted stencils with stencil width $q' + 1$ as shown

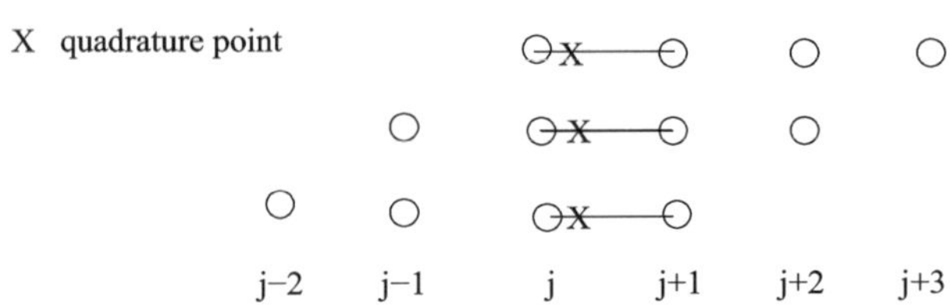

Using these stencils with nodal data $u_j$, $q'$ estimates of the $h_{j+1/2}(QP)$ can be obtained

$$\begin{cases} h^{(0)}_{j+1/2}(QP) = \sum_{i=j}^{j+3} c_i^{(0)} u_i + \mathcal{O}((\Delta\xi)^{q'+1}) \\ h^{(1)}_{j+1/2}(QP) = \sum_{i=j-1}^{j+2} c_i^{(1)} u_i + \mathcal{O}((\Delta\xi)^{q'+1}) \\ h^{(2)}_{j+1/2}(QP) = \sum_{i=j-2}^{j+1} c_i^{(2)} u_i + \mathcal{O}((\Delta\xi)^{q'+1}) \end{cases} . \tag{66}$$

A linear combination of these $q'$ stencils can then be calculated such that

$$h_{j+1/2}(QP) = \sum_{r=0}^{q'-1} d_r h^{(r)}_{j+1/2}(QP) + \mathcal{O}((\Delta\xi)^{2q'} \quad , \quad d_r > 0 \ . \tag{67}$$

To calculate the coefficients $d_r$, a $q$-order polynomial is fitted through all $q + 1$ points in the stencils. Coefficients of this polynomial are matched term-by-term to determine $d_r$. This completes the preprocessing phase of the reconstruction. In using these polynomials, the objective is to calculate modified coefficients $\tilde{d}_r$ such for smooth data

$$\tilde{d}_r = d_r + \mathcal{O}((\Delta\xi)^q) \ . \tag{68}$$

Otherwise, we would like to revert to the stencil of the candidates $h^{(r)}_{j+1/2}(QP)$, $r = 0, \ldots, q' - 1$ with smoothest polynomial approximation. Application of the reconstruction consists of the following steps:

- Calculate modified $\alpha_r$ coefficients,

$$\alpha_r = \frac{d_r}{(\beta_r + \epsilon)^2} \quad , \quad r = 0, \ldots, q' - 1 \ , \epsilon \approx 10^{-6} \ . \tag{69}$$

- Calculate the normalized coefficients $\tilde{d}_r$

$$\tilde{d}_r = \frac{\alpha_r}{\sum_{s=0}^{q'-1} \alpha_s} \tag{70}$$

where $\beta_r$ are smoothness coefficients estimated from a numerically approximated Sobolev semi-norm

$$\tilde{\beta}_r = \sum_{s=1}^{q'} \int_{\xi_i}^{\xi_{i+1}} (\Delta\xi_i)^{2s-1} \left( \frac{\partial^s p_r(\xi)}{\partial \xi^s} \right)^2 d\xi \tag{71}$$

and $p_r(\xi)$ an underlying high-order generating polynomial, i.e. the minimum order polynomial that interpolates all pointwise data $u_i$ in the r-th shifted WENO stencil such as in (66), see also Shu [37].

- Evaluate $h_{j+1/2}(QP)$ at the quadrature point $QP$,

$$h_{j+1/2}(QP) = \sum_{r=0}^{q'-1} \tilde{d}_r h^{(r)}_{j+1/2}(QP) \ .$$

### 4.3.3 Calculation of Statistics on the Response Surface

The mapping $\mu(\xi)$ enables a convenient procedure for calculating statistics. Let $\Delta\mu_{j+1/2} \equiv \mu_{j+1} - \mu_j$. Statistics integrals are then approximated as a sum of integrations on intervals $\Delta\xi_{j+1/2}$. For example on the response surface $u_h(x,t,\xi)$ in 1-D

$$\begin{aligned} E[u_h](x,t) &= \int_{\Xi} u_h(x,t,\xi) p(\xi) d\xi \\ &= \sum_{j=1}^{N-1} \int_{\Delta\xi_{j+1/2}} u_h(x,t,\xi)\, p(\xi)\, d\xi = \sum_{j=1}^{N-1} \int_{\Delta\mu_{j+1/2}} u_h(x,t,\xi(\mu))\, d\mu \ . \end{aligned} \tag{72}$$

On each interval, conventional $Q$-point Gauss-Legendre quadrature formulas with weights $w_m$ and locations $y_m$ are used

$$\int_{\Delta\mu_{j+1/2}} u_h(x,t,\xi(\mu))\, d\mu \approx \sum_{m=1}^{Q} w_m u_h(x,t,\xi(\mu(y_{j+1/2,m})))\Delta\mu_{j+1/2} \ . \tag{73}$$

Observe from (40) that $Q$-point Gauss-Legendre quadratures integrate $2Q - 1$ polynomials exactly. So that variances are accurately approximated using $q$-order piecewise polynomials, the number of quadrature points $Q$ is chosen such that (at a minimum) $q^2$-order piecewise polynomials are integrated exactly. The final quadrature formula is then given by

$$Q_N[E[u_h]](x,t) = \sum_{j=1}^{N-1} \sum_{m=1}^{Q} w_m u_h(x,t,\xi(\mu(y_{j+1/2,m})))\Delta\mu_{j+1/2} \ . \tag{74}$$

An estimate of the quadrature error is given by either the forward formula using $N$ and $2N - 1$ interpolation points for odd $N$

$$|R_N[E[u_h]]| = C_{q,N} |Q_{2N-1}[E[u_h]] - Q_N[E[u_h]]| \tag{75}$$

or the backward formula using $N$ and $N/2 + 1$ interpolation points

$$|R_N[E[u_h]]| = c_{q,N} |Q_N[E[u_h]] - Q_{N/2+1}[E[u_h]]| \tag{76}$$

for estimated constants $c_{q,N}$ and $C_{q,N}$.

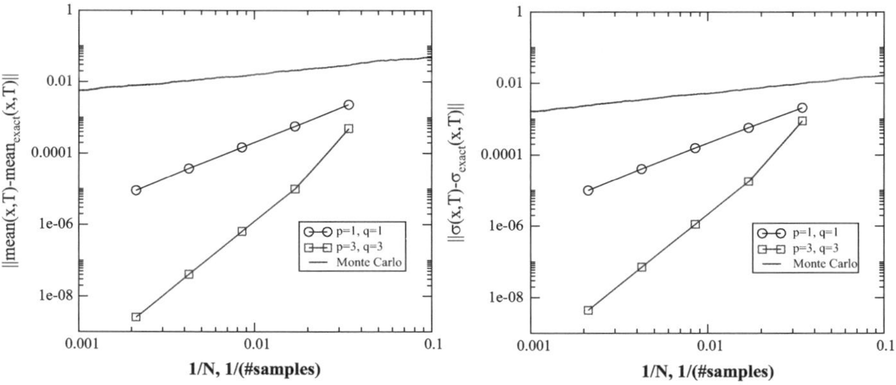


**Fig. 16** Graphs of the $L_2$ norm error in the mean$(x, T)$ (*left*) and $\sigma^2(x, T)$ (*right*) statistics of the numerical solution of the Burgers' equation problem (30) with phase uncertain initial data (29) at time $t = 0.15$ with comparison to Monte Carlo sampling (averaged over 100 simulations)

### 4.3.4 Smooth Solution Accuracy

The Burgers' equation problem (30) with phase uncertain initial data (29) described in Sect. 4.1.1 has been used to assess accuracy of the formulation using $p$-order piecewise polynomials in space and $q$-order piecewise polynomials in the phase coordinate. Nodal interpolation points $\{\xi^{(1)}, \ldots, \xi^{(N)}\}$ in the phase coordinate direction have been placed at the Clenshaw-Curtis quadrature point locations (50). So that accuracy of the methods can be accurately measured, calculations have been performed up to the time $t = 0.15$ such that a discontinuity has not yet formed in the domain. Figure 16 show graphs of the expectation and variance error for various mesh resolutions. The graphs show nearly optimal rates of convergence using linear ($p = q = 1$) and cubic ($p = q = 3$) polynomials. For reference, Monte Carlo sampling [24] using uniformly distributed samples (averaged over 100 simulations) shows approximately $(\#samples)^{-1/4}$ rate of convergence which is suboptimal with respect to the $(\#samples)^{-1/2}$ rate of convergence for classical Monte Carlo integration.

### 4.3.5 Resolution of Discontinuities

Figures 8 and 9 in Sect. 4.1.1 demonstrated the poor performance of the stochastic collocation method when discontinuities are present in random variable dimensions. Calculations for the Burgers' equation problem (30) with uniform probability density $\mathscr{U}(-0.25, 0.25)$ phase uncertain initial data at the time $t = 0.35$ are now repeated using a ($p = 2, q = 3$) piecewise polynomial approximation space with $N = 9$ points. These calculations use the same number of degrees of freedom in the physical and parameter coordinates as was used in Figs. 8 and 9. Solution

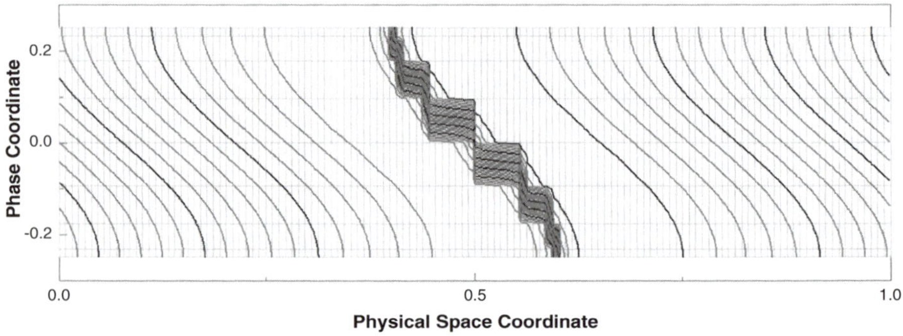


**Fig. 17** Solution contours obtained using piecewise polynomial approximation ($N = 9$) for the Burgers' equation solution (30) with phase uncertain initial data (29) at time $t = 0.35$

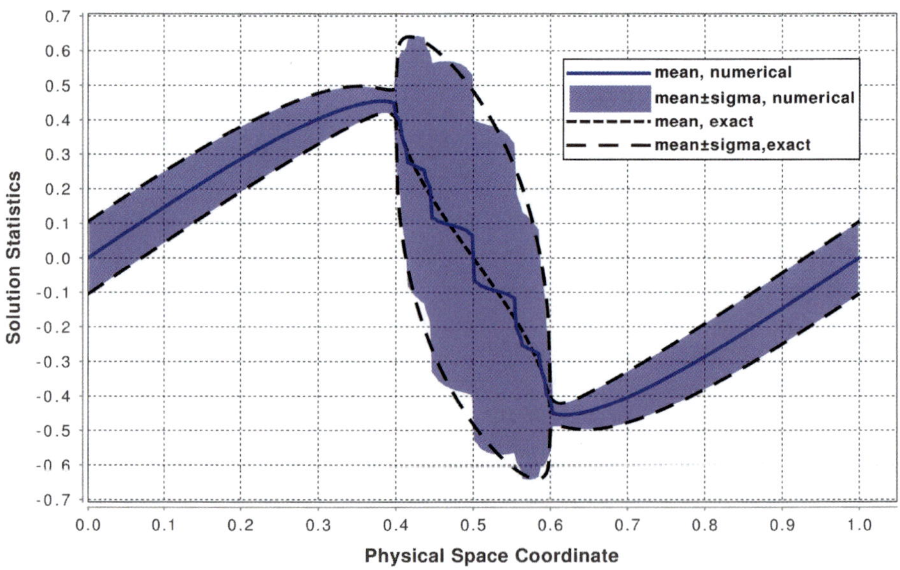


**Fig. 18** Expectation and standard deviation envelopes approximated using piecewise polynomial approximation ($N = 9$) for the Burgers' equation problem (30) with phase uncertain initial data (29) at time $t = 0.35$

contours and the resulting statistics calculated using piecewise polynomial approximation are given in Figs. 17 and 18. The piecewise polynomial results are notable improvements over the stochastic collocation method with regard to reducing or eliminating spurious oscillations in solution contours in Fig. 17. In particular, the use of piecewise polynomial approximation yields localization of non-smoothness to a region immediately surrounding the discontinuity in contrast to stochastic collocation which exhibits global oscillations. Unfortunately, the statistics graphed in Fig. 18 show almost no improvement over the stochastic collocation statistics graphed previously in Fig. 9. To understand this lack of improvement, note that the non-oscillatory piecewise polynomial approximation near the true discontinuity still suffers from $\mathscr{O}(1)$ error. This error is highly dependent on the discontinuity location within mesh spacing intervals and the choice of tensor-product interpolation on a cartesian mesh. Consequently, the calculation of statistics involving an integration over the parameter dimension exhibits a spurious stair-stepped behavior. To obtain further improvement without performing more realizations of Burgers' equation requires an improved estimation of information at scales smaller than the local mesh spacing when discontinuities are present.

### 4.3.6 Subscale Recovery

The poor approximation properties at or near discontinuities observed in Fig. 17 is a consequence of the choice of coordinate-dependent tensor-product interpolation and the lack of information at scales smaller than the mesh spacing. This missing information could be obtained by performing further realizations for output quantities of interest but these additional realizations may be prohibitively expensive or otherwise unavailable at the time uncertainty calculations are performed. An alternative approach is to use the realization information at hand but in a more meaningful way. To do this, we first revisit a related continuous problem. Consider a domain $\Omega$ embedded into a background field as depicted in Fig. 19 with prescribed boundary data on $\partial\Omega$ obtained from the background field. Without further information, there are an infinite number of ways to extend this boundary data into the interior of $\Omega$, e.g. harmonic extension, algebraic interpolation, etc. Therefore, a plausible selection principle is sought that yields a computationally efficient extension. This extension can be viewed as a form of interpolation. Extensions based on the minimization of function total variation have proven useful in image processing [7, 36] where the extension task is referred to as "inpainting". The minimization problem for total variation

$$\min_u \left( \int_\Omega |\nabla u| \, dx \right) \tag{77}$$

yields the well-known Euler-Lagrange curvature equation

$$\nabla \cdot \frac{\nabla u}{|\nabla u|} = 0 \ . \tag{78}$$

Thus, solving (78) in a domain $\Omega$ together with imposed Dirichlet boundary data on $\partial\Omega$ serves as our means for extending boundary data into the domain interior. Attractive features of this selection principle include

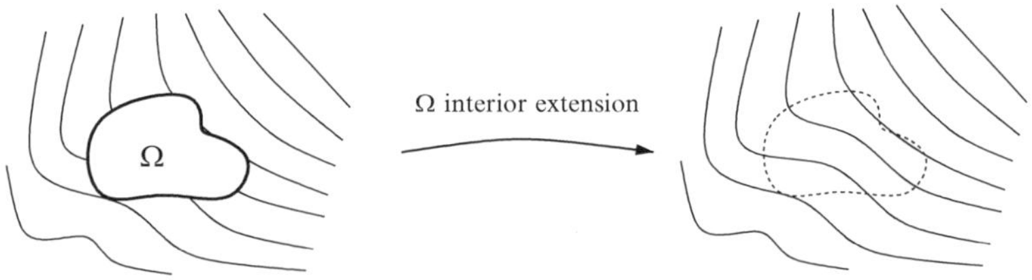


**Fig. 19** A domain $\Omega$ of missing information embedded in contours of a background field (*left*). Given boundary data on $\partial\Omega$, the interior of $\Omega$ is calculated and contoured by solving an auxiliary partial differential equation (*right*)

- Coordinate-free representation,
- Preserves discontinuities in data,
- Easily numerically approximated.

Because this technique will be used to fill in scales smaller that the mesh size, the technique is referred to herein as "subscale recovery" via the curvature equation (78).

Finite-difference formulas approximating (78) on cartesian meshes are readily constructed. Consider for example a 2-D cartesian mesh with index pairs $(i, j)$ and mesh points $(x_i, y_j)$, a second-order accurate finite-difference discretization of (78) takes the following form for all interior mesh points $(i, j) \in$ mesh

$$\begin{aligned} & c^{(x)}_{i+1/2,j}(u_{i+1,j} - u_{i,j}) + c^{(x)}_{i-1/2,j}(u_{i-1,j} - u_{i,j}) \\ & + c^{(y)}_{i,j+1/2}(u_{i,j+1} - u_{i,j}) + c^{(y)}_{i,j-1/2}(u_{i,j+1} - u_{i,j}) = 0 \end{aligned} \tag{79}$$

with solution dependent non-negative coefficients

$$c^{(x)}_{i\pm1/2,j} = \frac{1}{|\nabla u|_{i\pm1/2,j}|x_{i,j} - x_{i\pm1,j}|}, \quad c^{(y)}_{i,j\pm1/2} = \frac{1}{|\nabla u|_{i,j\pm1/2}|y_{i,j} - y_{i,j\pm1}|} \tag{80}$$

with the midpoint gradient magnitudes $|\nabla u|_{i\pm1/2,j}$ and $|\nabla u|_{i,j\pm1/2}$ approximated so that constant gradient data yields zero numerical curvature. When individual midpoint gradients vanish, the stencil is adjusted so that the remaining coefficients remain bounded, see also Sethian [35]. Numerical solutions obtained using schemes of the form (79) with bounded non-negative coefficients exhibit a discrete local minimum-maximum principle, i.e.

$$\min(u_{i\pm1,j}, u_{i,j\pm1}) \le u_{i,j} \le \max(u_{i\pm1,j}, u_{i,j\pm1}) \qquad \text{(min-max principle)}$$

and thus are non-oscillatory discrete interior extensions. Another consequence of the finite-difference form (79) with bounded non-negative coefficients is that the discrete equations are amenable to solution using standard relaxation iterative methods such as Gauss-Seidel iteration used in the present implementation.

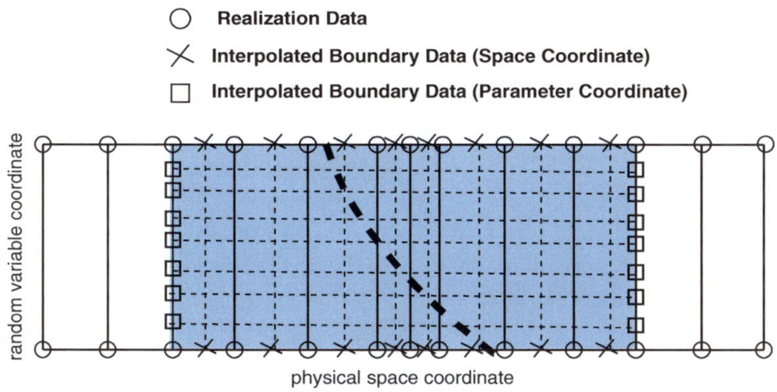


**Fig. 20** Locally refined region for use in local subscale recovery. *Circle* points denote boundary data from realizations, $x$ points denote boundary data obtained from spatial reconstruction, and *square* points denote from reconstruction in parameter dimensions. *Dashed lines* depict a locally refined mesh for use in solving the curvature equation (78)

The basic strategy is to replace tensor-product interpolation with subcell recovery via curvature equation whenever data in physical and/or uncertain parameter dimensions lack smoothness, e.g. near discontinuities and large gradients. Specifically, for each interval in a parameter dimension that is found to be non-smooth using the smoothness measure (71), a refinement domain surrounding this interval is formed as depicted in Fig. 20. The refinement domain is then slightly enlarged so that left and right boundary data (square points) may be interpolated from smooth data using the piecewise polynomial reconstructions in the parameter dimensions. Along the top and bottom boundaries of the refinement domain in Fig. 20, boundary data at circles is obtained from realization data. Data at $x$ points is obtained from the spatial data reconstruction used in the finite-volume discretization. Equation (78) is then discretized and solved using the finite-difference approximation (79). Note that this refined solution does not alter the value of the realization data at circles. The locally refined data calculated at mesh points is then used to enrich the pointwise data used in the piecewise polynomial reconstruction of Sect. 4.3.2 and the calculation of statistics quadratures. In summary, the subscale recovery algorithm consists of the following steps:

**Subscale Recovery Algorithm ($C_s$, $M_x$, $M_\xi$):**

1. Mark intervals $[\xi_i, \xi_{i+1}]$ in each parameter dimension for local refinement (shaded intervals in Fig. 21) based on the smoothness indicator (71)

$$\tilde{\beta}_i = \max_{\forall r} \sum_{s=1}^{q/2+1} \int_{\xi_i}^{\xi_{i+1}} (\Delta\xi_i)^{2s-1} \left( \frac{\partial^s p_{r,i}(\xi)}{\partial \xi^s} \right)^2 d\xi \ .$$

with $p_{r,i}(\xi)$ defined as in (71). An interval is deemed non-smooth if $\tilde{\beta}_i > C_s$.

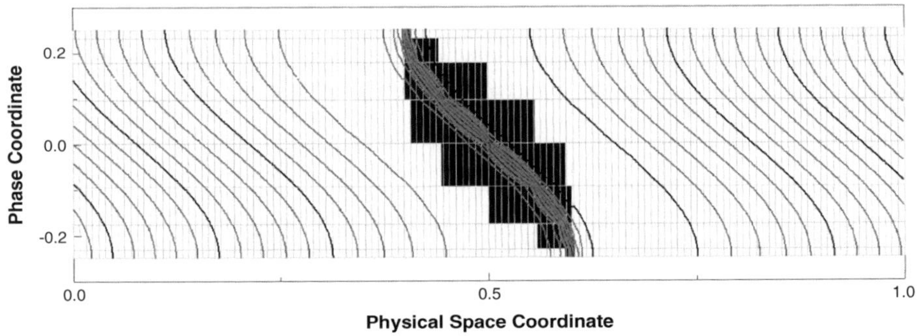


**Fig. 21** Solution contours obtained from the piecewise polynomial approximation ($N = 9$) with subscale recovery ($C_s = 1, M_x = 2, M_\xi = 8$) for the Burgers' equation problem (30) with phase uncertain initial data (29) at time $t = 0.35$. *Shaded horizontal strips* show the subdomain regions postprocessed using subscale recovery

2. Form local refinement domains by grouping together contiguously marked non-smooth intervals.
3. Enlarge the local refinement domain in physical space dimensions so that data along parameter coordinate boundaries (square points in Fig. 20) may be interpolated using piecewise polynomial reconstructions in the parameter dimensions from realization data with smoothness measure $< C_s$.
4. Locally enrich refinement domains by introducing $M_\xi$ new mesh points in each parameter interval and $M_x$ new mesh points in each physical dimension interval, e.g. $M_x = 1,\ M_\xi = 7$ in Fig. 20.
5. Interpolate and impose boundary data for each refinement domain.
6. Solve the curvature equation (78) with imposed boundary data in each refinement domain using the finite-difference scheme (79).
7. Evaluate statistics quadratures using the piecewise polynomial interpolation of Sect. 4.3.2 with enriched pointwise data coming from the locally refined domains.

Subscale recovery ($C_s = 1, M_x = 2, M_\xi = 8$) results using the curvature equation (78) are shown in Figs. 21 and 22 for the Burgers' equation problem (30) with uniform probability density $\mathscr{U}(-0.25, 0.25)$ phase uncertain initial data at the time $t = 0.35$. The resulting statistics are now nonoscillatory and agree well with the exact statistics. This is a significant advancement over the previous results in Figs. 17 and 18.

Observe that due to the use of dense product grids in subscale recovery, the computational cost of subscale recovery grows exponentially with the number of parameter dimensions. In the present applications, the number of uncertain parameters is rather small (typically $<5$). For practical engineering problems the cost of deterministic realizations requiring computational fluid dynamic (CFD) simulation is usually quite large thus making the cost of subscale recovery relatively insignificant given the substantial improvement in approximated statistics.

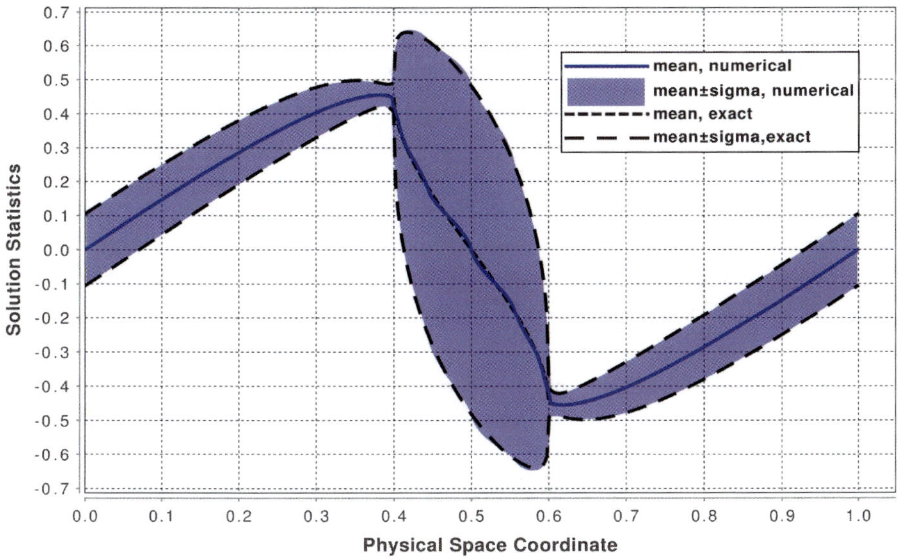


**Fig. 22** Expectation and standard deviation envelopes approximated using a piecewise polynomial approximation ($N = 9$) with subscale recovery ($C_s = 1, M_x = 2, M_\xi = 8$) for the Burgers' equation problem (30) with phase uncertain initial data (29) at time $t = 0.35$

## *4.4 The HYGAP Algorithm*

The excellent performance of the stochastic collocation method for smooth data and the superior performance of the piecewise adaptive polynomial approximation with subscale recovery for non-smooth data suggests a hybrid algorithm which uses either method depending on an assessment of smoothness of the data. We refer to this as the HYbrid Global and Adaptive Polynomial (HYGAP) algorithm.

**HYGAP Algorithm:**

1. Choose $N$ parameter evaluation points $\{\xi^{(1)}, \ldots, \xi^{(N)}\}$ for the stochastic collocation method at either (1) optimal quadrature points for specific probability densities or (2) node-nested quadratures (e.g. Gauss-Kronrod, Clenshaw-Curtis) when error bounds are sought.
2. Evaluate the outputs of interest $J(\boldsymbol{u}_h(\cdot;\xi);\xi)$ at collocation points

$$\{J(\boldsymbol{u}_h(\cdot;\xi^{(1)});\xi^{(1)}), \ldots, J(\boldsymbol{u}_h(\cdot;\xi^{(N)});\xi^{(N)})\}$$

and optionally estimates of the output of interest error for realizations

$$\{|J(\boldsymbol{u}(\cdot;\xi^{(1)});\xi^{(1)}) - J(\boldsymbol{u}_h(\cdot;\xi^{(1)});\xi^{(1)})|, \ldots, |J(\boldsymbol{u}(\cdot;\xi^{(N)});\xi^{(N)}) - J(\boldsymbol{u}_h(\cdot;\xi^{(N)});\xi^{(N)})|\} \ .$$

3. Determine smoothness of the outputs of interest using the Sobolev semi-norm estimate previously encountered in Eq. (71)

$$\beta = \max_{\forall r,i} \sum_{s=1}^{q} \int_{\xi_i}^{\xi_{i+1}} (\Delta \xi_i)^{2s-1} \left( \frac{\partial^s p_{r,i}(\xi)}{\partial \xi^s} \right)^2 d\xi$$

with $p_{r,i}(\xi)$ defined as in (71).
4. Depending on the smoothness estimate $\beta$, evaluate statistics integrals for that parameter dimension using either the stochastic collocation approximation or the piecewise adaptive polynomial approximation with subscale recovery $(C_s, M_x, M_\xi)$.

- $\beta > C_s$: Apply the adaptive piecewise polynomial reconstruction algorithm of Sect. 4.3 with local subscale recovery of Sect. 4.3.6 to evaluate output statistics integrals.
- $\beta < C_s$: Apply the stochastic collocation method of Sect. 4.1 to evaluate output statistics integrals.

The non-dimensional constant $C_s$ can be chosen $O(1)$ for problems containing sharp discontinuities in parameter dimensions. For problems not containing discontinuities but with very steep gradients not well resolved by stochastic collocation, choosing $C_s$ substantially smaller than unity activates the use of piecewise polynomials and subcell recovery in these high gradient regions.

### 4.4.1 HYGAP Calculations with Estimated Error Bounds

Revisiting the Burgers' equation problem (30) with uniform probability density $\mathscr{U}(-0.25, 0.25)$ phase uncertain initial data at the time $t = 0.35$, statistics have been approximated from $P_2$ polynomial finite-volume method realizations using a HYGAP approximation ($N = 9$) at Gauss-Kronrod quadrature points as graphed in Fig. 23. When the HYGAP algorithm adaptively chooses piecewise polynomial approximation, subscale recovery ($C_s = 1, M_x = 2, M_\xi = 8$) has been used. The statistics graphed in Fig. 23 are non-oscillatory with good agreement with the exact solution statistics. It is instructive to compare the accuracy of statistics approximated using the HYGAP and stochastic collocation methods as graphed in Fig. 24. In this figure, an interval is delimited wherein the underlying Burgers' equation solution with phase uncertainty is discontinuous in the uncertain phase coordinate. In this interval, piecewise polynomial approximation with subcell recovery is used in the HYGAP approximation. Exterior to this interval, the HYGAP and stochastic collocation methods are identical. In this interval but away from the interval boundaries, peak errors in expectation (mean) have been reduced approximated 1/2-order magnitude. Slightly less pronounced improvements in variance are also observed. At the boundary of this interval, the accuracy of the piecewise polynomial approximation is reduced due to the discontinuity intersecting the boundary as depicted in Fig. 21.

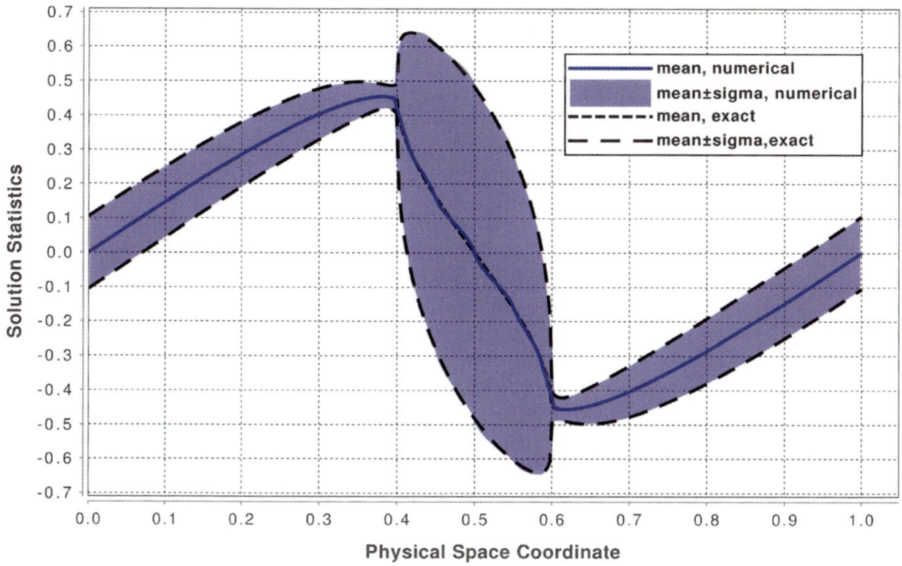


**Fig. 23** Expectation and standard deviation envelopes approximated using a HYGAP approximation ($N = 9$) at Gauss-Kronrod quadrature points for the Burgers' equation problem (30) with phase uncertain initial data (29) at time $t = 0.35$

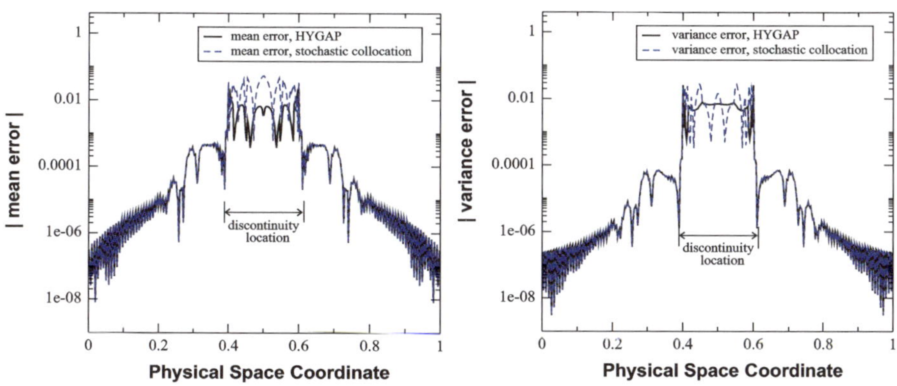


**Fig. 24** Comparison of expectation (mean) and variance errors calculated using HYGAP approximation ($N = 9$) at Gauss-Kronrod quadrature points and stochastic collocation approximation ($N = 9$) for the Burgers' equation problem (30) with phase uncertain initial data (29) at time $t = 0.35$

In addition, estimated error bounds using (24) and (25) have been included in the calculations. The use of Gauss-Kronrod mesh spacing in the phase coordinate permits an estimate of quadrature errors. Specifically, the backward formula (43) has been used with the associated constant estimated from (49) with constant

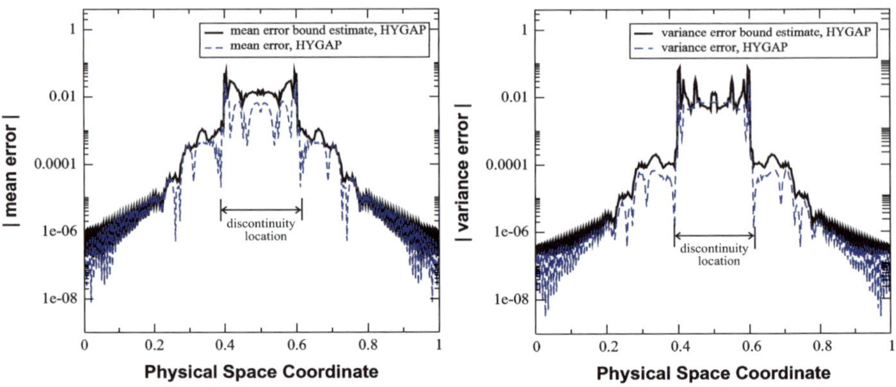


**Fig. 25** Exact and estimated error bounds for the variance statistic calculated using a HYGAP approximation ($N = 9$) at Gauss-Kronrod quadrature points for the Burgers' equation problem (30) with phase uncertain initial data (29) at time $t = 0.35$

$C_{BF} = 1.0$ resulting in an overall extrapolation constant of $\kappa^{GK} = 0.0065$ for $N = 9$ point quadrature. When the HYGAP algorithm adaptively chooses piecewise polynomial approximation, the piecewise polynomial quadrature estimate (76) has been used with a constant chosen for non-smooth data with value $c_{q,N} = 1.0$. For purposes of evaluating the derived error bounds for statistics, the error in realizations, $\{|u(x,t;\xi^{(1)}) - u_h(x,t;\xi^{(1)})|, \ldots, |u(x,t;\xi^{(N)}) - u_h(x,t;\xi^{(N)})|\}$, has been exactly specified in this example. In practice, these errors could be estimated using the methods described in Sect. 2.3. Figure 25 graphs error bound estimates for statistics obtained using (24) and (25) with comparison to the exact errors in statistics. In this figure, an interval wherein the underlying stochastic solution in the phase coordinate is discontinuous has been included. Exterior to this interval, the underlying stochastic solution is smooth and the estimated error bounds for approximated statistics appears to approximate the exact error in statistics quite well. In this smooth solution region, these approximate error bounds are reliable in the sense they approach true error bounds under uniform mesh refinement in physical and parameter dimensions. Inside this interval, the error bound estimates given herein remain, at this point, only approximate and are not provably reliable under uniform mesh refinement. Although the quadrature error estimates contain constants, the constants chosen appear suitable over the entire graph.

## 5 Further Selected Applications

Several example calculations are now presented to further verify the non-oscillatory behavior of the HYGAP algorithm. When the HYGAP algorithm adaptively selects piecewise polynomial approximation, $P_3$ cubic piecewise polynomials with subscale recovery ($C_s = 1, M_x = 2, M_\xi = 8$) are used.

## 5.1 The Propagation of Functional Uncertainty with Estimated Error Bounds for Subsonic Compressible Euler Flow

In Sect. 4.4.1, output statistics with estimated error bounds were previously calculated for the Burgers' equation problem with the error in deterministic realizations, $|u(x,t,\xi^{(i)}) - u_h(x,t,\xi^{(i)})|, i = 1,\ldots,N$, specified from knowledge of the exact solution. In this section, the error in outputs of interest representable as functionals will be calculated for compressible Euler flow using an a posteriori error estimation procedure.

### 5.1.1 Finite-Element Approximation of the Euler Equations

The Euler equations of compressible flow for mass, momentum, and energy in $d$ space dimensions are given by

$$\frac{\partial}{\partial t}\boldsymbol{u} + \sum_{i=1}^{d} \boldsymbol{f}_i = 0 \ , \quad \boldsymbol{u} = \begin{pmatrix} \rho \\ \rho u_j \\ E \end{pmatrix} \ , \qquad \boldsymbol{f}_i = \begin{pmatrix} \rho u_i \\ \rho u_i u_j + \delta_{ij} p \\ u_i(E+p) \end{pmatrix} \tag{81}$$

with $\gamma$-law gas equation of state

$$p = (\gamma - 1)\left(E - \frac{1}{2}\rho \sum_{i=1}^{d} u_i^2\right) \tag{82}$$

where $\boldsymbol{u}$ is the vector of mass, momentum, and energy, $\boldsymbol{f}$ is the inviscid flux. The fluid density is denoted by $\rho$, velocities by $u_i$, pressure by $p$, and total energy by $E$.

The two-dimensional flow equations have been approximated in space-time using the discontinuous Galerkin finite-element method [9, 21, 30] formulated in symmetrization variables $\boldsymbol{v}$ that symmetrize the Euler equations after the change of variable $\boldsymbol{u} \mapsto \boldsymbol{v}$. Consider a domain $\Omega$ with boundary $\Gamma$. Assume the domain has been tessellated with a mesh $\mathscr{T}$ composed of nonoverlapping spatial elements $K$ that are orthogonally extruded a time interval $I^n$ thus forming space-time prisms, $K \times I^n$. The union of $N_T$ time intervals discretizes the total time interval $[0,T]$, i.e. $\cap_{n=0}^{N_T-1} I^n = [0,T]$. The discontinuous Galerkin method in symmetrization variables utilizes a piecewise polynomial approximation space in the symmetrization variables given by

$$\mathscr{V}^h = \left\{\boldsymbol{v}_h \mid \boldsymbol{v}_h|_{K\times I^n} \in \left(\mathscr{P}_k(K) \times \mathscr{P}_k(I^n)\right)^m\right\} \ .$$

The space-time discontinuous Galerkin method using symmetrization variables [3] is then stated in the following weak formulation.

**Space-Time Discontinuous Galerkin Method:** Find $\boldsymbol{v}_h \in \mathcal{V}^h$ such that

$$B(\boldsymbol{v}_h, \boldsymbol{w}_h)_{\mathrm{DG}} \equiv \sum_{n=0}^{N_T-1} B^{(n)}(\boldsymbol{v}_h, \boldsymbol{w}_h)_{\mathrm{DG}} = 0 \ , \qquad \forall\, \boldsymbol{w}_h \in \mathcal{V}^h \tag{83}$$

with

$$\begin{aligned} B^{(n)}(\boldsymbol{v}, \boldsymbol{w})_{\mathrm{DG}} = & \int_{I^n} \sum_{K \subset \mathcal{T}} \int_K -(\boldsymbol{u}(\boldsymbol{v}) \cdot \boldsymbol{w}_{,t} + \sum_{i=1}^{d} \boldsymbol{f}_i(\boldsymbol{v}) \cdot \boldsymbol{w}_{,x_i})\, dx\, dt \\ & + \int_{I^n} \sum_{K \in \mathcal{T}} \int_{\partial K \cap \Gamma = 0} \boldsymbol{w}(x_-) \cdot \boldsymbol{h}(\boldsymbol{v}(x_-), \boldsymbol{v}(x_+); \boldsymbol{n})\, ds\, dt \\ & + \int_{I^n} \sum_{K \in \mathcal{T}} \int_{\partial K \cap \Gamma \neq 0} \boldsymbol{w}(x_-) \cdot \boldsymbol{h}_{bc}(\boldsymbol{v}(x_-)); \boldsymbol{n})\, ds\, dt \\ & + \int_{\Omega} \left( \boldsymbol{w}(t_-^{n+1}) \cdot \boldsymbol{u}(\boldsymbol{v}(t_-^{n+1})) - \boldsymbol{w}(t_+^n) \cdot \boldsymbol{u}(\boldsymbol{v}(t_-^n)) \right) dx \end{aligned} \tag{84}$$

where $\boldsymbol{h}(\cdot, \cdot; \boldsymbol{n})$ denotes a numerical flux and $\boldsymbol{h}_{bc}(\cdot; \boldsymbol{n})$ denotes a numerical flux including boundary conditions, see for example [3] when symmetrization variables are used.

### 5.1.2 A Posteriori Error Estimation of Functionals via Dual Problem

When a deterministic output of interest $J(\boldsymbol{u}_h(\cdot; \boldsymbol{\xi}); \boldsymbol{\xi})$ is a functional, the task of estimating the finite-dimensional approximation error is greatly simplified using the a posteriori error estimation theory developed by Eriksson et al. [11] and Becker and Rannacher [4]. This theory was applied to the discontinuous Galerkin finite-element method using symmetrization variables $\boldsymbol{v}$ in [18]. The abstract a posteriori error estimation theory for Galerkin approximations consists of the following steps:

1. Solve the primal numerical problem using finite-dimensional approximation spaces. In the abstract formulation, $F(\boldsymbol{v}_h) : \mathcal{V}^h \mapsto \mathbb{R}^m$ is a forcing term (equal to zero in the present calculations) and boundary conditions are assumed to be enforced weakly via fluxes.
   *Primal numerical problem:* Find $\boldsymbol{v}_h \in \mathcal{V}^h$ such that
   $$B(\boldsymbol{v}_h, \boldsymbol{w}_h) = F(\boldsymbol{w}_h) \ , \quad \forall\, \boldsymbol{w}_h \in \mathcal{V}^h \ .$$
2. Solve the mean-value linearized auxiliary dual problem $\overline{B}(\cdot, \cdot)$ using infinite-dimensional spaces given a mean-value linearized functional $\overline{J}$.

*Linearized auxiliary dual problem:* Find $\Phi \in \mathscr{V}$ such that

$$\overline{B}(\boldsymbol{w}, \Phi) = \overline{J}(\boldsymbol{w}) \ , \quad \forall\, \boldsymbol{w} \in \mathscr{V} \ .$$

3. Compute the error in a functional using the error representation formula derived from the following steps

$$\begin{aligned}
J(\boldsymbol{u}) - J(\boldsymbol{u}_h) &= \overline{J}(\boldsymbol{u} - \boldsymbol{u}_h) && \text{(mean value } J\text{)} \\
&= \overline{B}(\boldsymbol{u} - \boldsymbol{u}_h, \Phi) && \text{(dual problem)} \\
&= \overline{B}(\boldsymbol{u} - \boldsymbol{u}_h, \Phi - \pi_h \Phi) && \text{(Galerkin orthogonality)} \\
&= B(\boldsymbol{u}, \Phi - \pi_h \Phi) - B(\boldsymbol{u}_h, \Phi - \pi_h \Phi) && \text{(mean value } B\text{)} \\
&= F(\Phi - \pi_h \Phi) - B(\boldsymbol{u}_h, \Phi - \pi_h \Phi) && \text{(primal problem)}
\end{aligned}$$

where $\pi_h$ denotes any projection into the Galerkin test space (e.g. $L_2$ projection, interpolation). This yields the final error representation formula

$$J(\boldsymbol{u}) - J(\boldsymbol{u}_h) = F(\Phi - \pi_h \Phi) - B(\boldsymbol{u}_h, \Phi - \pi_h \Phi) \ . \tag{85}$$

The mean-value linearization given by the theory requires knowledge of the infinite-dimensional primal solution. In addition, solutions of the dual problem are posed in infinite-dimensional spaces. These solutions are generally not available and must be approximated. In the present computations, the mean-value linearization has been replaced by the Jacobian (tangent) linearization at the numerical solution state and the dual problem has been solved numerically using an approximation space that is one polynomial order higher than the approximation space of the primal numerical problem, i.e. $\mathscr{P}_{k+1}(K) \times \mathscr{P}_{k+1}(I^n)$. This permits the right-hand-side of (85) to be estimated.

The error representation formula (85) may be used to estimate the functional error. However, if the estimated error is too large, the formula does not provide information how the mesh or approximation space should be modified to further reduce the error. Element-wise decomposition of the error representation formula provides a pathway for deriving element refinement indicators and a systematic means for reducing the error by the refinement of elements. To simplify the notation, let $Q^n = K \times I^n$ denote a space-time prism. Observe that without further approximation, the error representation formula can be written as a sum over space-time elements

$$|J(\boldsymbol{u}) - J(\boldsymbol{u}_h)| = \left| \sum_{n=0}^{N_T - 1} \sum_{Q^n} F_{Q^n}(\Phi - \pi_h \Phi) - B_{Q^n}(\boldsymbol{v}_h, \Phi - \pi_h \Phi) \right| \tag{86}$$

where $B_{Q^n}(\cdot,\cdot)$ and $F_{Q^n}(\cdot)$ are the restriction of $B^{(n)}(\cdot,\cdot)$ and $F^{(n)}(\cdot)$ to a single space-time element. Application of the generalized triangle inequality provides a localized estimate of the contribution of each space-time element to the total error in the functional

$$|J(\boldsymbol{u}) - J(\boldsymbol{u}_h)| \leq \sum_{n=0}^{N_T-1} \sum_{Q^n} \underbrace{\left| F_{Q^n}(\Phi - \pi_h \Phi) - B_{Q^n}(\boldsymbol{v}_h, \Phi - \pi_h \Phi) \right|}_{\text{refinement indicator, } \eta(Q^n)} . \quad (87)$$

These localized estimates serve as *refinement indicators* for mesh adaptivity. A commonly used strategy in mesh adaptivity, used herein, is to refine a fixed fraction of element indicators $\eta(Q^n)$ that are too large and coarsen a fixed fraction of element indicators that are too small.

### 5.1.3 Euler Flow over a Multi-element Airfoil with Uncertainty Propagation and Estimated Error Bounds

Steady-state Euler flow with inflow Mach number 0.1 has been computed over a multi-element airfoil geometry using the discontinuous Galerkin finite-element method of Sect. 5.1.1. The inflow angle of attack (AOA) is assumed uncertain with truncated Gaussian probability density, AOA = Gaussian$_{4\sigma}(m = 5°, \sigma = 1°)$. In these calculations, the output of interest is the aerodynamic lift coefficient

$$\text{lift coefficient} = \frac{1}{\rho_\infty \sum_{i=1}^{d} u_{i,\infty}^2 \, L} \int_{\text{surface}} (\boldsymbol{n} \cdot \boldsymbol{l}) \, p \, ds$$

where $(\cdot)_\infty$ are reference inflow conditions, $L$ a reference airfoil chord length, and $\boldsymbol{l}$ is a unit vector orthogonal to the incoming free stream flow. The a posteriori error estimation procedure of Sect. 5.1.2 has been used to calculate the approximation error for realizations of the lift force coefficient thereby permitting error bounds for output statistics to be approximated. A single realization of the primal solution using $\mathcal{P}_2$ elements and the dual solution using $\mathcal{P}_3$ elements is shown in Fig. 26. In addition, the a posteriori error estimation procedure has been used to construct mesh refinement indicators for adaptive mesh refinement. Three levels of adaptive refinement have been used in the calculation of realizations with the 18,000 element mesh shown in Fig. 27. HYGAP uncertainty propagation using $N = 9$ Gauss-Kronrod quadrature points has been used to calculate uncertainty statistics for the aerodynamic lift coefficient functional with error bounds using (24) and (25). In these error bound formulas, Gauss-Kronrod quadrature error has been estimated using the estimation formula (43) with constant approximated from (49) with $C_{BF} = 1.0$. Due to the smooth behavior of the output functional with respect to the uncertain angle of attack parameter, no piecewise polynomial approximation was required for these calculations. For ease of presentation, let

$$\Delta_N E[J(\boldsymbol{u})] \equiv E[J(\boldsymbol{u})] - Q_N[E[J(\boldsymbol{u}_h)]] \ , \quad \Delta_N V[J(\boldsymbol{u})] \equiv V[J(\boldsymbol{u})] - Q_N[V[J(\boldsymbol{u}_h)]]$$

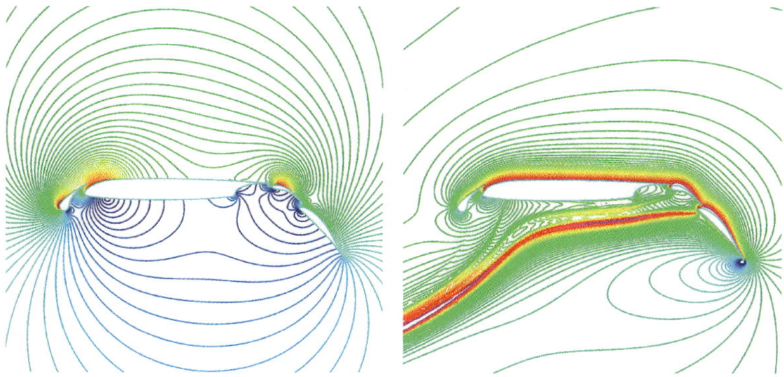

**Fig. 26** A single realization of subsonic Euler flow over a multi-element airfoil geometry with inflow Mach number 0.1 and 5° angle of attack. Shown are contours of Mach number associated with the primal solution (*left*) and contours of the $x$-momentum associated with the dual solution for a lift functional. *Blue color* denotes small values and *red color* denotes large values

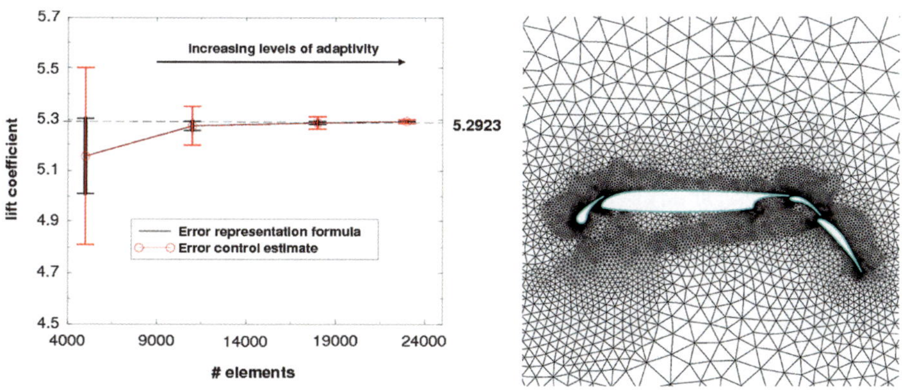


**Fig. 27** Adaptive mesh refinement for the 5° realization. The estimated error in lift coefficient functional using estimates (86) and (87) is graphed as vertical error bars for 3 levels of mesh refinement (*left*) and the resulting adapted mesh with 18,000 elements after 2 levels of adaptive refinement is plotted (*right*)

denote the errors in approximated expectation and variance, respectively. Using this notation, Table 1 summarizes lift coefficient statistics with error bounds during the adaptive mesh refinement process. Rapid reduction in the estimated error bounds is achieved during mesh refinement but eventually the fixed $N = 9$ Gauss-Kronrod quadratures should dominate and prevent further reduction unless $N$ is increased. This table shows the importance of an accurate estimation of realization errors for outputs of interest. The error indicator formula (87) is clearly not as accurate as the error representation formula (86) and consequently the approximated bounds on statistics are much larger.

**Table 1** Approximated statistics and error bounds for the aerodynamic lift coefficient functional. Tabulated are the computed estimates of expectation and variance together with error bounds using realization errors from either (86) or (87)

| level | # elements | $E[J(u_h)]$ | $V[J(u_h)]$ | Eq. (86) $\Delta_N E[J(u_h)]$ | Eq. (86) $\Delta_N V[J(u_h)]$ | Eq. (87) $\Delta_N E[J(u_h)]$ | Eq. (87) $\Delta_N V[J(u_h)]$ |
|---|---|---|---|---|---|---|---|
| 0 | 5,000 | 5.145 | 0.01157 | 0.147 | 0.05619 | 0.346 | 0.20112 |
| 1 | 11,000 | 5.274 | 0.01188 | 0.018 | 0.00462 | 0.076 | 0.02390 |
| 2 | 18,000 | 5.286 | 0.01191 | 0.006 | 0.00240 | 0.024 | 0.00630 |
| 3 | 32,000 | 5.292 | 0.01192 | 0.002 | 0.00048 | 0.007 | 0.00172 |

## 5.2 NACA0012 Airfoil Transonic Euler Flow

Transonic Euler flow past a NACA0012 airfoil with inflow Mach number 0.8 and inflow angle of attack of 2.26° has been calculated using a finite-volume approximation with $P_1$ polynomial MUSCL reconstruction [19, 20] in space. Flow field Mach number contours and the surface pressure coefficient distribution are shown in Fig. 28. A strong upper surface shock wave is clearly observed. Inflow Mach number uncertainty was then introduced into the calculation, $M_\infty = \text{Gaussian}_{4\sigma}(m = 0.8, \sigma = 0.01)$, and a series of computations were performed using $N = 4, 5$, and 6 collocation points. Field contours of Mach number statistics (mean and $\log_{10}$ variance) are shown in Figs. 29 and 30 using stochastic collocation (left) and HYGAP (right) with $N = 5$ collocation points. These contour plots show the dramatic improvement using HYGAP approximation. In regions of the flow field with smooth solution behavior, the HYGAP algorithm reverts to the stochastic collocation approximation so the resulting contours in both figures are identical. In the non-smooth region of the flow field, the collocation points have been reinterpreted as pointwise values for use in piecewise polynomial approximation with subscale recovery. Figure 31 shows graphs of surface pressure coefficient using stochastic collocation (left) and the HYGAP approximation (right) with $N = 4, 5$, and 6 Gauss-Hermite collocation points. The spurious "stair stepping" oscillations in the stochastic collocation results are a direct consequence of the shock discontinuity obliquely traversing through both physical and random variable dimensions as seen previously in Fig. 8 for Burgers' equation. Due to this effect, as $N$ increases the number of spurious "steps" increases in the stochastic collocation results. It is rather surprising to see that excellent results are obtained with HYGAP using as little as $N = 4$ evaluation points.

## 5.3 The Reynolds-Averaged Navier-Stokes Applications

The Reynolds-averaged Navier-Stokes (RANS) equations for a compressible ideal gas in $d$ space dimensions are given by

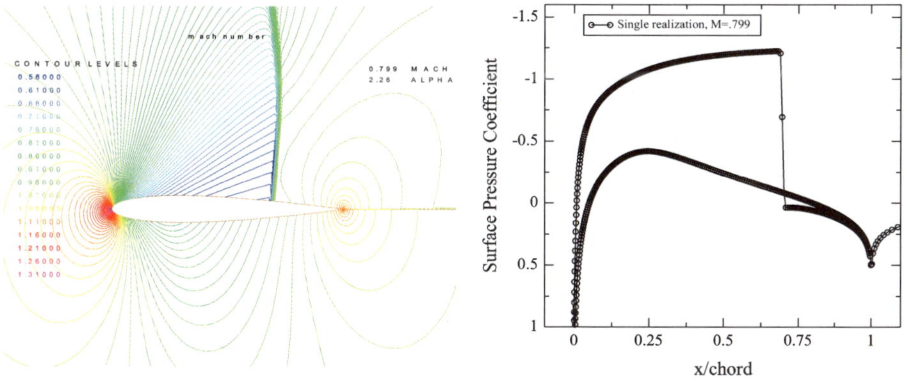


**Fig. 28** Transonic flow over a NACA airfoil using a single value of the free stream Mach number equal to 0.8. Shown are Mach number contours (*left*) and a graph of the surface pressure coefficient (*right*) exhibiting a strong upper surface shock wave

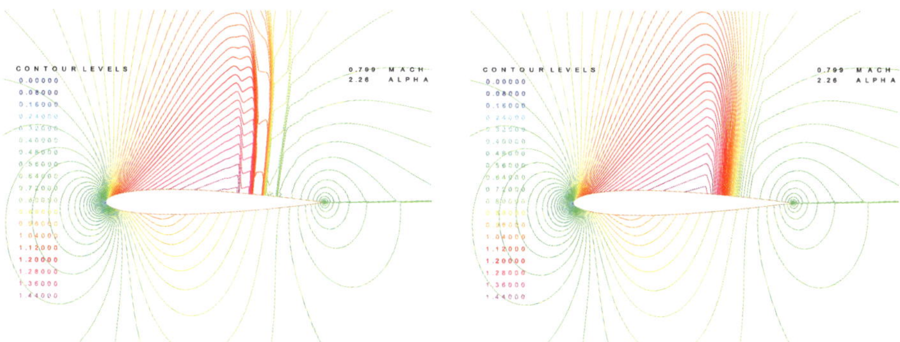


**Fig. 29** Transonic flow over a NACA0012 airfoil with $M_\infty$ = Gaussian$_{4\sigma}(m = 0.8, \sigma = 0.01)$ uncertainty. Shown are mean Mach number contours using $N = 5$ stochastic collocation approximation (*left*) and the HYGAP approximation (*right*)

$$\frac{\partial}{\partial t}\boldsymbol{u} + \sum_{i=1}^{d} \boldsymbol{F}_i - \frac{1}{\mathrm{Re}} \sum_{i=1}^{d} \boldsymbol{G}_i = 0 \tag{88}$$

$$\boldsymbol{u} = \begin{pmatrix} \rho \\ \rho u_j \\ E \end{pmatrix}, \quad \boldsymbol{F}_i = \begin{pmatrix} \rho u_i \\ \rho u_i u_j + \delta_{ij} p \\ u_i (E + p) \end{pmatrix}, \quad \boldsymbol{G}_i = \begin{pmatrix} 0 \\ \boldsymbol{\tau}_{ij} \\ u_k \tau_{ik} - q_i \end{pmatrix}$$

with Fourier heat flux and Newtonian fluid shear stress given by

$$q_i = -\frac{\gamma}{\gamma - 1}\left(\frac{\mu}{\mathrm{Pr}} + \frac{\mu_T}{\mathrm{Pr_T}}\right)\frac{\partial T}{\partial x_i}, \quad \boldsymbol{\tau}_{ij} = (\mu + \mu_T)\left(\left(\frac{\partial u_i}{\partial x_j} + \frac{\partial u_j}{\partial x_i}\right) - \frac{2}{3}\frac{\partial u_k}{\partial x_k}\delta_{ij}\right)$$

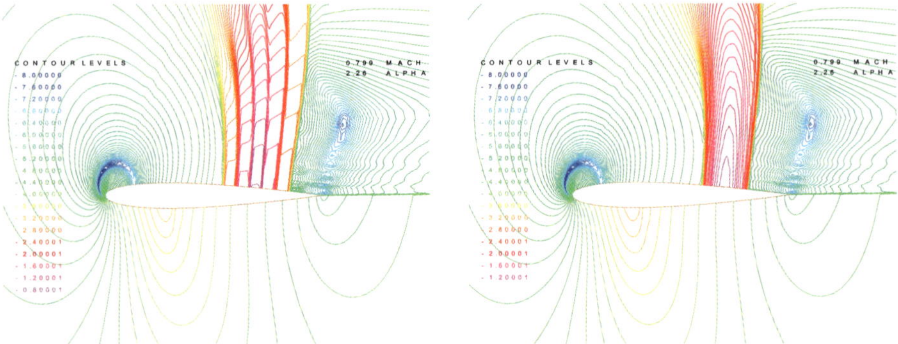

**Fig. 30** Transonic flow over a NACA0012 airfoil with $M_\infty = \text{Gaussian}_{4\sigma}(m = 0.8, \sigma = 0.01)$ uncertainty. Shown are $\log_{10}$ (variance Mach number) contours using $N = 5$ stochastic collocation approximation (*left*) and the HYGAP approximation (*right*)

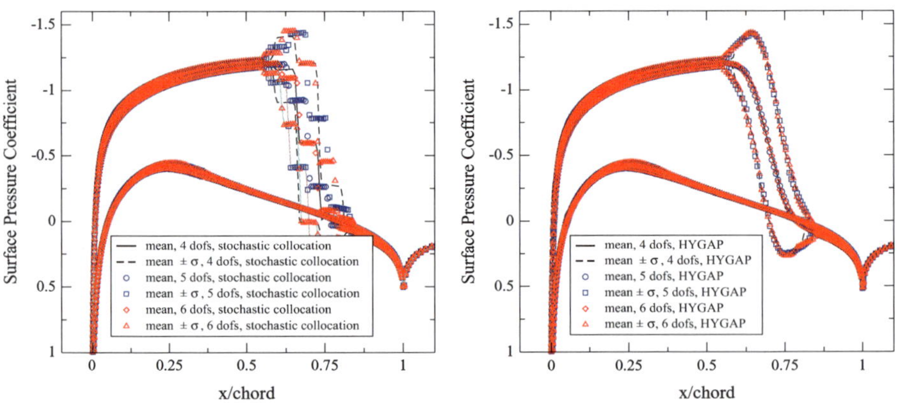


**Fig. 31** Transonic flow over a NACA0012 airfoil with $M_\infty = \text{Gaussian}_{4\sigma}$ $(m = 0.8, \sigma = 0.01)$ uncertainty. Shown are graphs of surface pressure coefficient mean and standard deviation envelopes approximated using $N = 4, 5$, and 6 stochastic collocation approximation (*left*) and the HYGAP approximation (*right*)

and $\gamma$-law equation of state

$$p = (\gamma - 1)\left(E - \frac{1}{2}\rho \sum_{i=1}^{d} u_i^2\right) .$$

In these equations, $\boldsymbol{u}$ is the vector of mass, momentum, and energy, $\boldsymbol{F}$ is the inviscid flux, and $\boldsymbol{G}$ the viscous flux. The fluid density is denoted by $\rho$, velocities by $u_i$, temperature by $T$, pressure by $p$, and total energy by $E$. In these equations, $\gamma$ is the ratio of specific heats, $R$ the ideal gas law constant, $Pr$ and $Pr_T$ the molecular and turbulent Prandtl numbers, and $\mu$ and $\mu_T$ the molecular and turbulent viscosities.

The non-dimensional molecular viscosity is calculated from Sutherland's law

$$\frac{\mu}{\mu_\infty} = \left(\frac{T}{T_\infty}\right)^{3/2} \frac{1 + S/T_\infty}{T/T_\infty + S/T_\infty}$$

where $S/T_\infty$ is a Sutherland's law parameter.

### 5.3.1 Baldwin-Barth Turbulence Model

To complete the Reynolds-averaged Navier-Stokes model, the turbulent viscosity $\mu_T$ must be provided. A simple one-equation turbulence model was introduced by Baldwin and Barth [2] in 1990. The Baldwin-Barth model is a one-equation PDE of the form

$$\frac{DR_T}{Dt} = (c_{\epsilon_2} f_2(y^+) - c_{\epsilon_1})\sqrt{R_T P} + (\nu + \frac{\nu_T}{\sigma})\nabla^2 R_T - \frac{1}{\sigma}(\nabla \nu_t) \cdot \nabla R_T$$

where

$$\mu_T = \rho \nu_T = \rho c_\mu D_1(y^+) D_2(y^+) R_T \ , \quad P = \nu_T \left(\frac{\partial u_i}{\partial x_j} + \frac{\partial u_j}{\partial x_i}\right)\frac{\partial u_i}{\partial x_j} - \frac{2}{3}\nu_T \left(\frac{\partial u_k}{\partial x_k}\right)^2$$

$$D_1 = 1 - \exp(-y^+/A^+) \ , \quad D_2 = 1 - \exp(-y^+/A_2^+), \quad \sigma^{-1} = (c_{\epsilon_2} - c_{\epsilon_1})\sqrt{c_\mu}/\kappa^2$$

and

$$f_2(y^+) = \frac{c_{\epsilon_1}}{c_{\epsilon_2}} + (1 - \frac{c_{\epsilon_1}}{c_{\epsilon_2}})(\frac{1}{\kappa y+} + D_1 D_2)$$
$$\times \left(\sqrt{D_1 D_2} + \frac{y+}{\sqrt{D_1 D_2}}(\frac{1}{A+}\exp(-y^+/A^+)D_2 + \frac{1}{A_2^+}\exp(-y^+/A_2^+)D_1)\right)$$

with model parameter values $\kappa = 0.41, c_\mu = 0.08, c_{\epsilon_1} = 1.2, c_{\epsilon_1} = 2.0, A^+ = 26$, and $A_2^+ = 10$ as given in [2].

### 5.3.2 NACA0012 Airfoil Transonic RANS Flow

Steady-state Reynolds-averaged Navier-Stokes flow at a Reynolds number of $9\times 10^6$ using a Baldwin-Barth turbulence model has been calculated over a NACA0012 airfoil geometry using a highly resolved $512 \times 64$ cell mesh. Experimental data by Harris [13] is available for comparison. Transonic flow with inflow Mach number 0.8 and angle of attack (AOA) 2.26° was chosen because previous calculations were

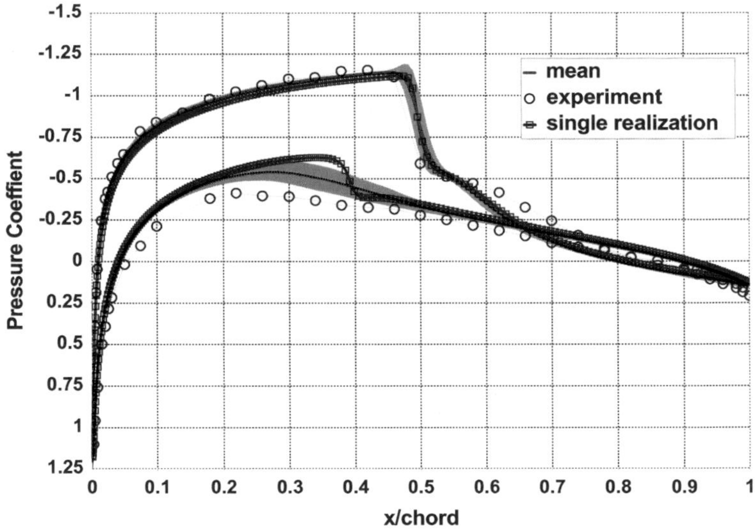


**Fig. 32** Transonic RANS flow over a NACA airfoil with inflow and turbulence model parameter uncertainty, $\{M_\infty, AOA, c_\mu, c_{\epsilon_2}\}$ parameter uncertainty. Surface pressure coefficient statistics are approximated using the HYGAP approximation ($N = 6 \times 6 \times 6 \times 6$) with subscale recovery. The *shaded band* denotes a one standard deviation envelope

anecdotally observed to be very sensitive to the choice of inflow and turbulence model parameters. Numerical calculations include

- The propagation of uncertain inflow data assuming truncated Gaussian probability density statistics
  - $M_\infty = \text{Gaussian}_{4\sigma}(m = 0.8, \sigma = 0.008)$,
  - $AOA = \text{Gaussian}_{4\sigma}(m = 2.26°, \sigma = 0.1°)$,
- The propagation of the most sensitive uncertain turbulence model parameters assuming uniform probability density statistics
  - $c_\mu = \mathscr{U}[0.0855, 0.0945]$,
  - $c_{\epsilon_2} = \mathscr{U}[1.9, 2.1]$.

The effects of inflow and turbulence model parameter uncertainty, $\{M_\infty, AOA, c_\mu, c_{\epsilon_2}\}$, have been simultaneously analyzed using a $N = 6 \times 6 \times 6 \times 6$ HYGAP approximation with subscale recovery. Pressures coefficient statistics on the airfoil surface are graphed in Fig. 32 in terms of mean surface pressure coefficient together with a shaded standard deviation envelope. Significant uncertainties are predicted near the upper surface shock wave location that extend further downstream of the shock wave. This is a very typical behavior for high speed compressible flows containing shock waves. For the particular flow conditions chosen, this figure also

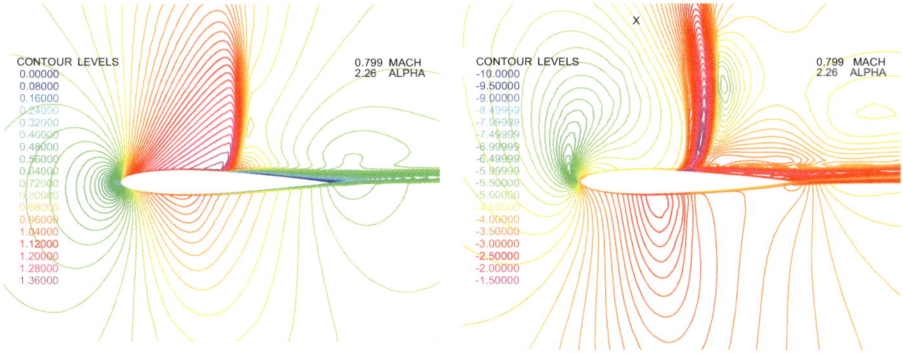


**Fig. 33** Contours of mean Mach number (*left*) and $\log_{10}$(variance Mach number) (*right*) calculated using the HYGAP approximation ($N = 6 \times 6 \times 6 \times 6$) with subscale recovery. *Blue color* denotes low values and *red color* denotes high values

shows the surprisingly large uncertainty in lower surface (lower curve) pressure coefficient values. To explain this lower surface uncertainty, Fig. 32 also graphs the surface pressure coefficient distribution for a single realization of the 4 uncertain parameters. This particular realization reveals a weak lower surface shock wave that is sometimes present when the uncertain parameters are varied. The presence of a lower surface shock wave significantly changes the solution and arguably explains the relatively large uncertainty in the lower surface pressure distribution. This uncertainty information may well explain the difficulties historically encountered in numerically matching the experimental data at these flow conditions. Contours of the solution Mach number statistics are shown in Fig. 33 (left). The thickening of the shock wave in this figure is a consequence of shock position uncertainty. In Fig. 33 (right) contours of $\log_{10}$ (variance Mach number) are also shown. These contours reveal that the regions near the upper surface shock wave, the lower surface near maximum airfoil thickness, and the flow inside the separation bubble down stream of the shock wave are all regions of relatively high uncertainty.

### 5.3.3 ONERA M6 Wing

Steady-state Reynolds-averaged Navier-Stokes flow past an ONERA M6 wing has been calculated using the OVERFLOW [16] Chimera grid Reynolds-averaged Navier-Stokes solver. The mesh system for this geometry consists of 81 overlapping 3D meshes containing approximately 6 million degrees of freedom. The inflow Mach number is 0.84 and the Reynolds number is $11.72 \times 10^6$. Calculations have been performed at 3.06° and 6° angle of attack (AOA). Experimental data by Schmidtt and Charpin [34] is available for comparison. Density and pressure coefficient contours of the numerical solution at $M_\infty = 0.84$, $AOA = 3.06°$

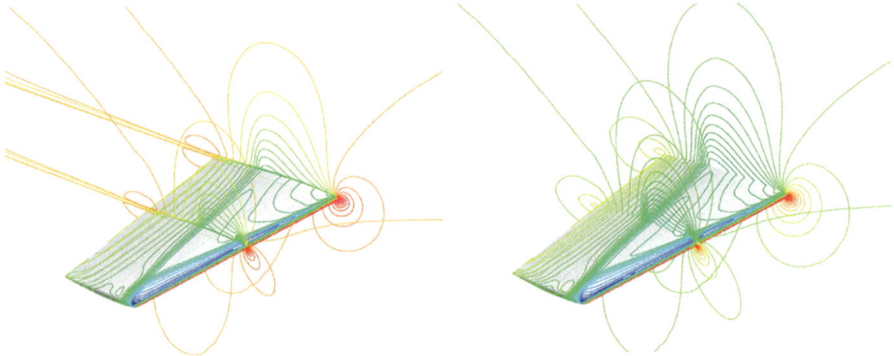

**Fig. 34** Transonic RANS flow over the ONERA M6 wing at Mach 0.84 and AOA = 3.06°. Shown are single realization density contours (*left*) and pressure coefficient contours (*right*). *Blue color* denotes low values and *red color* denotes high values

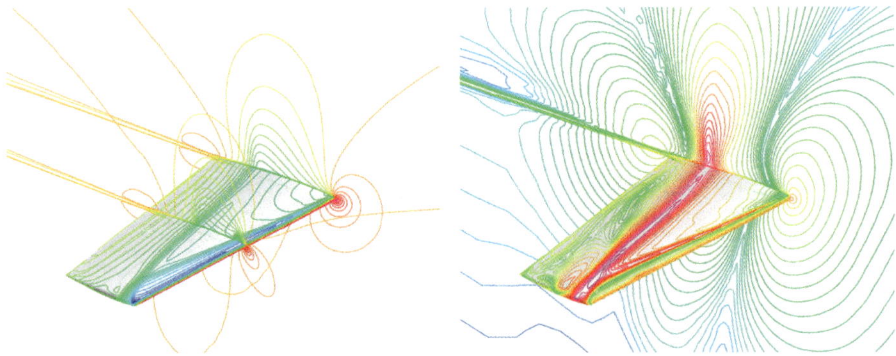

**Fig. 35** Transonic RANS flow over the ONERA M6 wing with inflow Mach number and angle of attack uncertainty, $M_\infty = \text{Gaussian}_{4\sigma}(m = 0.84, \sigma = 0.02)$ and $\text{AOA} = \text{Gaussian}_{4\sigma}(m = 3.06°, \sigma = 0.075°)$. Shown are surface and cutting plane contours of mean density (*left*) and $\log_{10}$(variance density) (*right*) calculated using a HYGAP approximation ($N = 6 \times 6$) with subscell recovery. *Blue color* denotes low values and *red color* denotes high values

(no uncertainty) are presented in Fig. 34. The well-known lambda-shock pattern is clearly seen on the upper wing surface.

**Case 1:** ONERA M6 wing RANS flow at 3.06° angle of attack

Uncertainty in the inflow Mach number and angle of attack (AOA) have been introduced into the analysis, $M_\infty = \text{Gaussian}_{4\sigma}(m = 0.84, \sigma = 0.02)$ and $\text{AOA} = \text{Gaussian}_{4\sigma}(m = 3.06°, \sigma = 0.075°)$. Uncertainty statistics have been estimated using a HYGAP approximation ($N = 6 \times 6$) with subscale recovery. Figure 35 shows contours of the mean and $\log_{10}$ (variance density) field on the surface of the wing and at cutting planes. Experimental data from Schmidtt and Charpin includes pressure coefficient distributions at various span stations on the

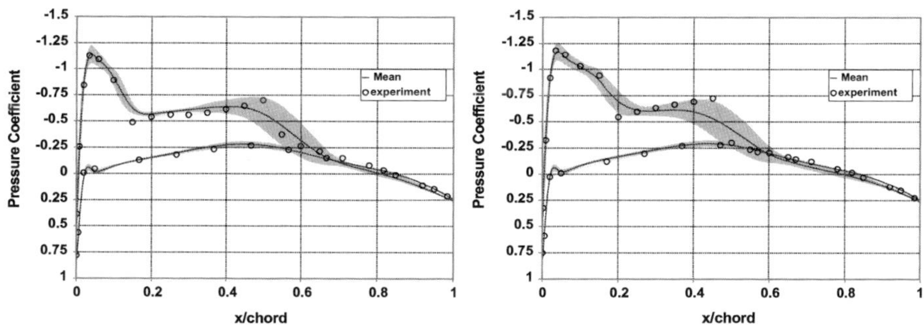


**Fig. 36** Transonic RANS flow over the ONERA M6 wing with free stream Mach number and angle of attack uncertainty, $M_\infty = \text{Gaussian}_{4\sigma}(m = 0.84, \sigma = 0.02)$ and $\text{AOA} = \text{Gaussian}_{4\sigma}(m = 3.06°, \sigma = 0.075°)$. Surface pressure coefficient statistics calculated using a HYGAP approximation ($N = 6 \times 6$) with subscell recovery are graphed at the 44 % span station (*left*) and the 65 % span station (*right*). The *shaded band* denotes a one standard deviation envelope

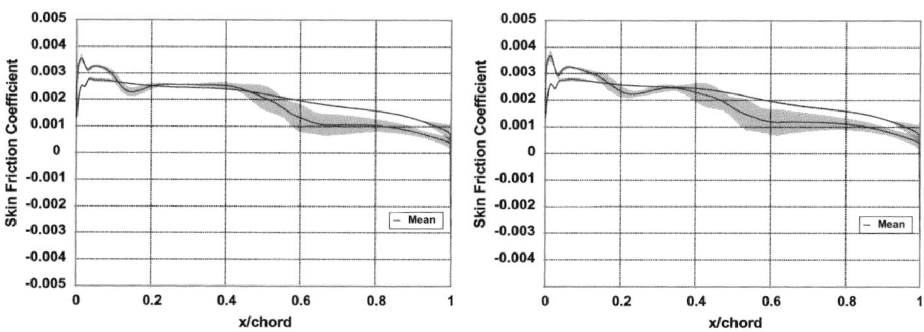


**Fig. 37** Transonic RANS flow over the ONERA M6 wing with free stream Mach number and angle of attack uncertainty, $M_\infty = \text{Gaussian}_{4\sigma}(m = 0.84, \sigma = 0.02)$ and $\text{AOA} = \text{Gaussian}_{4\sigma}(m = 3.06°, \sigma = 0.075°)$. Graphs of skin friction coefficient statistics calculated using a HYGAP approximation ($N = 6 \times 6$) with subscell recovery are graphed at the 44 % span station (*left*) and the 65 % span station (*right*). The *shaded band* denotes a one standard deviation envelope

M6 wing. This experimental data is included with uncertainty estimations in Fig. 36. Observe that the experimental data falls within one standard deviation of the approximated mean values. Skin friction coefficient statistics are also graphed at these wing span stations in Fig. 37. These graphs show significant uncertainty in skin friction coefficient at the lambda-shock locations with mild uncertainty downstream of the shock waves.

**Case 2:** ONERA M6 wing RANS flow at 6.0° angle of attack

For purposes of comparison, we have increased the mean angle of attack from 3.06° to 6°. Once again, uncertainty statistics have been estimated using a HYGAP

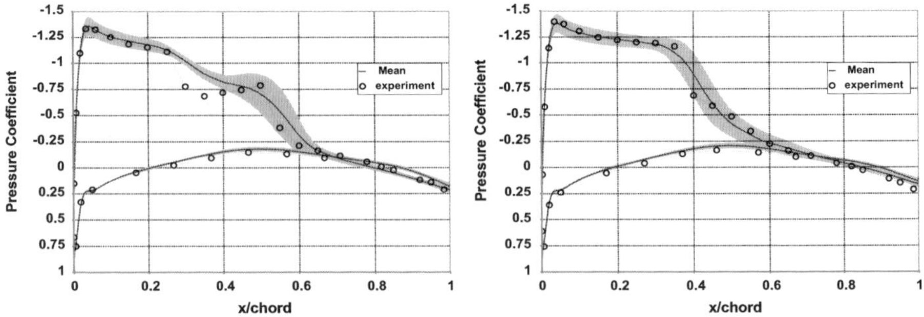


**Fig. 38** Transonic RANS flow over the ONERA M6 wing with free stream Mach number and angle of attack uncertainty, $M_\infty = \text{Gaussian}_{4\sigma}(m = 0.84, \sigma = 0.02)$ and $\text{AOA} = \text{Gaussian}_{4\sigma}(m = 6°, \sigma = 0.075°)$. Surface pressure coefficient statistics calculated using a HYGAP approximation ($N = 6 \times 6$) with subscell recovery are graphed at the 44 % span station (*left*) and the 65 % span station (*right*). The *shaded band* denotes a one standard deviation envelope

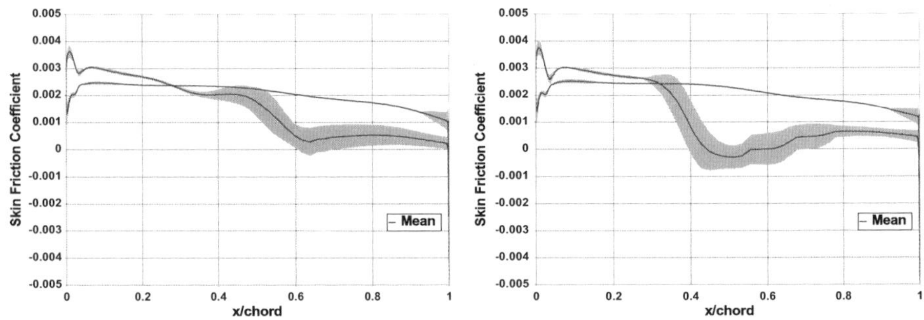


**Fig. 39** Transonic RANS flow over the ONERA M6 wing with free stream Mach number and angle of attack uncertainty, $M_\infty = \text{Gaussian}_{4\sigma}(m = 0.84, \sigma = 0.02)$ and $\text{AOA} = \text{Gaussian}_{4\sigma}(m = 6°, \sigma = 0.075°)$. Graphs of skin friction coefficient statistics calculated using a HYGAP approximation ($N = 6 \times 6$) with subscell recovery are graphed at the 44 % span station (*left*) and the 65 % span station (*right*). The *shaded band* denotes a one standard deviation envelope

approximation ($N = 6 \times 6$) with subscale recovery. Results for surface pressure coefficient and surface skin friction coefficient statistics at 44 and 65 % span stations are graphed in Figs. 38 and 39. These graphs show a noticeable increase in uncertainty on the entire upper surface as the angle of attack has been increased from 3.06° to 6°. The skin friction graphs indicate that a region of reversed flow exists at the 65 % span station. Observe that the experimental data no longer falls within one standard deviation of the mean value and the predictive capability of the turbulence model is questionable. Increasing the angle of attack further, the flow then becomes unsteady and uncertainty increases substantially.

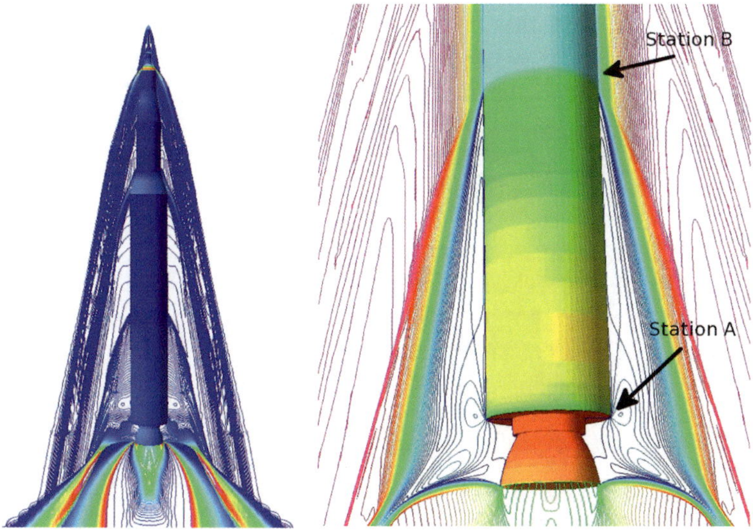


**Fig. 40** Mach 6.7 flow over launch vehicle configuration with rocket plume modeling. Pressure contours over the entire launch vehicle (*left*) and contouring of flow field Mach number and surface color shading of flow field density (*right*). *Blue color* denotes low values and *red color* denotes high values

### 5.3.4 Launch Vehicle Plume Analysis

As a final example, uncertainty propagation for a launch vehicle simulation with exhaust plume modeling is considered. In the design of launch vehicle systems, one must cope with very large forces, volatile chemicals, and extremely hot rocket plume gases. One phenomenon sometimes encountered in these designs is Plume Induced Flow Separation (PIFS). As an example of this phenomenon, Fig. 40 (left) shows Mach 6.7 flow over a launch vehicle with rocket plume modeling approximated using the OVERFLOW [16] Chimera grid Reynolds-averaged Navier-Stokes solver (courtesy Goetz Klopfer, NASA Ames). As the launch vehicle accelerates and ascends into the atmosphere, the rocket plume expands and eventually causes the flow to separate at Station B on the rocket body with reversed flow occurring between Stations A and B in Fig. 40 (right). This reversed flow may carry very hot plume gases in close proximity to the rocket body resulting in material failure unless additional thermal protection is provided. As a historical note, this phenomenon was encountered in the NASA Apollo Saturn-V launches during 1967–1973. Consequently, during those launches one of the five Saturn-V engines was intentionally powered off at a prescribed altitude to reduce the severity of the PIFS phenomenon. For the present simplified single engine configuration, we seek to quantify the extent of the PIFS flow reversal with respect to uncertain flow conditions. We have included into the analysis

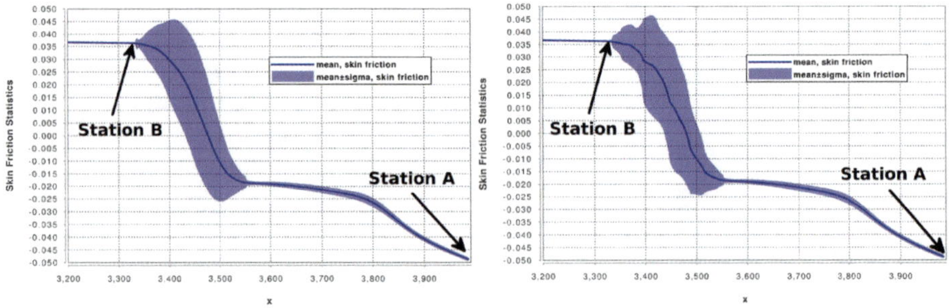


**Fig. 41** Mach 6.7 flow over a launch vehicle configuration. Uncertainty in skin friction coefficient approximated using $N = 6$ HYGAP approximation (*left*) and stochastic collocation approximation (*right*). The *shaded band* denotes a one standard deviation envelope

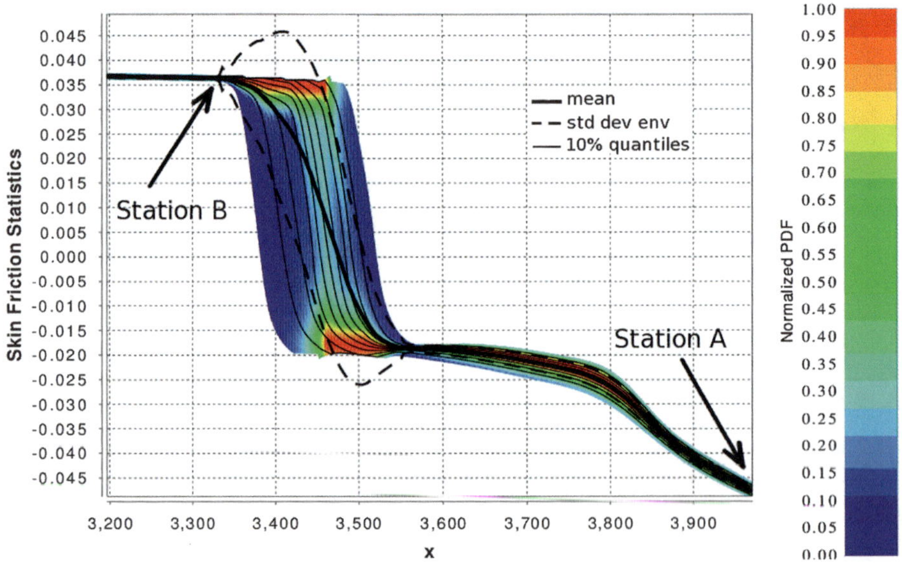


**Fig. 42** Mach 6.7 flow over a launch vehicle configuration. Uncertainty in skin friction coefficient approximated using the HYGAP approximation ($N = 6$). Graphed are the skin friction coefficient statistics and 10 % quantiles of probability. *Blue color* denote low values of the normalized probability density and *red color* denote high values

- Uncertainty in flight Mach number

$$M_\infty = \text{Gaussian}_{4\sigma}(m = 6.7, \sigma = 0.067)$$

- A simplified model of thrust uncertainty given two thrust settings

$$\text{thrust}(\zeta) = \text{thrust}_{80\,\%} + \zeta\,(\text{thrust}_{100\,\%} - \text{thrust}_{80\,\%})\ , \quad \zeta = \text{Gaussian}_{4\sigma}(m = 0.7, \sigma = 0.1)\ .$$

One way to characterize the extent of this reversed flow is via a skin friction coefficient that changes sign when the vertical component of velocity changes direction. Figure 41 contrasts uncertainty estimates for skin friction between Stations A and B using HYGAP approximation with subscale recovery (left) and conventional stochastic collocation (right) with $N = 6$ collocation points. The stochastic collocation results show spurious oscillations in the shock wave region for the reasons discussed at length in Sect. 4. The HYGAP results are non-oscillatory with negligible change as the number of quadrature points is increased. Finally, it should be noted that the stochastic collocation and HYGAP methods both accommodate the explicit construction of a response surface (see (34) and (63)). This can be valuable when the actual probability density of the output of interest is sought. After explicitly constructing the response surface in the HYGAP method, techniques such as kernel density estimation [27, 32] can be applied to obtain a continuous approximation of the output probability density such as shown in Fig. 42. In this figure, quantiles of 10 % probability are also graphed for reference. Looking at the normalized probability density at physical locations near $x = 3{,}460$, it becomes clear that the output probability density is bi-modal with two widely separated peaks. This explains the large standard deviation envelope and illustrates the rich information contained in the output probability density that may be lost in low order statistics.

**Acknowledgements** The author acknowledges the support of the NASA Fundamental Aeronautics Program for supporting this work. Computing resources have been provided by the NASA Ames Advanced Supercomputing Center.

## References

1. Babuska, I., Nobile, F., Tempone, R.: A stochastic collocation method for elliptic partial differential equations with random input data. SIAM J. Numer. Anal. pp. 1005–1034 (2007)
2. Baldwin, B.S., Barth, T.J.: A one-equation turbulence transport model for high Reynolds number wall-bounded flows. Tech. Rep. TM-102847, NASA Ames Research Center, Moffett Field, CA (1990)
3. Barth, T.J.: Numerical methods for gasdynamic systems on unstructured meshes. In: Kröner, Ohlberger, Rohde (eds.) An Introduction to Recent Developments in Theory and Numerics for Conservation Laws, *Lecture Notes in Computational Science and Engineering*, vol. 5, pp. 195–285. Springer-Verlag, Heidelberg (1998)
4. Becker, R., Rannacher, R.: Weighted a posteriori error control in FE methods. In: Proc. ENUMATH-97, Heidelberg. World Scientific Pub., Singapore (1998)
5. Brass, H., Föster, K.J.: On the estimation of linear functionals. Analysis **7**, 237–258 (1987)
6. Brezinski, C., Zaglia, M.R.: Extrapolation Methods. North Holland (1991)
7. Chan, T., Shen, J.: Image Processing and Analysis. SIAM (2005)
8. Clenshaw, C.W., Curtis, A.R.: A method for numerical integration on an automatic computer. Numer. Math. **2**, 197–205 (1960)
9. Cockburn, B., Hou, S., Shu, C.: TVB Runge-Kutta local projection discontinuous Galerkin finite element method for conservation laws IV: The multidimensional case. Math. Comp. **54**, 545–581 (1990)

10. Cockburn, B., Luskin, M., Shu, C.W., Süli, E.: Enhanced accuracy by postprocessing for finite element methods for hyperbolic equations. Math. Comp. **72**, 577–606 (2003)
11. Eriksson, K., Estep, D., Hansbo, P., Johnson, C.: Introduction to numerical methods for differential equations. Acta Numerica pp. 105–158 (1995)
12. Ghamen, R., Spanos, P.: Stochastic Finite Elements. Dover Pub. Inc., Mineola, New York (1991)
13. Harris, C.: Two-dimensional aerodynamic characteristics of the NACA 0012 airfoil in the Langley 8-foot transonic pressure tunnel. Tech. Rep. TM-81927, NASA Langley Research Center, Hampton, VA (1981)
14. Hildebrand, F.: Introduction to Numerical Analysis. McGraw-Hill, New York (1956)
15. Holtz, M.: Sparse Grid Quadrature in High Dimensions with Applications in Finance and Insurance, *Lecture Notes in Computational Science and Engineering*, vol. 77. Springer-Verlag, Heidelberg (2011)
16. Jespersen, D., Pulliam, T., Buning, P.: Recent enhancements to OVERFLOW. Tech. Rep. 97-0644, AIAA, Reno, NV (1997)
17. Jiang, G., Shu, C.W.: Efficient implementation of weighted ENO schemes. J. Comp. Phys. pp. 202–228 (1996)
18. Larson, M., Barth, T.: A posteriori error estimation for adaptive discontinuous Galerkin approximations of hyperbolic systems. In: B. Cockburn, C.W. Shu, G. Karniadakis (eds.) Discontinuous Galerkin methods. Theory, computation and applications, *Lecture Notes in Computational Science and Engineering*, vol. 11. Springer-Verlag, Heidelberg (2000)
19. van Leer, B.: Towards the ultimate conservative difference schemes V. A second order sequel to Godunov's method. J. Comp. Phys. **32**, 101–136 (1979)
20. van Leer, B.: Upwind-difference schemes for aerodynamics problems governed by the Euler equations. AMS Pub., Providence, Rhode Island (1985)
21. LeSaint, P., Raviart, P.: On a finite element method for solving the neutron transport equation. In: C. de Boor (ed.) Mathematical Aspects of Finite Elements in Partial Differential Equations, pp. 89–145. Academic Press (1974)
22. Loeven, G., Bijl, H.: Probabilistic collocation used in a two-step approach for efficient uncertainty quantification in computational fluid dynamics. Comp. Modeling in Engrg. Sci. **36**(3), 193–212 (2008)
23. Mathelin, L., Hussaini, M.Y., Zang, T.: Stochastic approaches to uncertainty quantification in CFD simulations. Num. Alg. pp. 209–236 (2005)
24. Metropolis, N., Ulam, S.: The Monte Carlo method. J. Amer. Stat. Assoc. **44**(247), 335–341 (1949)
25. Nobile, F., Tempone, R., Webster, C.: A sparse grid stochastic collocation method for partial differential equations with random input data. SIAM J. Numer. Anal. **46**, 2309–2345 (2008)
26. Novak, E., Ritter, K.: High dimensional integration of smooth functions over cubes. Numer. Math. **75**(1), 79–97 (1996)
27. Parzen, E.: On estimation of a probability density function and mode. Annals of Mathematical Statistics **33**, 1065–1076 (1962)
28. Piessens, R., de Doncker-Kapenga, E., Überhuber, C., Kahaner, D.: QUADPACK: A Subroutine Package for Automatic Integration. Springer Series in Computational Mathematics. Springer-Verlag (1983)
29. Prudhomme, S., Oden, J.: On goal-oriented error estimation for elliptic problems: application to the control of pointwise errors. Comp. Meth. Appl. Mech. and Eng. pp. 313–331 (1999)
30. Reed, W.H., Hill, T.R.: Triangular mesh methods for the neutron transport equation. Tech. Rep. LA-UR-73-479, Los Alamos National Laboratory, Los Alamos, New Mexico (1973)
31. Richardson, L., Gaunt, J.: The deferred approach to the limit. Trans. Royal Soc. London, Series A **226**, 299–361 (1927)
32. Rosenblatt, M.: Remarks on some nonparametric estimates of a density function. Annals of Mathematical Statistics **27**, 832–837 (1956)
33. Sack, R., Donovan, A.: An algorithm for gaussian quadrature given modified moments. Num. Math. pp. 465–478 (1971)

34. Schmidtt, V., Charpin, F.: Pressure distributions on the ONERA M6 wing at transonic mach numbers. Tech. Rep. AGARD AR-138, Advisory Group for Aerospace Research and Development (1979)
35. Sethian, J.: Level Set Methods and Fast Marching Methods. Cambridge University Press, Cambridge (1999)
36. Sethian, J.A.: Image processing via level set curvature flow. Proc. Natl. Acad. Sci. USA **92**, 7045–7050 (1995)
37. Shu, C.W.: High order ENO and WENO schemes for computational fluid dynamics. In: Barth, Deconinck (eds.) High-Order Discretization Methods in Computational Physics, *Lecture Notes in Computational Science and Engineering*, vol. 9, pp. 439–582. Springer-Verlag, Heidelberg (1999)
38. Smolyak, S.: Quadrature and interpolation formulas for tensor products of centain classes of functions. Dok. Akad. Nauk SSSR **4**, 240–243 (1993)
39. Tatang, M.A.: Direct incorporation of uncertainty in chemical and environmental engineering systems. Ph.D. thesis, MIT (1994)
40. Wan, X., Karniadakis, G.: An adaptive multi-element generalized polynomial chaos method for stochastic differential equations. J. Comput. Phys. pp. 617–642 (2005)
41. Wheeler, J.: Modified moments and gaussian quadratures. Rocky Mtn. J. Math. pp. 287–296 (1974)
42. Wiener, N.: The homogeneous chaos. Am. J. Math. pp. 897–936 (1938)
43. Xiu, D., Karniadakis, G.: The Wiener-Askey polynomial chaos for stochastic differential equations. SIAM J. Sci. Comput. pp. 619–644 (2002)
44. Zienkiewicz, O.C., Zhu, J.Z.: The superconvergent patch recovery and a posteriori error estimates. Part I: the recovery technique. Int. J. Numer. Meth. Engrg. **33**, 1331–1364 (1992)

# Uncertainty Quantification in Aeroelasticity

**Philip Beran and Bret Stanford**

**Abstract** It is important to account for uncertainties in aeroelastic response when designing and certifying aircraft. However, aeroelastic uncertainties are particularly challenging to quantify, since dynamic stability is a binary property (stable or unstable) that may be sensitive to small variations in system parameters. To correctly discern stability, the interactions between fluid and structure must be accurately captured. Such interactions involve an energy flow through the interface, which if unbalanced, can destablize the structure. With conventional computational techniques, the consequences of imbalance may require large simulation times to discern, and evaluating the dependence of stability on numerous system parameters can become intractable. In this chapter, the challenges in quantifying aeroelastic uncertainties will be explored and numerical methods will be described to decrease the difficulty of quantifying aeroelastic uncertainties and increase the reliability of aircraft structures subjected to airloads. A series of aeroelastic analyses and reliability studies will be carried out to illustrate key concepts.

## 1 Introduction

*Aeroelasticity* is the discipline that concerns itself with interactions between an aircraft structure and the surrounding airflow, which arise through some combination of aerodynamic, elastic, and inertial forces [15]. The discipline includes practitioners stretching from the research to the regulatory communities, owing to safety risks

P. Beran (✉)
Air Force Research Laboratory, 2210 Eighth St, Wright-Patterson AFB, OH 45433, USA
e-mail: philip.beran@wpafb.af.mil

B. Stanford
Universal Technology Corporation, 2210 Eighth St, Wright-Patterson AFB, OH 45433, USA
e-mail: bretkennedystanford@gmail.com

H. Bijl et al. (eds.), *Uncertainty Quantification in Computational Fluid Dynamics*, Lecture Notes in Computational Science and Engineering 92,
DOI 10.1007/978-3-319-00885-1_2, 

arising from adverse aeroelastic phenomena that must be considered in existing and future commercial and military aircraft. In the United States, certification of aircraft for safe aeroelastic behavior is currently a deterministic enterprise. However, with new understanding of uncertainty quantification principles and aeroelastic prediction methodologies, it is useful to view aeroelasticity from an uncertainty perspective as a means to improve reliability of new aircraft designs.

Aeroelasticity can be divided into the sub-disciplines *steady aeroelasticity* and *dynamic aeroelasticity*. Steady aeroelasticity considers only steady interactions between aircraft and flowfield, whereas dynamic aeroelasticity also considers time-dependent responses. This chapter focuses on uncertainty quantification of dynamic aeroelastic behaviors, and describes components of an uncertainty framework for problems involving stability change and nonlinear dynamics. Two dynamic aeroelastic phenomena will be discussed: *flutter* and *limit-cycle oscillation*. Flutter is classically regarded as an oscillation between aircraft and flowfield that grows in time in an unbounded manner, i.e., a dynamic instability. Typically, this situation occurs at a certain flight speed, called the *flutter speed*, such that flight at a speed larger than the flutter speed causes a structural component (e.g., wing or control surface) to fail. Limit-cycle oscillation [21], or LCO, is a sustained oscillation between aircraft and flowfield representative of an underlying aeroelastic instability quenched by some aerodynamic or structural dynamic nonlinearity. While not always true, the onset of LCO first occurs at the flutter speed for the stiffening nonlinearities considered herein.

It should be recognized that the use of the term "flutter" to describe an aeroelastic response is sometimes ambiguous, which makes a clear discussion on uncertainty quantification for aeroelasticity no easier. For example, the term *transonic flutter* is often used in the literature to describe LCO, since it is recognized that the physical condition that leads to LCO in the transonic regime, i.e., the underlying instability, is essentially flutter. However, the unbounded character of the instability is not manifested owing to the action of a different, growth attenuating, physical mechanism (e.g., a shock). As the ability to predict this mechanism will depend on model fidelity, prediction of flutter or LCO can be method dependent.

Much is said about the science of uncertainty quantification in this book, so a broad development of uncertainty quantification principles is not repeated in this chapter. Instead, these principles are applied to problems in aeroelasticity, under the assumption that the reader has the requisite background in uncertainty quantification. Also, attention is restricted here to aleatoric uncertainties, i.e., statistical uncertainties either in the vehicle structure (e.g., variations in material properties or geometric shape) or the airstream (e.g., turbulence intensity or even flight speed), although epistemic uncertainties (e.g., modeling uncertainties) can be much more fundamental to a discussion of aeroelasticity, as just described in the previous paragraph. Why consider the intersection of uncertainty quantification and aeroelasticity? Possible reasons are organized into two broad categories: (1) assessing aircraft safety from a non-deterministic standpoint, and (2) designing the vehicle to be risk-minimized.

## 1.1 Assessing Flutter Safety

Safety assessment culminates in aircraft certification; aircraft must be certified to operate in a manner that is free of flutter. In many cases, the onset of flutter is abrupt, and so a safety margin, the *flutter margin*, is imposed. In the United States, this margin is 15 %, i.e., the vehicle is allowed to accelerate to a speed 15 % below that of the anticipated flutter speed. Since flutter (as opposed to LCO) can quickly destroy the aircraft, the speed at which a particular vehicle would experience flutter is in general not known, and has to be estimated from test and analysis.

Deriving flutter predictions from test and analysis can be problematic, and testing can be very costly. Various challenges exist in carrying out flutter tests in wind tunnels; flutter speeds are extrapolated estimates. Debris from a scaled model that fails may damage the wind tunnel, so conditions leading to flutter are avoided. Scaled models and boundary conditions do not exactly, or sometimes cannot, represent the actual flight configuration and support conditions. Analytical models enable the theoretical simulation of flutter and LCO, but the accuracy and reliability of these tools is still a matter of debate in the community, particularly in the absence of supporting test data. Thus, defining bounds of safety for an event that no one wishes to observe instills great caution amongst aeroelasticians.

It is interesting that the flutter margin is deterministic (i.e., a fixed safety factor), since so many ingredients in determining flutter speed are uncertain. Flutter safety is not currently expressed as *reliability*, the probability the vehicle flies free of flutter or dangerous LCO over the intended flight envelope. However, an argument can be made in favor of using a non-deterministic approach to prioritize tests forming a basis for clearance recommendations. Uncertainty quantification can be used to identify which are the most critical tests; i.e., the tests that carry the most risk, and which serve to reduce uncertainty to the greatest extent. These same statements apply to placing limits (known as *placards*) on the operation of existing aircraft that are susceptible to LCO when externally carrying munitions and fuel tanks [19]. Like flutter, a deterministic 15 % margin is employed for LCO, with a frequency-scheduled, g-loading limit used to quantify the failure state. Probabilistic techniques are useful tools to quantify those store configurations that pose the greatest risk when testing the aircraft system.

## 1.2 Designing for Flutter Safety

Another motivation for quantifying uncertainties in aeroelastic responses is to promote reliability in aircraft designs. The premise is that while flutter certification does not explicitly account for reliability, a reliability-based approach to aircraft design will more effectively and inexpensively mitigate aeroelastic problems prior to the first testing of an aircraft and its base components. In this sense, if flutter or large-amplitude LCO are predicted to occur within the vehicle's intended operational envelope, then design changes (e.g., increased skin thickness) should target the most probable failure modes and the failure modes that are most catastrophic in their impact. In this way, increases in vehicle mass can be more prudently allocated. This is the perspective taken in this manuscript: *mitigate the effects of uncertainty on aeroelastic constraints early during design to reduce the downstream costs of developing a high-performance aircraft.*

It should be noted that algorithms used to promote reliability in design may impact certification in the future if certification becomes more reliant on computational methods.

Another important comment is that aircraft offer a variety of aeroelastic failure modes [8, 9]. Mignolet has noted [31] that the weakened and heavy Goland wing may fail through two modes, flutter and divergence, and that the particular form of failure is sensitive to structural parameters (divergence is not a dynamic phenomenon like flutter, but is a static phenomenon that results in unbounded structural response beyond a critical dynamic pressure). Furthermore, different flutter modes can compete as the influence of different physical mechanisms changes with flight conditions and vehicle properties. Missoum and his students have developed a variety of techniques to characterize joint failure surfaces with relevance to aeroelastic problems [6, 20]. *In this study, attention will be limited to the uncertainty quantification of individual failure modes.*

## 1.3 Big Picture

The goal of this chapter will be to describe enabling methodologies needed to predict the probability that an aeroelastic system will fail owing to large amplitude flutter or LCO. Ultimately, this probability prediction can be used to avoid designs that offer little reliability in a desired region of the flight envelope. The main concept is shown in Fig. 1, where large amplitude flutter or LCO is treated as a probabilistic constraint boundary in a design optimization context. Deterministically, it is assumed that at speeds below a critical value, the system is acceptable, and above this speed, the system is unacceptable. In reality, the critical failure state depends on variations in the environment (e.g., fighter aircraft flying in lower density air in cold weather have greater safety margins than those flying in higher density air at the same flight speed) or in the vehicle (e.g., fatigue may alter an aircraft's structural properties over time [40], and different tail numbers of the same vehicle type may have different structural characteristics). In this way, the constraint boundary does not occur with certainty at a point, but instead is distributed over a range of flight conditions. Thus, at speeds approaching the deterministic critical value, there may exist a probability of failure that is unacceptable.

*One significant way in which aeroelasticity diverges from Computational Fluid Dynamics (CFD) in the treatment of uncertainty is that flutter is generally viewed as an inverse problem.* Instead of simply trying to quantify uncertainties in aerodynamic responses, the aeroelastician attempts to quantify uncertainties in

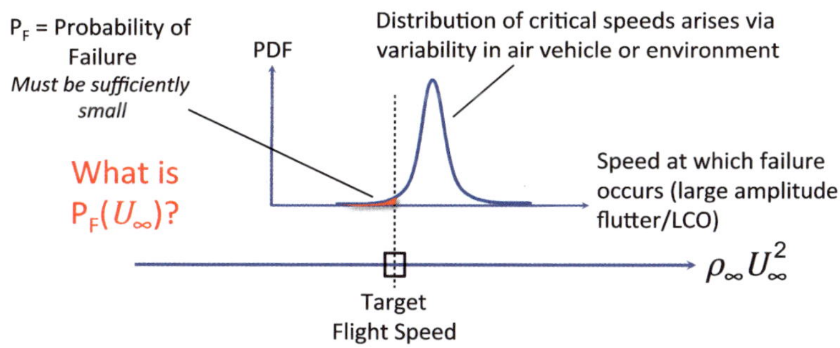


**Fig. 1** Notional diagram interpreting onset of aeroelastic "failure" as a probability distribution with respect to the flight speed, $U_\infty$, for a given atmospheric density, $\rho_\infty$. The probability of failure corresponds to the *red area* for which flight speeds exceed the failure onset speed

the aeroelastic response *and* the location in the design space where aeroelastic responses can be dangerous. However, as the reader shall see, prediction of the location of dangerous aeroelastic responses naturally introduces uncertainty through the inverse nature of the flutter problem. *Thus, particular effort is given in this chapter to (1) orient the reader to fundamentals in aeroelasticity, and (2) describe numerical techniques formulated to minimize numerical sources of uncertainty in flutter prediction while promoting efficiency.*

## 1.4 Chapter Organization and Scope

The rest of this chapter is organized as follows. In Sect. 2, various fundamental mathematical concepts in aeroelasticity are described, as guided by some very simple examples. This section will highlight aeroelastic stability and differences between time-linearized and nonlinear aeroelastic behaviors, and will minimize details in the formulation of the governing equations. In Sect. 3, a bifurcation procedure is described that computes flutter points and their sensitivities to parameters. In Sect. 4, the flutter procedure is extended to nonlinear behaviors and LCO through a perturbation approach, again including provision for the computation of sensitivities. Finally, in Sect. 5, the preceding material is revisited from an uncertainty standpoint to yield probabilistic measures of flutter speed and LCO. These notes are complemented with a practical problem that exemplifies aeroelastic and uncertainty quantification concepts. In this manuscript, attention is focused on the structure, and the analysis of the airloads is usually simplified (some comments will be added concerning extension to CFD). With this limitation, uncertainties of greatest interest herein will pertain to the structure and are parametric in nature. Important discussions related to model-form uncertainty are omitted. As stated above, this exposition applies principles in uncertainty quantification to aeroelasticity. Thus, it is assumed that the reader possesses greater familiarity with the underpinnings of uncertainty quantification than aeroelasticity.

## 2 Fundamental Mathematical Concepts in Aeroelasticity

In this section, a mathematical approach to aeroelasticity is reviewed, and certain key concepts related to flutter and LCO are described. The framework puts focus on the structural equations, and leaves unexplored, except for one class of problems, the nature and mathematical character of the aerodynamic forces acting on the structure and the means by which these loads are transmitted to the structural degrees of freedom.

### *2.1 Equations of Motion*

The linear structural equations of motion are written as

$$\mathbf{M}\ddot{\mathbf{u}} + \mathbf{C}\dot{\mathbf{u}} + \mathbf{K}\mathbf{u} = \mathbf{F}(\ddot{\mathbf{u}}, \dot{\mathbf{u}}, \lambda), \tag{1}$$

where $\mathbf{u}$ is a vector of structural degrees of freedom ($N_{\mathrm{FOM}}$ displacements and/or rotations, where the subscript associates this quantity with a Full-Order Model) from some spatially discrete (e.g., finite element) model, the over-dot implies time differentiation, and vectors and matrices are typeset in boldface. The coefficient matrices, $\mathbf{M}$, $\mathbf{C}$, and $\mathbf{K}$, are the mass, damping, and stiffness matrices, respectively. The aerodynamic force vector, $\mathbf{F}$, is assumed to be interpolated in a work-conserving way from the flowfield domain to the structure as an expression of the airloads, and $\lambda$ is a parameter, such as Mach number (in general, $\lambda$ may be a set of parameters).

The airloads, which may depend nonlinearly on the structural state, may be modeled in any number of different ways, analytically to CFD, although examples will rely on the former. The aerodynamic forces balance the elastic ($\mathbf{Ku}$) and inertial ($\mathbf{M}\ddot{\mathbf{u}}$) forces, in addition to the contributions from damping. Certainly, structural nonlinearities can be present. For example, when the stiffness is nonlinear in $\mathbf{u}$, the structural equations are re-written as

$$\mathbf{M}\ddot{\mathbf{u}} + \mathbf{C}\dot{\mathbf{u}} + \mathbf{P}(\mathbf{u}) = \mathbf{F}(\mathbf{u}, \dot{\mathbf{u}}, \lambda), \tag{2}$$

where $\mathbf{P}$ is a vector of internal elastic forces, and the tangent stiffness matrix $d\mathbf{P}/d\mathbf{u}$ defines the sensitivities of these forces with respect to $\mathbf{u}$.

The linear structural equation is placed in first-order form by expanding the set of dependent variables, $\mathbf{q} \equiv (\mathbf{u}^T, \dot{\mathbf{u}}^T)^T$:

$$\dot{\mathbf{q}} = \begin{pmatrix} \dot{\mathbf{u}} \\ -\mathbf{M}^{-1}\mathbf{C}\dot{\mathbf{u}} - \mathbf{M}^{-1}\mathbf{K}\mathbf{u} + \mathbf{M}^{-1}\mathbf{F}(\mathbf{q};\lambda) \end{pmatrix} \equiv \mathbf{H}(\mathbf{q};\lambda) \tag{3}$$

$$\dot{\mathbf{q}} = -\begin{pmatrix} \mathbf{I} & \mathbf{0} \\ \mathbf{0} & \mathbf{M} \end{pmatrix}^{-1} \begin{pmatrix} \mathbf{0} & -\mathbf{I} \\ \mathbf{K} & \mathbf{C} \end{pmatrix} \mathbf{q} + \begin{pmatrix} \mathbf{0} \\ \mathbf{M}^{-1}\mathbf{F} \end{pmatrix}, \tag{4}$$

where $\mathbf{I}$ is an identity matrix of rank $N_{\text{FOM}}$.

It is conventional to express structural responses as a linear combination of free vibration modes (here interest is in the dynamic response about a static solution, which is now neglected):

$$\mathbf{u} \approx \sum_{i=1}^{N_{\text{ROM}}} \eta_i \boldsymbol{\phi}_i = \boldsymbol{\Phi}\eta \tag{5}$$

where $\boldsymbol{\phi}_i$ is the $i$th mode, $\eta_i$ is the $i$th modal amplitude, $N_{\text{ROM}}$ is the number of modes permitted (a *Reduced Order Model*), and $\boldsymbol{\Phi}$ is the modal matrix whose columns are the individual modes. Defining a new set of dependent variables makes sense when the number of modes retained in (5) is smaller than the original number of degrees of freedom: i.e., $N_{\text{ROM}} \ll N_{\text{FOM}}$. The use of such a reduced order model is not described beyond this section, but is a common aspect of aircraft aeroelastic analysis, owing to the numerous degrees of freedom typically present in aircraft models.

The free-vibration modes are obtained from an airload-free, eigen-analysis of the structural equations:

$$\dot{\mathbf{q}} = -\begin{pmatrix} \mathbf{I} & \mathbf{0} \\ \mathbf{0} & \mathbf{M} \end{pmatrix}^{-1} \begin{pmatrix} \mathbf{0} & -\mathbf{I} \\ \mathbf{K} & \mathbf{C} \end{pmatrix} \mathbf{q} \equiv \mathbf{J}_0\mathbf{q}. \tag{6}$$

The structural response $\mathbf{q}$ is assumed to take the form $\boldsymbol{\phi} e^{\beta t}$, which when inserted into (6) yields $\beta\boldsymbol{\phi} e^{\beta t} = \mathbf{J}_0\boldsymbol{\phi} e^{\beta t}$, or

$$\beta\boldsymbol{\phi} = \mathbf{J}_0\boldsymbol{\phi}. \tag{7}$$

Solution of (7) yields the $N_{\text{ROM}}$ eigenvectors $\boldsymbol{\phi}_i$ and the associated imaginary eigenvalues, $\beta_i$, the natural frequencies. Substitution of (5) into the dynamical equation provides a set of generalized equations:

$$\mathbf{M}\boldsymbol{\Phi}\ddot{\eta} + \mathbf{C}\boldsymbol{\Phi}\dot{\eta} + \mathbf{K}\boldsymbol{\Phi}\eta = \mathbf{F}(\boldsymbol{\Phi}\eta, \boldsymbol{\Phi}\dot{\eta}, \lambda), \tag{8}$$

$$\hat{\mathbf{M}}\ddot{\eta} + \hat{\mathbf{C}}\dot{\eta} + \hat{\mathbf{K}}\eta = \boldsymbol{\Phi}^T\mathbf{F}(\boldsymbol{\Phi}\eta, \boldsymbol{\Phi}\dot{\eta}, \lambda), \tag{9}$$

where $\hat{\mathbf{C}} \equiv \boldsymbol{\Phi}^T\mathbf{C}\boldsymbol{\Phi}$, $\hat{\mathbf{K}} \equiv \boldsymbol{\Phi}^T\mathbf{K}\boldsymbol{\Phi}$, and $\hat{\mathbf{M}} \equiv \boldsymbol{\Phi}^T\mathbf{M}\boldsymbol{\Phi}$. In aeroelastic analysis of aircraft, it is common for the eigenvectors to be normalized in a manner that leads to $\hat{\mathbf{M}} = \hat{\mathbf{I}}$, where $\hat{\mathbf{I}}$ is an identity matrix of rank $N_{\text{ROM}}$.

## 2.2 Condition of Flutter

Flutter is considered to be a loss of linear dynamic stability of the structural equations about an equilibrium solution, which for the sake of brevity is assumed to be the trivial state: $\mathbf{u} = \mathbf{0}$ with $\mathbf{F} = \mathbf{0}$. When the transient solution is of small amplitude, the airloads may be linearized, resulting in

$$\frac{d}{dt}\begin{pmatrix}\mathbf{u}\\ \dot{\mathbf{u}}\end{pmatrix} = \begin{pmatrix}\mathbf{I} & \mathbf{0}\\ \mathbf{0} & \mathbf{M}\end{pmatrix}^{-1}\begin{pmatrix}\mathbf{0} & \mathbf{I}\\ \frac{\partial \mathbf{F}}{\partial \mathbf{u}} - \mathbf{K} & \frac{\partial \mathbf{F}}{\partial \dot{\mathbf{u}}} - \mathbf{C}\end{pmatrix}\begin{pmatrix}\mathbf{u}\\ \dot{\mathbf{u}}\end{pmatrix} \equiv \mathbf{J}\begin{pmatrix}\mathbf{u}\\ \dot{\mathbf{u}}\end{pmatrix}. \tag{10}$$

The dynamical behavior local to the equilibrium solution is discerned by again assuming that the response takes the form $\mathbf{q} = \boldsymbol{\phi} e^{\beta t}$, leading to the eigen-problem

$$\beta\boldsymbol{\phi} = \mathbf{J}\boldsymbol{\phi}, \tag{11}$$

where $\mathbf{J} = \mathbf{J}(\lambda)$ and the influence of aerodynamics is accounted for in the linearization. The real and imaginary parts of $\beta_i = g_i + i\Theta_i$ correspond to the damping and frequency of the eigenresponse, respectively:

$$\boldsymbol{\phi}_i e^{\beta_i t} = \boldsymbol{\phi}_i e^{g_i t}\left(\cos(\Theta_i t) + i\,\sin(\Theta_i t)\right),$$

where the damping parameter for each mode is $g_i = Re(\beta_i)$, and the frequency of each mode is $\Theta_i$. Thus, the aeroelastic damping is defined to be positive when the mode is unstable and grows exponentially, counter to the physical sense of damping. Also, the damping of the composite response, $G$, is driven by the eigenvalue with the largest real part (i.e., the most unstable or least stable mode), $G \equiv \overset{max}{i}\,(g_i)$. *The flutter point is the value of* $\lambda$ *at which* $G = 0$.

This definition of flutter is consistent with the appearance of a Hopf bifurcation point (supercritical) on the solution path of trivial solutions. The eigenvalues $\{\beta_i\}$ of $\mathbf{J}$ at the Hopf point are characterized in Fig. 2a, while the solution diagrams representative of the flutter point and LCO (post-bifurcation) are shown in Fig. 2b. It should be noted that Fig. 2 depicts a situation only involving one free parameter. When there are two parameters, the flutter point becomes a flutter boundary, and in higher dimensions a flutter surface exists. For now, $\lambda$ is taken as a scalar, and flutter occurs at $\lambda = \lambda^*$, such that $G(\lambda^*) = 0$, where the superscript * is used to denote a variable evaluated at the flutter point.

When cast in generalized form (9), the eigenvalues of the retained modes well approximate a subset of $\beta_i$. The modes that are not retained correspond to high-frequency behavior, and can usually be safely omitted from an aeroelastic analysis, since the associated aerodynamic damping is typically strongly negative. When $\lambda$ is varied in a manner that moves the critical, conjugate pair of eigenvalues into the right-half of the complex plane, LCO may develop as shown in Fig. 2b.

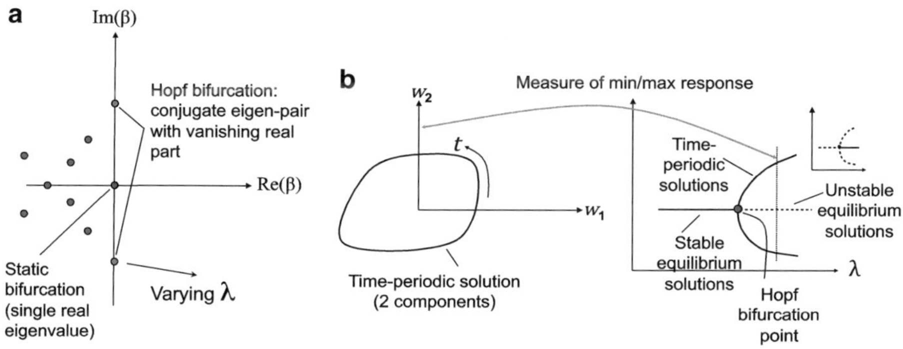


**Fig. 2** (**a**) Notional root-locus diagram of flutter ($G = 0$) when a conjugate pair of eigenvalues crosses the imaginary axis, from left to right, as $\lambda$ is varied; (**b**) schematics showing a LCO at $\lambda > \lambda^*$ that develops from a supercritical Hopf bifurcation, and whose converged periodic behavior is notionally characterized by two response variables, $w_1$ and $w_2$

The conditions $(\mathbf{q}^*, \boldsymbol{\phi}^*, \Theta^*, \lambda^*)$ satisfied at the Hopf bifurcation point can be expressed as a solution to an expanded system of equations, $\mathbf{H}_{\text{exp}} = 0$:

$$\mathbf{H}_{\text{exp}}(\mathbf{q}, \boldsymbol{\phi}, \Theta, \lambda) \equiv \begin{pmatrix} \mathbf{H}(\mathbf{q}, \lambda) \\ \mathbf{J}\boldsymbol{\phi} - i\Theta\boldsymbol{\phi} \\ \mathbf{b}^T\boldsymbol{\phi} - 1 \end{pmatrix} = 0. \tag{12}$$

The first equation corresponds to the dynamical equation, the second equation represents the condition of an eigenvalue on the imaginary axis, and the third equation is a normalization condition using a constant vector **b**. Two strategies for solving this system are discussed in the next section.

## 2.3 *Description of Panel Problem*

A sample problem is now considered that embodies some of the concepts discussed above. This is the problem of a pinned, flexible panel in supersonic flow, which has rich behavior for a simple configuration (see Fig. 3). The aerodynamic loads are algebraically modeled with linear piston theory [3], and the structural dynamics are modeled with von Karman's large-deflection plate theory. The aeroelastic interaction is simply enforced by directly applying aerodynamic loads at structural grid points. This problem has been examined by many, but reference is given to the seminal work of Dowell [22]. The equations of motion (without structural damping) are

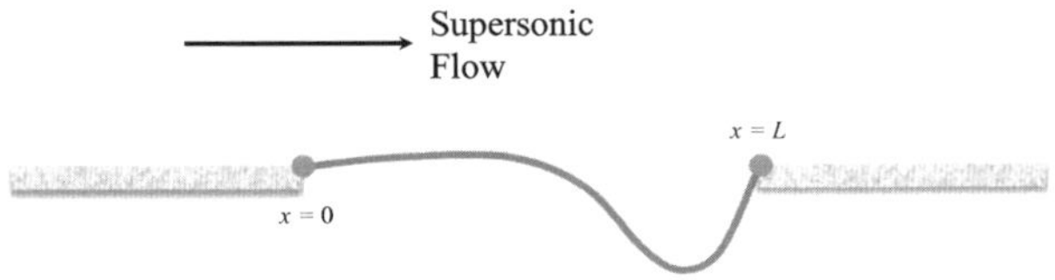


**Fig. 3** Schematic of a flexible panel of length $L$ in supersonic flow (infinite in spanwise direction)

$$D\frac{\partial^4 w}{\partial x^4} - \left(N_x + N_x^{(a)}\right)\frac{\partial^2 w}{\partial x^2} + \rho_s h\frac{\partial^2 w}{\partial t^2} = \Delta p - (p - p_\infty), \quad (13)$$

$$N_x \equiv \frac{Eh}{2L}\int_0^L \left(\frac{\partial w}{\partial x}\right)^2 dx, \quad (14)$$

$$p - p_\infty = \frac{\rho_\infty U_\infty^2}{\sqrt{M_\infty^2 - 1}}\left[\frac{\partial w}{\partial x} + \left(\frac{M_\infty^2 - 2}{M_\infty^2 - 1}\right)\frac{1}{U_\infty}\frac{\partial w}{\partial t}\right], \quad (15)$$

where $x$ and $t$ are the spatial and time coordinates, $N_x$ is the resultant in-plane force, and $N_x^{(a)}$ is the applied in-plane force. Additional parameters are defined in Table 1. The equations of motion are placed in non-dimensional form by scaling lengths by $L$, velocities by $U_\infty$, and time by $L/U_\infty$:

$$\frac{\mu}{\lambda}\frac{\partial^4 \hat{w}}{\partial \hat{x}^4} - \left(\hat{N}_x + \pi^2 R_x\right)\frac{\partial^2 \hat{w}}{\partial \hat{x}^2} + \frac{\partial^2 \hat{w}}{\partial \hat{t}^2} = \Delta\hat{p} - F\left(\frac{\partial \hat{w}}{\partial \hat{x}}, \frac{\partial \hat{w}}{\partial \hat{t}}\right), \quad (16)$$

$$\hat{N}_x \equiv \frac{6\mu}{\lambda}\left(\frac{h}{L}\right)^{-2}(1 - \nu^2)\int_0^1 \left(\frac{\partial \hat{w}}{\partial \hat{x}}\right)^2 d\hat{x}, \quad (17)$$

$$F = \frac{\mu}{\sqrt{M_\infty^2 - 1}}\left[\frac{\partial \hat{w}}{\partial \hat{x}} + \left(\frac{M_\infty^2 - 2}{M_\infty^2 - 1}\right)\frac{\partial \hat{w}}{\partial \hat{t}}\right], \quad (18)$$

where an over-hat denotes a non-dimensional dependent or independent variable. For a pinned panel, the boundary conditions are $\hat{w} = \hat{w}_{\hat{x}\hat{x}} = 0$ at the panel endpoints. Non-dimensional panel responses are computed, assuming non-dimensional values of $\lambda$, $M_\infty$, $\mu$, $h/L$, $\nu$, $\Delta\hat{p}$, and $R_x$. When the (non-dimensional) pressure differential, $\Delta\hat{p}$, is non-zero, equilibrium solutions of the governing equations are non-trivial, but when this parameter vanishes, as is assumed here, a trivial equilibrium solution, $\hat{w} = 0$, exists. For the sake of interpreting some results, it is useful to extract physical quantities in dimensional form. Given $E$, $L$, and $\rho_s$ in a consistent dimensional form, along with the values of the non-dimensional parameters, then the expressions in Table 1 can be used to quantify $\rho_\infty$, $D$, and $U_\infty$.

**Table 1** Dimensional and non-dimensional parameters associated with the panel problem

| Symbol | Parameter | Formula | Value[a] |
|---|---|---|---|
| $U_\infty$ | Freestream velocity | $\sqrt{D\lambda/(\rho_\infty L^3)}$ | – |
| $\rho_\infty$ | Freestream density | $\mu\rho_s h/L$ | – |
| $M_\infty$ | Freestream Mach number | – | 10 |
| $p_\infty$ | Freestream pressure | – | – |
| $p$ | Local static pressure | – | – |
| $\Delta p$ | Pressure differential | – | 0 |
| $L$ | Plate length | – | – |
| $h$ | Plate thickness | – | $0.001L$ |
| $\rho_s$ | Plate density | – | – |
| $E$ | Young's (tensile) modulus | – | – |
| $D$ | Plate stiffness | $\frac{1}{12}Eh^3/(1-\nu^2)$ | – |
| $\mu$ | Mass ratio | $(\rho_\infty L)/(\rho_s h)$ | 0.1 |
| $\nu$ | Poisson's ratio | – | 0.3 |
| $\lambda$ | Dynamic pressure parameter | $\rho_\infty U_\infty^2 L^3/D$ | varied |
| $R_x$ | In-plane parameter | – | 0 |

[a]Baseline values assumed unless otherwise specified

## *2.4 Simulation of LCO*

Using the conditions specified in Table 1, panel LCOs are simulated with time integration for a variety of values of $\lambda$. Details of the procedure reported by Ref. [10] are summarized here. The equations are discretized in space with 33 evenly distributed grid points and 2nd-order-accurate, finite-difference approximations. They are marched in time using a 2nd-order-accurate, 3-time-level, backward-difference (implicit) approximation assuming a constant time step of 0.01 (leading to about 3,600 time steps per cycle). The calculations are initiated with a very small velocity distribution to accelerate convergence to LCO. Peak deflections of $\hat{w}$ are measured at the 3/4-chord location and then post-processed with high-order fits near the peaks for more precise assessment of LCO amplitude and frequency.

Converged LCO amplitudes are plotted in Fig. 4 as a function of $\lambda$. A Hopf bifurcation point is evident with $\lambda^* > 3{,}400$, above which LCO amplitude is observed to grow. Time histories are shown in Fig. 5 for simulations at three different values of $\lambda$ in the neighborhood of $\lambda^*$: 4,000, 3,400, and 3,420. At $\lambda = 4{,}000$, well above the Hopf point, the unsteady solution strongly converges to LCO (grows and then saturates). At $\lambda = 3{,}400$, below the Hopf point, the panel returns to a stable equilibrium. At the third value, $\lambda = 3{,}420$, panel LCO develops, but much more slowly than at $\lambda = 4{,}000$. Of the data shown, $\lambda = 3{,}400$ and $\lambda = 3{,}420$ *bracket* the Hopf point: i.e., $3{,}400 < \lambda^* < 3{,}420$. But what is $\lambda^*$? Two observations pertain to this question. First, it is not usually practical to precisely determine $\lambda^*$ with time integration, since the time needed to assess stability grows as $\lambda \to \lambda^*$. This occurs because the damping vanishes at the flutter point. *The second observation is that with this imprecision in flutter point computation, it is even more difficult to compute*

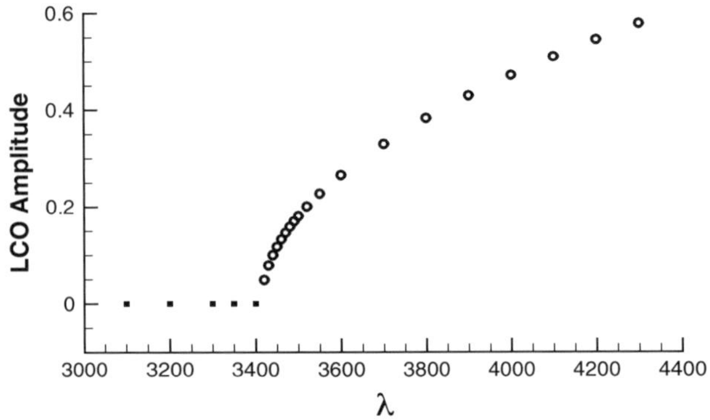


**Fig. 4** LCO amplitude as a function of the dynamic pressure parameter (*hollow circles* are converged LCO's; *filled squares* are stable equilibria)

*with time integration the sensitivity of the flutter speed to relevant parameters.* This result makes uncertainty quantification in aeroelasticity much more challenging.

## 2.5 Time-Linearized Behavior

The onset of classical flutter can also be predicted by linearized analysis, since the loss of stability for $\lambda > \lambda^*$ reflects the dynamics of infinitesimal perturbations governed by (10). For the panel problem described above, linearized analysis amounts to neglecting the nonlinear term (but when $\Delta\hat{p} \neq 0$, the nonlinear term must be retained to obtain the correct static response, about which the dynamics are linearized). Linearized simulations are shown in Fig. 6. In this figure, nonlinear and linearized analysis predict the same bracket of the Hopf point: $3{,}400 < \lambda^* < 3{,}420$. The effect of nonlinearity becomes significant when the dynamic response grows in amplitude. When $\lambda = 4{,}000$ the growth is rapid, and the linearized solution quickly diverges from the nonlinear solution, while at $\lambda = 3{,}420$ the linearized solution well approximates the nonlinear solution until peak deflections reach about 0.02 at the $3/4$-chord location, or approximately 20 panel thicknesses. In some problems, the inclusion of the nonlinear term greatly increases the computational cost of the calculation, which can be avoided if only an estimate of $\lambda^*$ is required.

## 2.6 Eigen-Analysis

The eigenvalues and eigenvectors of the linearized system $\mathbf{J}$ can be interrogated to gain more understanding of the panel dynamics. Also, the use of eigenvectors as modes (5) to develop a low-order representation of the dynamics can be useful

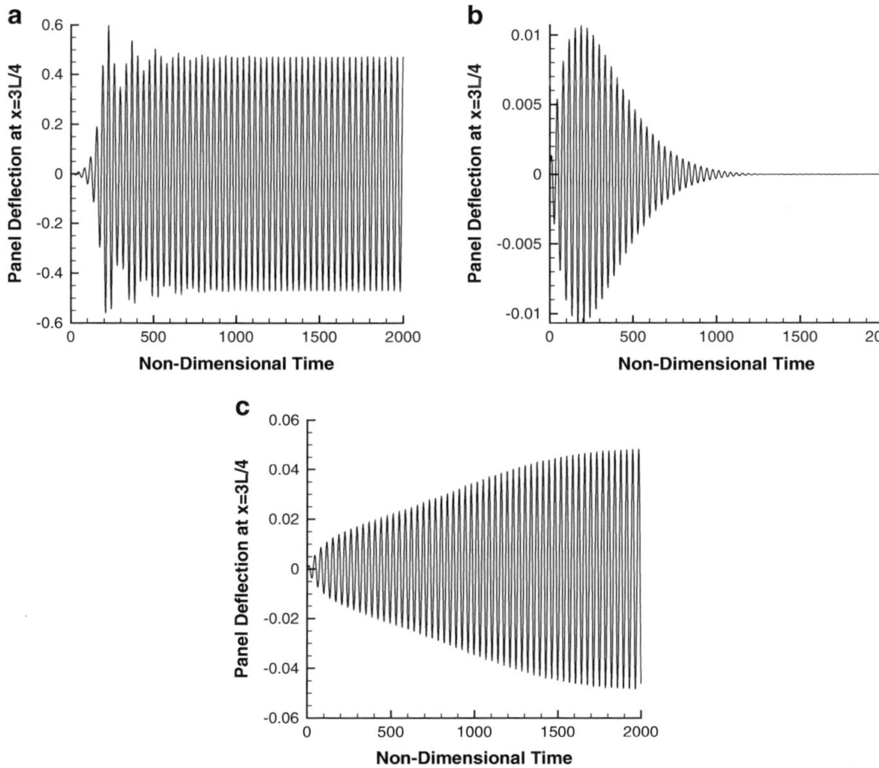


**Fig. 5** Dynamics in neighborhood of Hopf point: (**a**) converged LCO above Hopf point at $\lambda = 4{,}000$; (**b**) stable equilibrium below Hopf point at $\lambda = 3{,}400$, and (**c**) LCO slowly developing just above Hopf point at $\lambda = 3{,}420$

when the number of structural degrees of freedom becomes large. Eigenvectors are shown in Fig. 7 for a number of different values of $\lambda$, starting with 10 and extending to 3,413.64, the flutter point (this precise value is obtained in Sect. 3).

Generally, free-vibration modes (corresponding to a vacuum, $\lambda = 0$) are used in expansions (5) for all values of $\lambda$ of interest, since these modes can be computed without knowledge of the airloads and can be experimentally quantified using ground vibration testing (which is subject to measurement uncertainty). If all possible vibration modes are used, i.e., $N_{\text{ROM}} = N_{\text{FOM}}$, then the basis is complete and (5) is exact, but generally $N_{\text{ROM}} \ll N_{\text{FOM}}$ is desired, in which case, (5) is an approximation whose accuracy depends on $N_{\text{ROM}}$ and the degree to which the selected modes represent the linearized dynamics at a selected value of $\lambda$. In this work, the equations are non-dimensionalized in such a way that $\lambda$ appears in the denominator of the stiffness terms and, thus, $\lambda = 0$ cannot be specified as a condition at which **J** is evaluated. Instead, modes are computed at $\lambda = 10$, which are nearly identical to the vibration modes. It should also be noted that the modes used

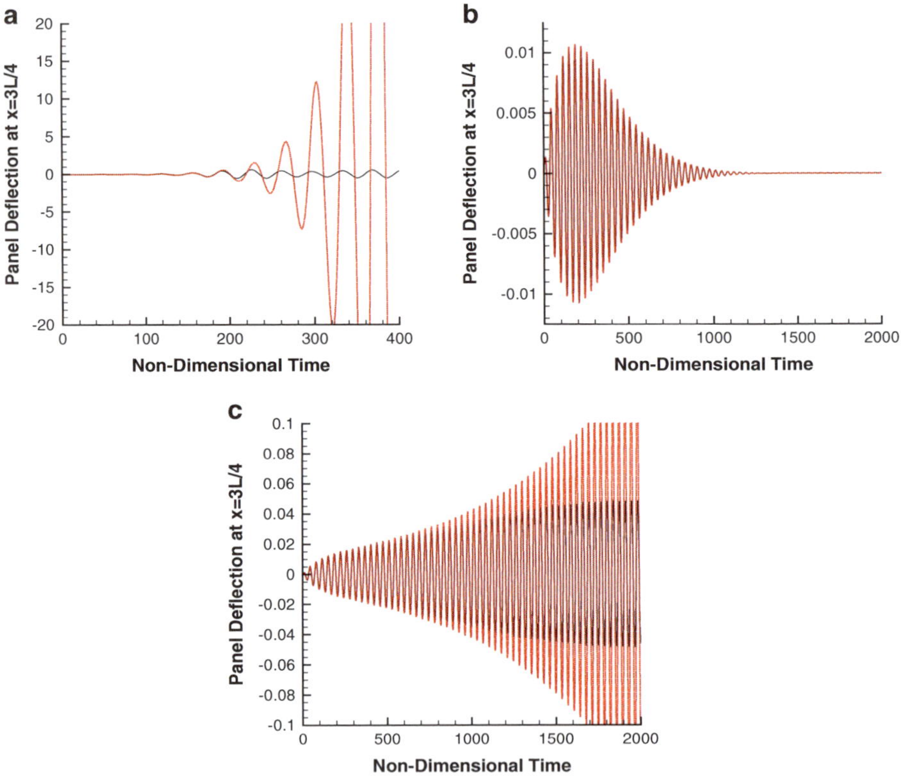


**Fig. 6** Linearized dynamics in neighborhood of Hopf point: (**a**) converged LCO above Hopf point at $\lambda = 4{,}000$; (**b**) stable equilibrium below Hopf point at $\lambda = 3{,}400$, and (**c**) LCO slowly developing just above Hopf point at $\lambda = 3{,}420$. Nonlinear response is shown in *black*; linearized in *red dotted*

in (5) are the portions of the eigenvectors of **J** that correspond to deflection (i.e., not velocity). Also, modes are normalized to have Euclidian norms of 1, including the components corresponding to velocity.

The non-dimensional frequencies of the modes, $\hat{\Theta}_i$, are simply the imaginary components of the computed eigenvalues. For $\lambda = 10$, $\hat{\Theta}_i$ are 0.986, 3.94, 8.82, and 15.6 rad/TU for the first four modes, respectively, where "TU" is an abbreviation of non-dimensional time unit. Generally, higher frequencies are ignored, since, typically, aerodynamic damping of these modes is relatively rapid and strong, and since the distribution of aerodynamic loading is often characterized by a length scale $L$ that doesn't drive response in higher order modes. The computed modes correctly match the analytical vibration modes, $\hat{u} \sim \sin(n\pi\hat{x})$, and are symmetric (odd modes) or anti-symmetric (even modes) about the mid-chord of the panel. At higher values of $\lambda$ (1,000, 2,000, and 3,000) the symmetry is progressively lost as the peak of Mode 1 moves aft towards the 3/4-chord location. At the bifurcation

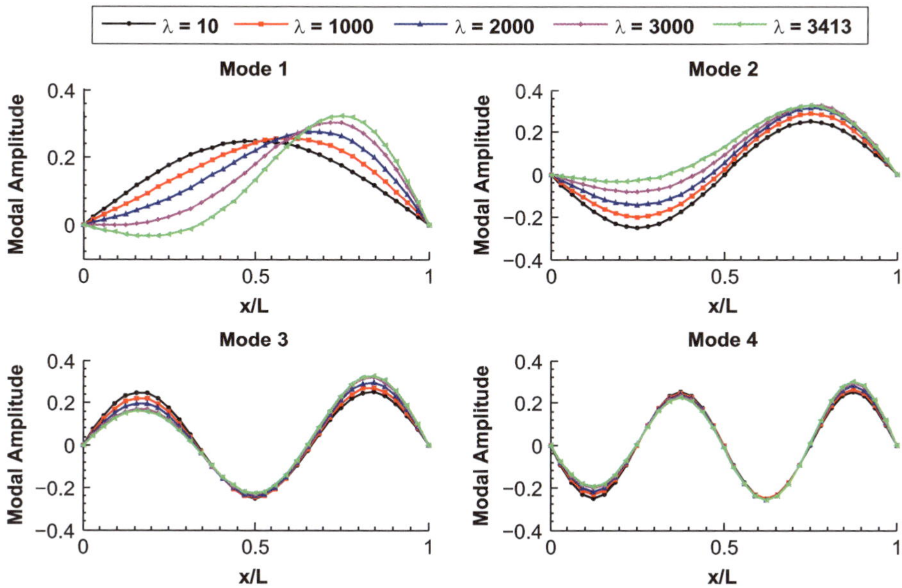


**Fig. 7** The first four eigenvectors of **J** (modes) at selected values of $\lambda$, starting at a nearly unloaded condition and extending to the flutter point

point, $\lambda \approx 3{,}413$, Modes 1 and 2 are nearly reflections of one another, and the frequencies become 0.175, 0.175, 0.474, and 0.843 rad/TU, considerably lower than at $\lambda = 10$.

These frequencies correspond to non-dimensional periods of oscillation, $\hat{T}_i$:

$$\hat{\Theta}_i \hat{T}_i = 2\pi \quad \rightarrow \quad \hat{T}_i = 2\pi / \hat{\Theta}_i. \tag{19}$$

The dimensional period of oscillation, $T_i$, and frequency, $\Theta_i$, for each mode are, respectively:

$$T_i = \frac{L}{U_\infty}\hat{T}_i = \frac{2\pi L}{U_\infty \hat{\Theta}_i}, \qquad \Theta_i = \frac{U_\infty}{L}\hat{\Theta}_i = \sqrt{\frac{D\lambda}{\rho_\infty L^5}}\hat{\Theta}_i. \tag{20}$$

For unloaded structures (vanishing freestream density), the period of oscillation should be invariant with respect to velocity, which is achieved when $U_\infty \hat{\Theta}_i$ is a constant. Thus, the non-dimensional frequency varies inversely with flight speed, owing to the chosen scale factors. As a result, frequency will be reported using dimensional values (in units of radians per second) to avoid skewing results by scaling. It should be noted that the scale factor $\sqrt{\frac{D\lambda}{\rho_\infty L^5}}$ is also used to derive dimensional damping values from the non-dimensionally computed values.

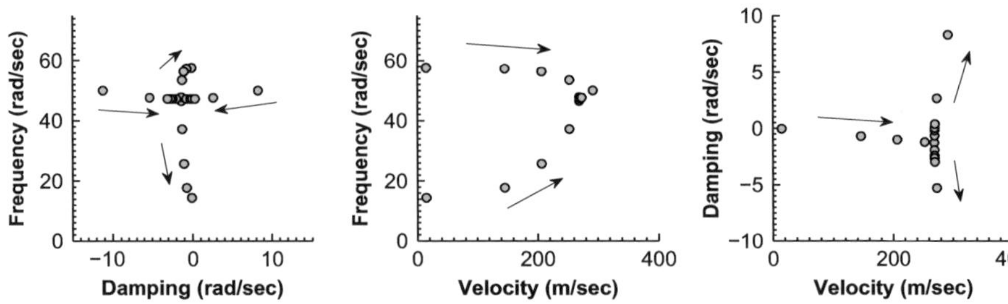


**Fig. 8** First two eigenvalues in dimensional form at selected values of $\lambda$ from 10 to 4,000: (*left*) root-locus diagram showing real (damping) and imaginary (frequency) eigenvalue parts; (*center*) $U_\infty$ versus $\Theta$, and (*right*) velocity versus $g$. *Arrows* indicate direction of increasing $\lambda$

For a pinned panel, the natural frequencies can be easily derived analytically, and are $\Theta_i = \sqrt{D/\rho_s h}\,(\pi i)^2$. Assuming the panel to be made of titanium ($E$ = 105 GPa and $\rho_s$ = 4,500 kg/m$^3$), the first four frequencies are 14.427, 57.708, 129.84, and 230.83 rad/s, respectively.

The eigenvalues corresponding to the first two modes shown in Fig. 7 are provided in Fig. 8 for $\lambda$ from 10 to 4,000, along with eigenvalues at other dynamic pressures for completeness. Results are shown in three ways: first, as a root-locus diagram showing damping, $g$, and frequency; second, as velocity-frequency diagram, and third, as a velocity-damping diagram (traditionally known as a "$V-g$" diagram). Note: in Fig. 8 (left) and (center), only positive frequencies are shown, and so only two eigenvalues per value of $\lambda$ are included; in Fig. 8 (right), all four eigenvalues for the two modes appear. As recorded in Fig. 8 (left), the eigenvalues start ($\lambda = 10$) very near the imaginary axis at frequencies of 14.42 and 57.52 rad/s (these asymptote to the natural frequencies as $\lambda$ decreases). As $\lambda$ is increased, modal frequencies approach (but don't ultimately reach) a common value ($\approx 47.5$) while the system gains stability. As $\lambda$ is increased further, the eigenvalues separate, with mode 1 going in the direction of increasing $g$ (de-stabilizing) while mode 2 moves in the direction of decreasing $g$ (stabilizing). The *coalescence* of the two modes is also seen in Fig. 8 (center), now plotting dimensional velocity on the abscissa. Finally, Fig. 8 (right) records the onset of flutter in a traditional manner by plotting velocity against damping.

Together, the figures show that at relatively low speeds, $g$ decreases with increased flight speed; i.e., stronger aerodynamics increases the physical damping. This effect is provided by the $\partial\hat{w}/\partial\hat{t}$-term in the aerodynamic force, which modifies the damping matrix as seen in (10). Likewise, aerodynamics has the effect of stiffening the panel from the perspective of the first mode, whose frequency increases nearly three-fold. This effect is provided by the $\partial\hat{w}/\partial\hat{x}$-term in the aerodynamic force term, which modifies the stiffness matrix, also as seen in (10). However, with continued increase of flight speed and as the frequencies of the two modes coalesce, the damping characteristics of the system change very rapidly, and damping of the first mode changes from negative to positive in less than 1 m/s. Not only is the change in damping rapid, but also represents a transition to levels of strong instability that would be quickly destructive without arresting nonlinearities. At this critical juncture, the interaction of the two modes enables mode 1 to begin to extract energy from the flowfield in an unstable manner (positive and growing $g$) while $g$ diminishes for mode 2 (negative and declining).

## 3 Computation of Flutter Points and Their Sensitivities

In the previous section, a condition was derived for flutter, and time-domain simulations were performed to observe the behavior of the aeroelastic system and bracket the approximate location of the flutter point. In this section, a precise location of the bifurcation point is computed, and with this alternative formulation, sensitivities of flutter speed with respect to various parameters are obtained.

A precise location of the bifurcation point may be computed with a class of techniques that track the position of the least stable eigenmode of the system Jacobian, **J**. Three methods fall within this class: (1) direct eigen-analysis for the most unstable mode; (2) direct analysis of the expanded system of equations for the Hopf bifurcation point, and (3) an inverse power method with shifting. Emphasis is given here to the first two approaches, since they are simple to describe and apply in this limited space focused on low-order problems. The interested reader should study the third approach [4, 5, 44], which has been successfully applied to practical problems in aeroelasticity.

### 3.1 Direct Evaluation of System Damping

As described above, the stability of an aeroelastic system can be characterized by the system damping, $G(\lambda) = \overset{max}{i} \, (g_i(\lambda))$, where $g_i$ is the damping associated with the $i$th mode of the aeroelastic system. The simplest approach to finding the flutter point at $\lambda^*$ is to iteratively solve for $G(\lambda^*) = 0$ with Newton's method, which is practical when the problem size is not too large. The basic elements of this approach are:

- Solve $dG/d\lambda|_{\lambda=\lambda^n} \, \Delta^n\lambda = -G(\lambda^n)$ until $G = 0$
- Evaluate $G(\lambda^n) \equiv G(\mathbf{J}(\lambda^n))$ by computing the eigenvalues of **J** and finding their maximum real part (e.g., using MATLAB, LAPACK or some other math library)
- Approximate the term $dG/d\lambda$ (via finite differences or analytical expressions)
- Relax the correction and update approximation to $\lambda^*$: $\lambda^n + \omega\Delta^n\lambda \rightarrow \lambda^{n+1}$

When applied to the discretized equations, the region of convergence for the scheme may be narrow below the bifurcation point, but wide above (where the unstable mode is clearly distinguished from all other modes). Use of an under-relaxation parameter $\omega < 1$ is generally necessary to avoid overshoots that drive the

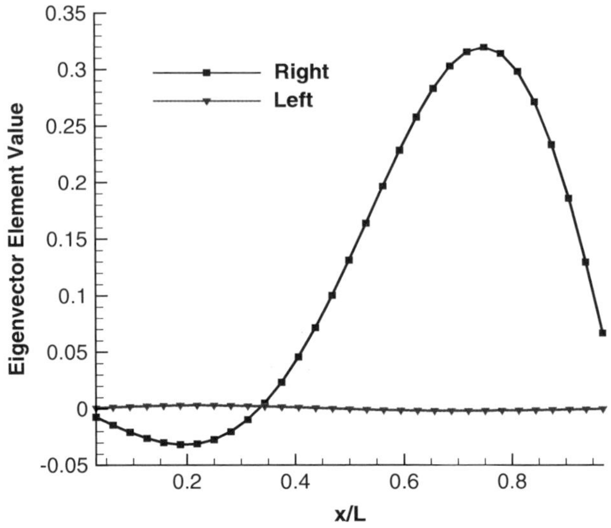


**Fig. 9** Left and right eigenvectors for the critical mode at $\lambda^* \approx 3{,}413.64$ (displacement component shown only)

approximation well below the bifurcation point. For the panel problem, a converged flutter point is obtained at $\lambda^* \approx 3{,}413.64$ using 30 iterations, a starting guess of $\lambda = 4{,}000$, and $\omega = 0.3$ (achieving a tolerance of $|G| \leq 10^{-6}$).

As will be seen, it is very useful to compute both the left and right eigenvectors of $\mathbf{J}(\lambda^*)$ corresponding to the neutral mode, i.e., the mode that satisfies the Hopf bifurcation criteria given in the previous section. We will denote these eigenvectors as $\boldsymbol{\phi}_L^*$ and $\boldsymbol{\phi}_R^*$, respectively (and normalize them to have Euclidean norms of value equal to 1). This step of computing eigenvectors can be performed at the end of the process just described. For the bifurcation point computed at $\lambda^* \approx 3{,}413.64$, the displacement components of $\boldsymbol{\phi}_L^*$ and $\boldsymbol{\phi}_R^*$ are shown in Fig. 9. It should be noted that the deflection patterns seen during simulation of LCO are very similar to $\boldsymbol{\phi}_R^*$. This is true because the nonlinearities that quench the growth of the flutter mode, which is $\boldsymbol{\phi}_R^*$, are very weak and serve mainly to constrain amplitude.

## *3.2 Sensitivities of Flutter Speed to Parameters via Perturbation Analysis*

The procedure described above is a practical tool for computing flutter speed for modestly sized problems. It is also a practical means for computing a sensitivity of the flutter speed to a single parameter, since the flutter point can be computed precisely and a finite-difference approximation can be employed to evaluate the derivative. For example, flutter points can be computed for two nearly equal values of the mass ratio, 0.1 and 0.101, to enable a sensitivity with respect to this parameter to be approximated. With greater precision, the flutter speeds at these points are, respectively, 3,413.64040143 and 3,414.75910765, and the rather large sensitivity is $\Delta\lambda^*/\Delta\mu \approx 111.87$ (a 2nd-order central difference approximation yields 111.84). Clearly, it would be very difficult to estimate this sensitivity using the bracketing approach of Sect. 2.

However, the need to compute sensitivities with respect to a large number of parameters would erode the practicality of the approach. If there are $N_{\text{par}}$ parameters, then computing sensitivities with respect to every parameter would require $O\left(N_{\text{par}}\right)$ flutter calculations. Since thousands of parameters may typically define a wing structure (e.g., skin, rib and spar thicknesses, etc.), it is very costly to compute flutter solutions for changes in each parameter. A different approach is needed whose cost grows more slowly with $N_{\text{par}}$.

One such approach is based on the perturbation analysis of eigensystems [24]. The flutter eigensystem is

$$(\mathbf{J}-\beta\mathbf{I})\,\boldsymbol{\phi}_R=0, \tag{21}$$

where the eigenvalue $\beta$ is constrained to be imaginary, $\beta = i\Theta$, and $\boldsymbol{\phi}_R$ is the critical right eigenvector. The goal is to use this relation to link changes in a parameter (such as $\mu$) to changes in $\Theta$, $\boldsymbol{\phi}_R$, and $\mathbf{J}$, and then connect these changes to movement of the flutter point. These changes are written as

$$\boldsymbol{\phi}_R=\boldsymbol{\phi}_R^*+\delta\boldsymbol{\phi}_R,\qquad \beta=\beta^*+\delta\beta,\qquad \mathbf{J}=\mathbf{J}^*+\delta\mathbf{J},\qquad \lambda=\lambda^*+\delta\lambda. \tag{22}$$

In the neighborhood of the baseline flutter point, denoted with a "*", the right eigenvector can be normalized via $\mathbf{b}^{\dagger}\boldsymbol{\phi}_R-1=0$ (please note that with "*" reserved for the designation of critical points, the superscript "†" implies inner product with the complex conjugate in the subsequent development of the formulation). Furthermore, a multiple of the left eigenvector, $\boldsymbol{\phi}_L^*$, can be assigned to $\mathbf{b}$ to yield $\left(\boldsymbol{\phi}_L^*\right)^{\dagger}\boldsymbol{\phi}_R-\left(\boldsymbol{\phi}_L^*\right)^{\dagger}\boldsymbol{\phi}_R^*=0$, since it can be shown that $\left(\boldsymbol{\phi}_L^*\right)^{\dagger}\boldsymbol{\phi}_R^*\neq 0$. The left eigenvector satisfies the equation $(\boldsymbol{\phi}_L)^{\dagger}(\mathbf{J}-\beta\mathbf{I})=0$, and the normalization condition yields

$$\begin{aligned}\left(\boldsymbol{\phi}_L^*\right)^{\dagger}\left(\boldsymbol{\phi}_R^*+\delta\boldsymbol{\phi}_R\right)-\left(\boldsymbol{\phi}_L^*\right)^{\dagger}\boldsymbol{\phi}_R^*&=0,\\ \left(\boldsymbol{\phi}_L^*\right)^{\dagger}\delta\boldsymbol{\phi}_R&=0.\end{aligned} \tag{23}$$

Substitution of the perturbation quantities into the eigen-system, followed by neglect of 2nd-order terms and cancelation of 0th-order terms provides

$$\left(\mathbf{J}^*+\delta\mathbf{J}-(\beta^*+\delta\beta)\mathbf{I}\right)\left(\boldsymbol{\phi}_R^*+\delta\boldsymbol{\phi}_R\right)=0, \tag{24}$$

$$(\delta\mathbf{J}-\delta\beta\mathbf{I})\,\boldsymbol{\phi}_R^*+\left(\mathbf{J}^*-\beta^*\mathbf{I}\right)\delta\boldsymbol{\phi}_R=0. \tag{25}$$

Now, the dot product of this equation is taken with the complex conjugate of $\boldsymbol{\phi}_L^*$ to yield, after applying the perturbation normalization condition, an equation for $\delta\beta$:

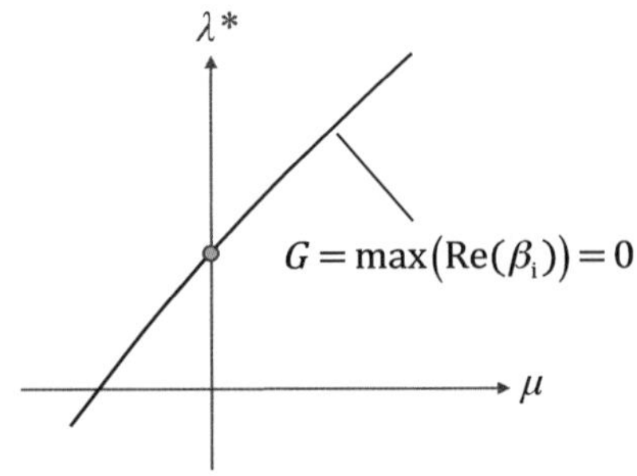


**Fig. 10** Notional relationship between flutter speed and a parameter

$$\boldsymbol{\phi}_L^{*\dagger}(\delta\mathbf{J} - \delta\beta\mathbf{I})\,\boldsymbol{\phi}_R^* + \boldsymbol{\phi}_L^{*\dagger}(\mathbf{J}^* - \beta^*\mathbf{I})\,\delta\boldsymbol{\phi}_R = 0, \tag{26}$$

$$\boldsymbol{\phi}_L^{*\dagger}\delta\mathbf{J}\boldsymbol{\phi}_R^* - \boldsymbol{\phi}_L^{*\dagger}\delta\beta\boldsymbol{\phi}_R^* + \boldsymbol{\phi}_L^{*\dagger}\mathbf{J}^*\delta\boldsymbol{\phi}_R - \boldsymbol{\phi}_L^{*\dagger}\beta^*\delta\boldsymbol{\phi}_R = 0,$$

$$\boldsymbol{\phi}_L^{*\dagger}\delta\mathbf{J}\boldsymbol{\phi}_R^* - \boldsymbol{\phi}_L^{*\dagger}\boldsymbol{\phi}_R^*\delta\beta + \boldsymbol{\phi}_L^{*\dagger}\beta^*\delta\boldsymbol{\phi}_R = \boldsymbol{\phi}_L^{*\dagger}\delta\mathbf{J}\boldsymbol{\phi}_R^* - \boldsymbol{\phi}_L^{*\dagger}\boldsymbol{\phi}_R^*\delta\beta = 0,$$

$$\delta\beta = \left(\boldsymbol{\phi}_L^{*\dagger}\delta\mathbf{J}\boldsymbol{\phi}_R^*\right) / \left(\boldsymbol{\phi}_L^{*\dagger}\boldsymbol{\phi}_R^*\right). \tag{27}$$

Equation (27) is a very interesting and powerful result [33]. By evaluating (once) the left and right eigenvectors associated with the critical mode at $\lambda^*$, and $\delta\mathbf{J}$ corresponding to any particular parameter variation, the associated value of $\delta\beta$ can be predicted. Stanford and Beran have used this approach for the topological optimization of wing structures [41]. This calculation is very efficient, since $\delta\mathbf{J}$ is evaluated and not a result of a flutter solution. The matrix perturbation must be recomputed for each parameter of interest, but the cost is far lower than the first calculation of the flutter point, especially since variations of interest may sometimes be confined to the structure. When applied to the critical mode, (27) may be expressed as a sensitivity with respect to a second parameter $\mu$:

$$\frac{\partial G}{\partial \mu} = Re\left[\left(\boldsymbol{\phi}_L^{*\dagger}\frac{\partial \mathbf{J}}{\partial \mu}\boldsymbol{\phi}_R^*\right) / \left(\boldsymbol{\phi}_L^{*\dagger}\boldsymbol{\phi}_R^*\right)\right]. \tag{28}$$

Equation (28) describes a damping sensitivity and can be used to estimate the change in flutter speed, $\delta\lambda^*$, resulting from variation of $\mu$. As $\mu$ is varied, the variation of $\lambda^*$ forms a *flutter boundary*, $G(\lambda, \mu) = 0$, in the $\mu$-$\lambda^*$ parameter space. See Fig. 10. On the flutter boundary:

$$dG = \frac{\partial G}{\partial \lambda^*}d\lambda^* + \frac{\partial G}{\partial \mu}d\mu = 0 \quad \rightarrow \quad \frac{\partial G}{\partial \lambda^*}d\lambda^* = -\frac{\partial G}{\partial \mu}d\mu,$$

$$\frac{d\lambda^*}{d\mu} = -\left(\frac{\partial G}{\partial \lambda^*}\right)^{-1}\frac{\partial G}{\partial \mu}. \tag{29}$$

Equation (29) is efficiently solved for any number of parameters, since $\partial G/\partial\lambda^*$ need only be computed once, independent of any parameters being varied. For the test case discussed at the beginning of this section, the value of $d\lambda^*/d\mu$ predicted with perturbation analysis is found to be 111.86, which compares very favorably with the difference result directly computed.

## 3.3 Computation of Flutter Points Through a Bifurcation Approach

Flutter points are now computed by directly solving the expanded system of equations (12), $\mathbf{H}_{\text{exp}} = \mathbf{0}$, for the Hopf bifurcation point, $(\mathbf{q}^*, \boldsymbol{\phi}^*, \Theta^*, \lambda^*)$ [27, 32]. At the critical point, $\lambda = \lambda^*$, the $N_{\text{FOM}}$ equations of motion, $\mathbf{H} = \mathbf{0}$, are satisfied with equilibrium solutions $\mathbf{q}^*$, and the $2N_{\text{FOM}}$ linearized equations $\mathbf{J}\boldsymbol{\phi} - i\Theta\boldsymbol{\phi} = 0$ are solved by a complex, right eigenvector $\boldsymbol{\phi}^*$ corresponding to a response frequency of $\Theta^*$, where $\mathbf{J}(\mathbf{q}, \lambda) = \partial\mathbf{H}/\partial\mathbf{q}$.

The expanded system is comprised of $3N_{\text{FOM}} + 2$ unknowns collocated into the array $\mathbf{Q}_{\text{exp}}$. The system can be solved in complex form using Newton's method after defining a Jacobian for the expanded system of equations, $\mathbf{J}_{\text{exp}}$ [14]:

$$\mathbf{J}^{\nu}_{\text{exp}} \left( \mathbf{Q}^{\nu+1}_{\text{exp}} - \mathbf{Q}^{\nu}_{\text{exp}} \right) = -\mathbf{H}_{\text{exp}}(\mathbf{q}^{\nu}, \boldsymbol{\phi}^{\nu}, \Theta^{\nu}, \lambda^{\nu}), \tag{30}$$

$$\mathbf{J}_{\text{exp}} \equiv \begin{pmatrix} \mathbf{J} & \mathbf{0} & \left(0, \frac{\partial \mathbf{H}}{\partial \lambda}\right) \\ \frac{\partial(\mathbf{J}\boldsymbol{\phi})}{\partial \mathbf{q}} & \mathbf{J} - i\Theta\mathbf{I} & \left(-i\boldsymbol{\phi}, \frac{\partial(\mathbf{J}\boldsymbol{\phi})}{\partial \lambda}\right) \\ 0 & \mathbf{b}^T & (0, 0) \end{pmatrix}, \tag{31}$$

$$\mathbf{Q}_{\text{exp}} \equiv (\mathbf{q}, \boldsymbol{\phi}, \Theta, \lambda)^T. \tag{32}$$

Attention will not be given to how this equation is solved; this information can be found elsewhere. For the application being studied here, $\mathbf{J}_{\text{exp}}$ is of modest size and can be explicitly computed and LU-decomposed.

In the same way as employed in the perturbation analysis, the condition that $\mathbf{H}$ vanishes on branches of flutter points can be used to compute the sensitivity of flutter speed to a free parameter, $\mu$ [11]:

$$d\mathbf{H}_{\text{exp}} = \left( \frac{\partial \mathbf{H}_{\text{exp}}}{\partial \mathbf{Q}_{\text{exp}}} \right)^* d\mathbf{Q}_{\text{exp}} + \left( \frac{\partial \mathbf{H}_{\text{exp}}}{\partial \mu} \right)^* d\mu = \mathbf{0}, \tag{33}$$

$$\mathbf{J}^*_{\text{exp}} d\mathbf{Q}_{\text{exp}} + \left( \frac{\partial \mathbf{H}_{\text{exp}}}{\partial \mu} \right)^* d\mu = \mathbf{0},$$

$$\mathbf{J}^*_{\text{exp}} \left( \frac{d\mathbf{Q}_{\text{exp}}}{d\mu} \right)^* = -\left( \frac{\partial \mathbf{H}_{\text{exp}}}{\partial \mu} \right)^*, \tag{34}$$

$$\left( \frac{d\mathbf{Q}_{\text{exp}}}{d\mu} \right)^* = \left( \left( \frac{d\mathbf{q}}{d\mu} \right)^*, \left( \frac{d\boldsymbol{\phi}}{d\mu} \right)^*, \left( \frac{d\Theta}{d\mu} \right)^*, \left( \frac{d\lambda}{d\mu} \right)^* \right)^T. \tag{35}$$

Equation (34) takes the same form as (30); flutter sensitivities can thus be computed with only one additional solve per parameter of interest using a frozen Jacobian, $\mathbf{J}_{\text{exp}} = \mathbf{J}^*_{\text{exp}}$. If $\mathbf{J}^*_{\text{exp}}$ is LU-decomposed, then each of these computations costs much less than computing the flutter point itself.

Computations of flutter points with the bifurcation method, hereafter referred to as the "direct B-method", yielded results essentially identical to that found through direct evaluation of system damping, hereafter referred to as the "direct G-method". Both techniques rely on interrogation of a linearized time-domain formulation of the governing equations. However, the roots of aeroelastic analysis lie in the frequency domain, and many effective frequency-domain techniques have been developed for flutter analysis, such as Chen's g-method [17]. For problems of small size, the direct B- and G-methods are fast and relatively easy to implement. As problem size grows, both methods become increasingly expensive, and the sparsity pattern of $\mathbf{J}_{\text{exp}}$ should be taken advantage of to improve efficiency. It is also found that the LU-decomposed form of $\mathbf{J}^*_{\text{exp}}$ can be used to compute the flutter hyper-surface in the neighborhood of the flutter point at $\lambda^*$. On the other hand, it was shown that the direct G-method offers a very attractive means for computing sensitivities of flutter speed to a large number of parameters. Thus, an efficient process for uncertainty quantification of flutter points is to identify a key flutter point with the direct B-method, assess flutter sensitivity with the direct G-method, and conduct further exploration around this flutter point with the B-method (e.g., a local Monte-Carlo simulation).

# 4 Computation of LCOs and Their Sensitivities

A methodology is now considered for the direct computation of limit-cycle oscillations based on perturbation analysis through the method of multiple scales [34] (hereafter referred to as MMS). Beran used this approach to study the character of Hopf bifurcations in the transonic regime for simply supported airfoils (where MMS is applied to the structural and CFD equation sets) [12]. This work is contrasted with that of Beran et al., who developed stochastic projections of fully developed LCOs [13]. There are many other significant studies applying MMS to aeroelasticity [16, 26, 35, 37, 43].

## *4.1 LCO Formulation*

Having located the flutter point using either of the direct tools described above, further perturbation schemes may be considered in order to assess the nature of the concomitant limit cycle oscillation. A typical supercritical LCO is seen in Fig. 4, though the presence of destabilizing leading-order nonlinearities can lead to dangerous subcritical LCOs, with stable high-amplitude behavior at flight speeds lower than the linear flutter speed. Similar to complications with flutter points, time integration is typically an inefficient tool for quantifying the limit cycle. The method of multiple scales, described next, distills the complex nonlinear dynamics to a two degree-of-freedom system, the parameters of which dictate the leading-order nature of the LCO emanating directly from the flutter point (i.e., subcritical or supercritical?), and the strength of the branch. The method may not reliably predict system dynamics far removed from the Hopf bifurcation point, but does provide highly accurate, local information, which is amenable to sensitivity and uncertainty analyses.

Perturbations to the equilibrium solution at the flutter point are written as:

$$\mathbf{q} = \mathbf{q}^* + \epsilon\hat{\mathbf{q}}, \tag{36}$$

$$\lambda = \lambda^* + \epsilon\hat{\lambda}. \tag{37}$$

Substituting this perturbation into the equations of motion ($\mathbf{H} = \mathbf{0}$), and expanding via a Taylor series (with the assumption that $d\mathbf{H}/d\lambda$ is zero, which is entirely true if the equilibrium solution $\mathbf{q}^*$ is trivial) provides:

$$\frac{d\hat{\mathbf{q}}}{dt} = \mathbf{J}^*\hat{\mathbf{q}} + \epsilon^2\hat{\lambda}\left(\frac{\partial \mathbf{J}}{\partial \lambda}\right)^* \hat{\mathbf{q}} + \epsilon\mathbf{C}(\hat{\mathbf{q}}, \hat{\mathbf{q}}) + \epsilon^2\mathbf{D}(\hat{\mathbf{q}}, \hat{\mathbf{q}}, \hat{\mathbf{q}}) + \ldots \tag{38}$$

where $\mathbf{C}(\hat{\mathbf{q}}, \hat{\mathbf{q}})$ and $\mathbf{D}(\hat{\mathbf{q}}, \hat{\mathbf{q}}, \hat{\mathbf{q}})$ are vector-valued symmetric bilinear and trilinear directional derivative operators. These may be easily computed with finite differences, for example:

$$\mathbf{C}(\mathbf{W}_1, \mathbf{W}_2) \approx \frac{1}{8\delta_C}\{\mathbf{H}(\mathbf{q}^* + \delta_C\mathbf{W}_1 + \delta_C\mathbf{W}_2; \lambda^*) - \mathbf{H}(\mathbf{q}^* - \delta_C\mathbf{W}_1 + \delta_C\mathbf{W}_2; \lambda^*)$$
$$- \mathbf{H}(\mathbf{q}^* + \delta_C\mathbf{W}_1 - \delta_C\mathbf{W}_2; \lambda^*) + \mathbf{H}(\mathbf{q}^* - \delta_C\mathbf{W}_1 - \delta_C\mathbf{W}_2; \lambda^*)\}.$$

The appropriate size of $\delta_C$ (and $\delta_D$) is arrived at empirically. In general, $\mathbf{W}_1$ and $\mathbf{W}_2$ will be complex-valued vectors, and the aeroelastic routines used to compute $\mathbf{H}$ must be modified to receive complex variables. The local solution is expanded as:

$$\hat{\mathbf{q}} = \hat{\mathbf{q}}_1(T_0, T_2) + \epsilon\hat{\mathbf{q}}_2(T_0, T_2) + \epsilon^2\hat{\mathbf{q}}_3(T_0, T_2) + \ldots \tag{39}$$

The multiple time scales are given as $T_0 = t$ and $T_2 = \epsilon^2 t$. Substituting this expression into (38), equating like powers of $\epsilon$, and enforcing a normalization condition on $\boldsymbol{\phi}_L^*$, ${\boldsymbol{\phi}_L^*}^T\boldsymbol{\phi}_R^* = 1$, provides solutions for the different terms comprising $\hat{\mathbf{q}}$. For like powers of $\epsilon^0$, the solution is:

$$O(1): \quad \hat{\mathbf{q}}_1 = A\left(\boldsymbol{\phi}_R^* e^{i\Theta^* T_0}\right) + \overline{A}\left(\overline{\boldsymbol{\phi}_R^*} e^{-i\Theta^* T_0}\right) \tag{40}$$

Substituting (40) into the $O(\epsilon)$ equation yields:

$$O(\epsilon): \quad \hat{\mathbf{q}}_2 = 2\mathbf{Z}_0 A\overline{A} + 2\mathbf{Z}_2 A^2 e^{2i\Theta^* T_0} + cc, \tag{41}$$

where "$cc$" refers to a complex conjugate, and $\mathbf{Z}_0$ and $\mathbf{Z}_2$ are computed by solving the following systems:

$$\mathbf{J}^*\mathbf{Z}_0 = -\frac{1}{2}\mathbf{C}(\boldsymbol{\phi}_R^*, \overline{\boldsymbol{\phi}_R^*}), \tag{42}$$

$$(2i\Theta^*\mathbf{I} - \mathbf{J}^*)\mathbf{Z}_2 = \frac{1}{2}\mathbf{C}(\boldsymbol{\phi}_R^*, \boldsymbol{\phi}_R^*). \tag{43}$$

Finally, the expressions for $\hat{\mathbf{q}}_1$ and $\hat{\mathbf{q}}_2$ are substituted into the equation for like powers of $\epsilon^2$. Rather than explicitly solve for $\hat{\mathbf{q}}_3$, secular terms are removed from the equation (solvability) to obtain the characteristic equations for $A = \frac{1}{2}ae^{i\Theta}$:

$$O(\epsilon^2): \qquad \dot{a} = \hat{\lambda}\beta_{1r}a + \beta_{2r}a^3, \qquad \dot{\Theta} = \hat{\lambda}\beta_{1i} + \beta_{2i}a^2. \tag{44}$$

The complex-valued $\beta$ coefficients are computed as:

$$\beta_1 = \beta_{1r} + i\beta_{1i} = \boldsymbol{\phi}_L^{*T}\mathbf{J}_\lambda^*\boldsymbol{\phi}_R^*, \tag{45}$$

$$\beta_2 = \beta_{2r} + i\beta_{2i} = \boldsymbol{\phi}_L^{*T}\left(2\mathbf{C}(\mathbf{P}, \mathbf{Z}_0) + \mathbf{C}(\overline{\mathbf{P}}, \mathbf{Z}_2) + \frac{3}{4}\mathbf{D}(\mathbf{P}, \mathbf{P}, \overline{\mathbf{P}})\right). \tag{46}$$

The steady-state solution to the characteristic equation provides the LCO amplitude:

$$a = \sqrt{-\hat{\lambda}\beta_{1r}/\beta_{2r}}. \tag{47}$$

If $\beta_{1r}$ and $\beta_{2r}$ have the same sign, limit cycles will exist for negative values of $\hat{\lambda}$ (i.e., $\lambda < \lambda^*$), which is a "subcritical" case. Otherwise, a supercritical limit cycle exists. Furthermore, if $\beta_{1r}$ is positive (which is the definition of a dynamically-unstable flutter point), then the supercritical LCO will be very benign for large values of $|\beta_{2r}|$. The complete solution is found by $\mathbf{q} = \mathbf{q}^* + \hat{\mathbf{q}}_1 + \hat{\mathbf{q}}_2$ (having set $\epsilon$ to unity). This can be done for a range of $\hat{\lambda}$ values at little cost, once the nonlinear system parameters $\beta_1$ and $\beta_2$ are computed. The cost of computing $\beta_1$ and $\beta_2$ is driven by four sources: (1) the cost of computing the bifurcation point satisfying $\mathbf{H}_{\text{exp}} = 0$, which yields $\boldsymbol{\phi}_R^*$; (2) the cost of computing a left eigenvector satisfying $\boldsymbol{\phi}_L^{*T}\boldsymbol{\phi}_R^* = 1$, which requires an analysis of the form $\mathbf{Ax} = \mathbf{b}$, where $\mathbf{A}$ is the same rank as $\mathbf{J}$, (3) two more analyses of the same form $\mathbf{Ax} = \mathbf{b}$, to find $\mathbf{Z}_0$ and $\mathbf{Z}_2$, and (4) the cost of approximating the bilinear and trilinear operators $\mathbf{C}$ and $\mathbf{D}$. The cost of steps (2)–(4) is on par with that of step (1), and thus MMS represents an efficient means for extracting much additional information about the dynamics local to a bifurcation point.

## 4.2 LCO Results

The MMS methodology just outlined is now applied to the panel problem, and comparisons are made to limit-cycles computed through simulation of the original equations. As the approach is based on a perturbation expansion, results are anticipated to degrade as distance from the bifurcation point increases (i.e., with increasing $\hat{\lambda}$, where $\lambda = \lambda^* + \epsilon^2\hat{\lambda}$). It is already known that the LCOs are supercritical, therefore: $\hat{\lambda} > 0$, $\beta_{1r} > 0$, and $\beta_{2r} < 0$. For this problem, the ratio $|\beta_{1r}/\beta_{2r}|$ is quite small:

$$\beta_1 = 2.22 \times 10^{-4} - 8.23 \times 10^{-6} i, \quad \beta_2 = -6.01 \times 10^4 + 3.39 \times 10^3 i. \tag{48}$$

As described in Sect. 2, damping decreases near bifurcation points; at small distances away from the bifurcation care must be taken to insure that the solution is fully converged. Results are compared in Fig. 11 for three values of $\lambda$: 3,420, 4,000, and 4,600. In these phase portraits of displacement vs. velocity at the 3/4-chord location, it can be seen that LCO amplitude grows with distance away from the bifurcation point at $\lambda^* = 3{,}413.64$, and that accuracy of the MMS approximation is quite good, although degrades slowly as the pressure parameter increases.

The variation in accuracy with changing distance from the bifurcation point is easily seen by comparing predicted values of LCO period, as provided in Fig. 12. Here it is observed that LCO period decreases with $\lambda$, which is expected, since the geometric nonlinearity present in the panel equation provides a stiffening influence whose effect is stronger as LCO amplitude grows. It is also apparent that MMS slightly under-predicts LCO period, but provides exceedingly accurate predictions near the bifurcation point (it should be noted that subcritical bifurcations [21], whose LCOs are arrested by stronger nonlinearities, may not be nearly so well approximated).

## 4.3 LCO Sensitivities

Computing sensitivities of LCO characteristics with respect to various parameters is challenging. In Sect. 3.2, a perturbation analysis was employed to directly obtain sensitivities of the flutter location with respect to a parameter of interest, $\mu$. While still relevant, one must now additionally propagate parametric variation through the expressions just developed to predict updated LCO characteristics about the new location of the bifurcation point. This idea can be captured in the following notional equation for the LCO amplitude, $A_{\text{LCO}}$, extracted from (47):

$$A_{\text{LCO}}(\lambda;\mu) = \Pi(\mu)\hat{\lambda}^{1/2} = \Pi(\mu)\sqrt{\lambda - \lambda^*(\mu)}, \tag{49}$$

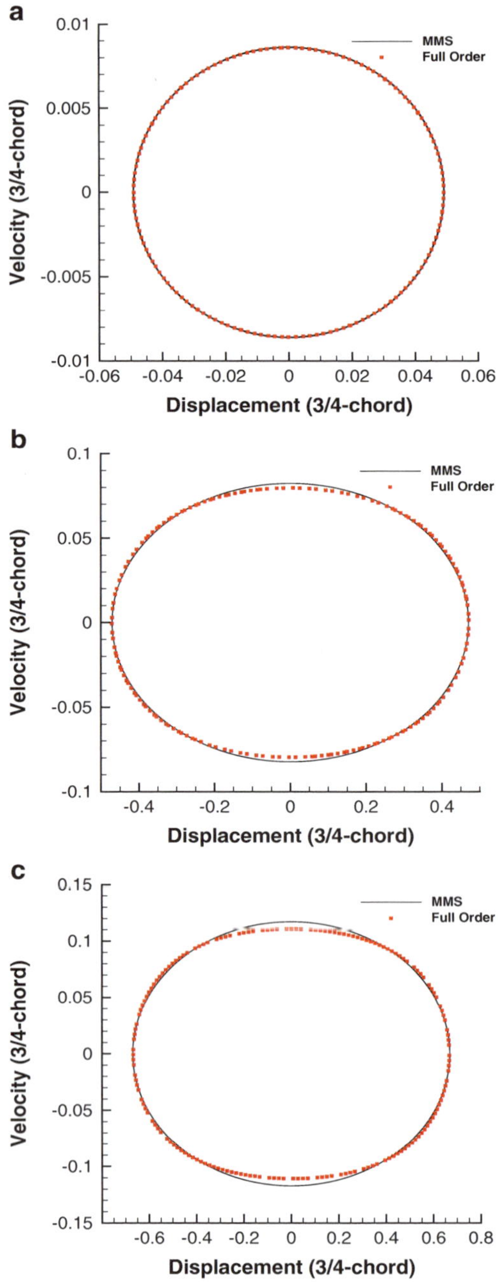


**Fig. 11** Comparisons of full-order simulation of panel LCO response with MMS predictions at selected values of $\lambda$: (**a**) $\lambda = 3{,}420$, (**b**) $\lambda = 4{,}000$, and (**c**) $\lambda = 4{,}600$

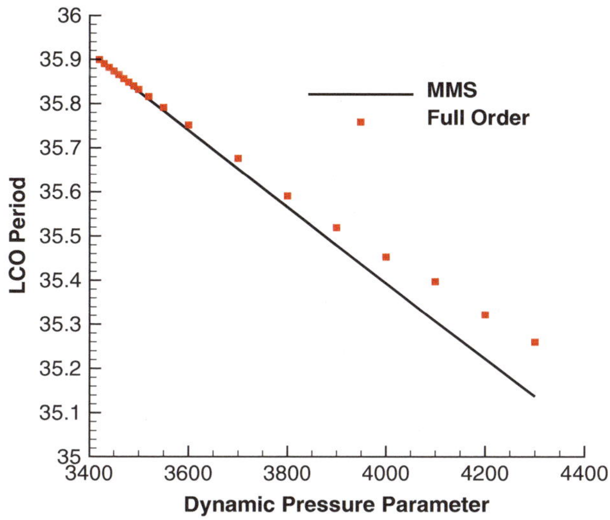


**Fig. 12** Comparison of LCO periods predicted by full-order simulation and MMS predictions

where $\Pi = \Pi(\mu)$ is a function dependent on $\mu$ that casts $a$ as a deflection at 3/4-chord (where deflection is assumed maximal). Taking the partial derivative with respect to $\mu$ yields (i.e., at a fixed value of $\lambda$):

$$\frac{\partial A_{\text{LCO}}}{\partial \mu} = -\frac{\Pi}{2\sqrt{\lambda - \lambda^*}}\left(\frac{\partial \lambda^*}{\partial \mu}\right) + \frac{d\Pi}{d\mu}\sqrt{\lambda - \lambda^*} \tag{50}$$

The first term of the right-hand side of (50) captures the dependence of LCO amplitude on the location of the bifurcation point (i.e., influence of the sensitivity of the linearized equations to $\mu$). This information is easily obtained using the formulation described in Sect. 3.2. The second term describes the variation of the inherently nonlinear dynamics on $\mu$. Computing the second term is challenging analytically, since there are many terms arising in (45) and (46) that require specialized treatment. For the sake of brevity, finite-differences will be used here to capture these variations.

A trial case is examined where $\partial A_{\text{LCO}}/\partial \mu$ is computed about the LCO at $\lambda = 3{,}420$ and $\mu = 0.1$, with $\mu$ specified to be the mass ratio (the convention defined in Table 1). Using the full-order model, LCO amplitudes are found at two close values of mass ratio: 0.1 ($A_{\text{LCO}} = 0.049116$) and 0.101 ($A_{\text{LCO}} = 0.048682$), leading to $\partial A_{\text{LCO}}/\partial \mu \approx -0.434$. This result makes sense, since the effect of increasing mass ratio is to raise the flutter speed. Evaluation of $A_{\text{LCO}}$ using MMS at $\mu = 0.10$ and $\lambda = 3{,}420$ yields $A_{\text{LCO}} = 0.04900344$; (49) then implies $\Pi(0.1) = 0.019431$. We also found previously that $\partial \lambda^*/\partial \mu = 111.864$. The contribution to the sensitivity from movement of the bifurcation point is

$$\left.\frac{\partial A_{\text{LCO}}}{\partial \mu}\right|_{\text{moving bif pt}} = -\frac{\Pi(\mu)}{2\sqrt{\lambda - \lambda^*}}\left(\frac{\partial \lambda^*}{\partial \mu}\right) \approx -0.4310. \tag{51}$$

This result is almost identical to the sensitivity of the full-order amplitude. Thus, most of the contribution to the observed sensitivity arises from the movement of the bifurcation point in response to changes in $\mu$. To verify this, the contribution to changing bifurcation conditions is evaluated. Further analysis yields $\dot{\Pi}(0.1) = 0.00019$, which when inserted into the last term of (50) yields

$$\left.\frac{\partial A_{\text{LCO}}}{\partial \mu}\right|_{\text{changing bif conditions}} = \dot{\Pi}(\mu)\sqrt{\lambda - \lambda^*(\mu)} \approx 0.000479. \tag{52}$$

Put differently, the amplitude of LCO *in this problem with its assumed parameter values* derives its sensitivity to parameters from the sensitivity of bifurcation location to these same parameters, not from higher order dependencies of the nonlinear dynamics on these parameters. Thus, for parameter values not too different from that assumed here, it should be adequate to assess variability in panel LCO amplitude from the perspective of assessing variability in bifurcation location. Other problems, or this problem with dramatically different parameter values, may exhibit relatively larger contributions from the nonlinear terms.

## 5 Uncertainty Quantification of Aeroelastic Responses

Flutter and LCO are now considered from an uncertainty quantification perspective. Sections 5.1–5.4 examine the uncertainty quantification of flutter, whose onset is governed by the behavior of linearized dynamical equations. In contrast, Sect. 5.5 is dedicated to the uncertainty quantification of LCO, whose response characteristics are fundamentally nonlinear. While the panel problem is used herein as the general basis for the uncertainty quantification of both flutter and LCO, the problems are individually formulated in ways that render the flutter and LCO results not directly comparable. Indeed, for LCO, the panel problem is crafted in a manner intended to cause LCO amplitude, but not flutter speed, to be sensitive to certain structural parameters.

An ensemble of flutter solutions is first computed for a family of panels, whose thicknesses are random, using both sampling and sensitivity-based techniques. Then, the probability that a panel will fail to flutter, subject to variability in a bulk parameter and a boundary condition parameter (selecting between a pinned or clamped condition), is computed using sampling and a First-Order Reliability Method. *Beyond the idea that thickness of components will always vary during the manufacturing process, or that the panel is an idealization of a more complicated structure whose properties are essentially random in a manner that is convenient to represent as thickness or boundary-condition variation, no motivation is given for the way in which the panels are randomized. Emphasis is instead given to how these uncertainties propagate through the aeroelastic analysis*. For LCO, uncertainty is again introduced in the boundary conditions, but through parametric uncertainty in nonlinear torsional springs and linear in-plane strings that affix the panel to a supporting structure.

While not described here, but of relevance to the manuscript topic, Stanford and Beran studied the reliability of plate-like wings in supersonic flow, subject to an LCO constraint [42]. They modeled aerodynamics with piston theory and accurately captured LCO without using a perturbation approach. One particularly noteworthy paper is that of Ghommem et al., who united MMS, aeroelasticity and uncertainty quantification [25].

## *5.1 Assessment of Flutter for Panels of Random Thickness*

The equations of motion for a panel of variable thickness, in non-dimensional form, are idealized as

$$\frac{\mu}{\lambda}\frac{\partial^4 \hat{w}}{\partial \hat{x}^4} + \frac{3\mu}{\lambda}\left[\frac{2}{\hat{h}^2}\left(\frac{\partial \hat{h}}{\partial \hat{x}}\right)^2 + \frac{1}{\hat{h}}\frac{\partial^2 \hat{h}}{\partial \hat{x}^2}\right]\frac{\partial^2 \hat{w}}{\partial \hat{x}^2} - \hat{N}_x\frac{\partial^2 \hat{w}}{\partial \hat{x}^2} + \frac{\partial^2 \hat{w}}{\partial \hat{t}^2} = -F, \tag{53}$$

$$\hat{N}_x = \frac{6\mu}{\lambda}\left(\frac{h_0}{L}\right)^{-2}\left(1-\nu^2\right)\int_0^1 \frac{1}{\hat{h}^2}\left(\frac{\partial \hat{w}}{\partial \hat{x}}\right)^2 d\hat{x}, \tag{54}$$

where $F$ is defined above (it is assumed that variations in thickness only influence the structural response and not the aerodynamics; removing this restriction is a very interesting departure point for linking with uncertainty quantification in computational fluid dynamics). The dimensional thickness distribution is given by

$$h(\hat{x}) = h_0\hat{h}(\hat{x}), \quad \hat{h}(\hat{x}) = 1 + r(\hat{x}), \tag{55}$$

where $r(\hat{x})$ is a random field over the panel. The realization of spatially correlated random fields is in itself an important discipline. Here, a 1D random field is desired with a specified correlation length. Pettit and Beran [38] considered the propagation of uncertainty for a model like that studied here, except that: (1) variability was injected into the coefficient of the nonlinear term, and (2) only static responses to a uniform load were recorded stochastically. Their main finding was that to capture the correct behavior in the tails of the stochastic response, more refined grids were required. While significant, this observation was not utilized or studied here to preserve the simplicity of the presentation.

Realizations of random panels are constructed using the approach of Grigoriu [28]. The thickness distribution of the panel at each grid point, $\hat{x}_n$ $\left(n = 1, \ldots, N_{\text{grid}}\right)$, is given by

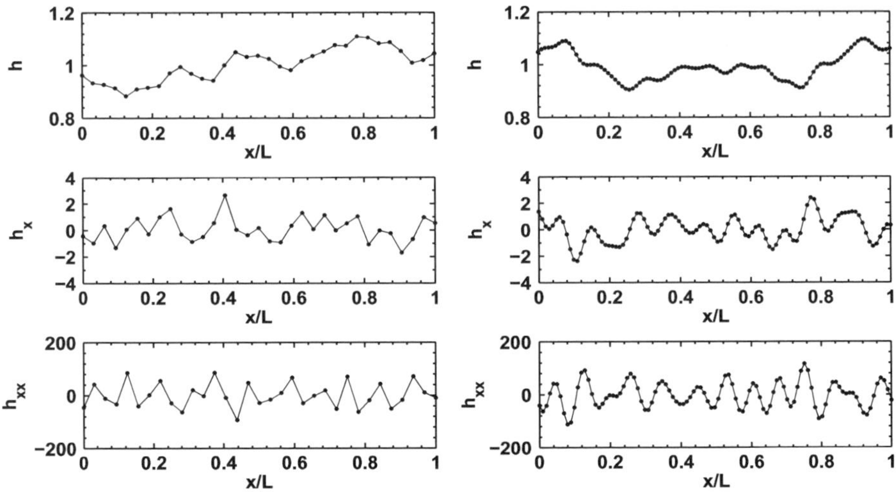


**Fig. 13** (*Left*) Realization of random panel and thickness distribution with first and second derivatives for the baseline grid; (*Right*) Realization of random panel with derivatives for a panel discretized by 101 grid points. For both figures, a large coefficient of variation of 0.1 is assumed

$$\hat{h}(\hat{x}_n) = 1 + \sum_{k=1}^{N_\nu} \left[ A_k \cos(\nu_k \hat{x}_n) + B_k \sin(\nu_k \hat{x}_n) \right], \tag{56}$$

where the amplitudes $A_k$ and $B_k$ are random variables corresponding to the $k$th frequency, $N_\nu = 10$, and the correlation length is 2/3. The amplitudes are independent normal random variables with zero mean and variance proportional to $\alpha_{\mathrm{CL}}^2/(\nu^2 + \alpha_{\mathrm{CL}}^2)$, where $\alpha_{\mathrm{CL}}$ is the inverse of the correlation length. Random numbers are generated with the MATLAB random number generator. To avoid egregious discretization errors in (53), $\partial\hat{h}/\partial\hat{x}$ and $\partial^2\hat{h}/\partial\hat{x}^2$ are computed analytically using (56).

An ensemble of 1,000 panel realizations was constructed assuming the coefficient of variation in the thickness to be 1 % (smaller than that shown in Fig. 13). The flutter speed of each panel was computed using the B-method. Values of $\lambda$ at flutter ranged from about 3,300 to about 3,500 in a fairly linear fashion, with a mean of 3,413, a standard deviation of 43 (resulting in a coefficient of variation of 1.3 %), a skewness of 0.04, and a kurtosis of 2.9. All realizations are shown in Fig. 14 in terms of flutter pressure parameter and frequency (all panels exhibited flutter). In a similar manner, the 1,000 panel realizations were analyzed using the linearized form of the G-method. The linearized approach well predicted the statistical characteristics of the ensemble studied nonlinearly, indicating that for the perturbations considered the flutter point varies quite linearly. The statistics of the linearized approach is as follows: mean = 3,416 (a slight positive shift), standard deviation = 43, skewness = $8 \times 10^{-4}$ (essentially linear), and a kurtosis = 2.8. Changes in flutter

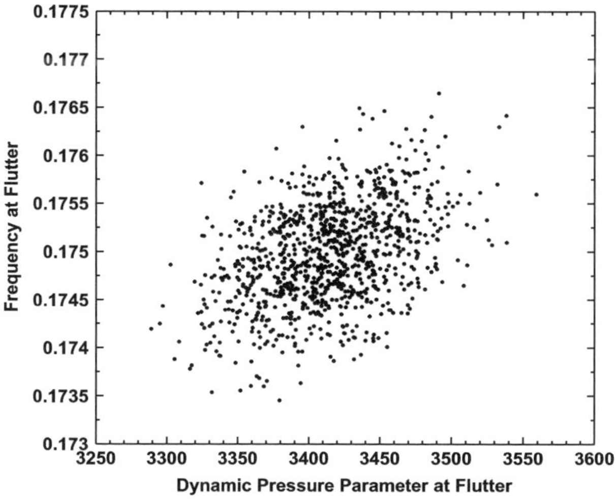


**Fig. 14** Distribution of flutter points for 1,000 random panels computed with direct B-method in terms of the dynamic pressure parameter and frequency

speed are predicted by pre-computing flutter sensitivities to thickness and the first two spatial derivatives of thickness and then multiplying each of these sensitivities by the perturbations assigned to each panel realization:

$$(\lambda^*)^k \approx (\lambda^*)^0 + \sum_{i=1}^{max(i)} \left[ \left(\frac{\partial \lambda^*}{\partial \hat{h}}\right)_i^0 \delta\hat{h} + \left(\frac{\partial \lambda^*}{\partial \hat{h}_{\hat{x}}}\right)_i^0 \delta\hat{h}_{\hat{x}} + \left(\frac{\partial \lambda^*}{\partial \hat{h}_{\hat{x}\hat{x}}}\right)_i^0 \delta\hat{h}_{\hat{x}\hat{x}} \right] \tag{57}$$

Here the subscript "$i$ " denotes the node of interest and the superscript "0" denotes the baseline panel. Panel realizations are shown in Fig. 14 as predicted by the B-method (Sect. 3.3) and the perturbation method (Sect. 3.2). It can be seen that the two methods are in close agreement. A systematic study of the errors was not carried out, but it was seen that the largest contributions to flutter deviation by grid point corresponded to the second-order terms, for which the large values of $\delta\hat{h}_{\hat{x}\hat{x}}$ are driven by the highest frequency component of the panel thickness variation. These locally large changes may stress the assumption of linearity, but may be more or less significant depending on how contributions sum across the panel.

The nonlinear results of all 1,000 panels were obtained with about 1 min of CPU time on a laptop, which indicates the efficiency of the B-method, particularly when solutions are grouped near to each other. In these calculations, each flutter point was used as an initial condition for the next flutter calculation. Typically, only a few Newton iterates were used per flutter point, and the remaining iterates were carried out with a modified-Newton method using a frozen and decomposed Jacobian of the expanded system. The linearized results were obtained by first computing the flutter speed of the baseline panel with the B-method and then computing sensitivities

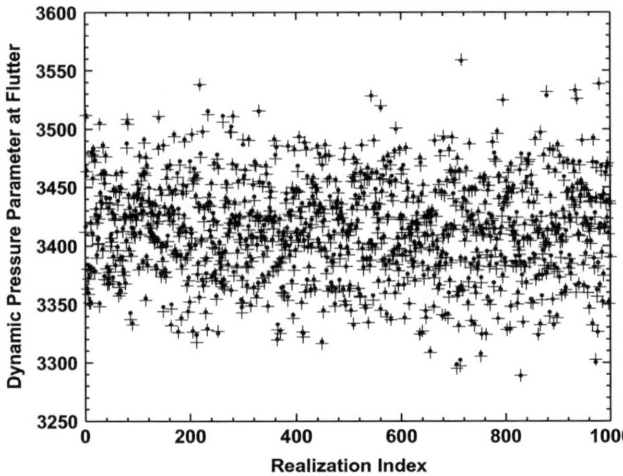


**Fig. 15** Comparison of flutter point ensembles for 1,000 random panels computed linearly via perturbation analysis (*dots*) and nonlinearly via the B-method (*crosses*) in terms of the realization index and flutter pressure parameter

about this point to the thickness and the first two derivatives of thickness at each of the 31 grid points using perturbation analysis. With this approach, all 1,000 panels were assessed in about 1 s (see Fig. 15).

Finally, a larger ensemble of 10,000 panels is studied by generating random thickness assuming a large coefficient of variation (COV) equal to 5 %. While perhaps unrealistic, it can be seen that this large value of COV generates panels that are unsafe, even when a flutter margin is imposed. "Failure" is defined to mean exceeding 85 % of the flutter "speed" of the baseline panel (3,416), or about $\lambda = 2{,}900$ (the dynamic pressure parameter is proportional to the square of velocity; satisfying a 15 % margin on velocity implies a larger margin on dynamic pressure, but 15 % is used here for illustrative purposes). 82 panels have flutter speeds below 2,900, leading to a probability of failure of slightly less than 1 % (Fig. 16).

## 5.2 Impact of Other Structural Nonlinearities on Flutter

Many other structural parameters can, and should, be considered uncertain. For example, the pre-load $R_x$ in (16) is naturally uncertain, with variation occurring through assembly, thermal effects, and aircraft aging. Also, the nature of the boundary condition is uncertain. The boundary conditions studied herein are that of a pinned panel. A clamped condition has also been extensively studied and better resembles a panel fixed to a supporting structure. However, it is not unreasonable to realize variability between the extremes of clamped and pinned structures [7] that can be captured through linear interpolation:

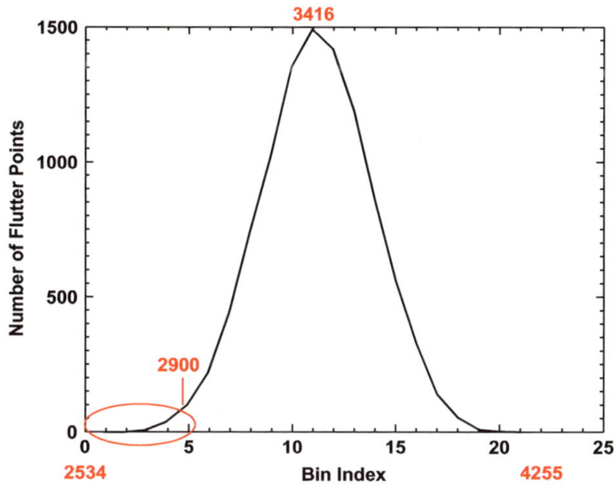


**Fig. 16** Unscaled probability distribution (21 bins) of flutter points generated from 10,000 random panels assuming a COV = 5 %. Eighty-two panels fail by fluttering below a "safe" value of $\lambda = 2{,}900$ (a probability of failure <1 %)

$$BC = \alpha \times (BC)_{\text{clamped}} + (1 - \alpha) \times (BC)_{\text{pinned}}, \tag{58}$$

where $\alpha$ is a real parameter between 0 and 1, and $BC$ is a symbolic representation of the type of boundary condition enforced. Here, $\alpha$ is 0 when the panel is pinned and 1 when the panel is clamped, and can be considered random with certain statistical properties. It should be noted that negative values of $R_x$ rapidly diminish flutter speed (a compressive load increasing susceptibility to buckling) while positive values of $\alpha$ increases flutter speed in a weakly nonlinear fashion. This latter result is a reminder that while linearized, the appearance of flutter is governed by a stability exchange whose character is potentially nonlinear.

Simultaneous variation of $R_x$ and $\alpha$ is now considered, where $\alpha$ is associated with the boundary condition at $x = 0$; the other end of the panel is assumed to be pinned. The parameters are first considered to be deterministic, and flutter boundaries (curves comprised of flutter points) are computed for variations in these parameters for specified values of $\lambda$. These curves in the 2D $\alpha - R_x$ space are slices of a flutter surface lying in the 3D $\lambda - \alpha - R_x$ space. From these curves, a "design" point in this space is selected in Sect. 5.3 that deterministically is flutter-free. Once the design point is selected, $\alpha$ and $R_x$ are treated as stochastic. Pseudo-random (a sequence of numbers generated by machine algorithm intended to approximate the properties of random numbers, and generally distinguished from low-discrepancy sequences often called quasi-random) samples of $\alpha$ and $R_x$ are collected, centered about the design point with $\lambda$ held fixed ($\lambda = \lambda_D$). As in Sect. 5.1, the flutter speeds of the members of the ensemble are computed, and those members for which $\lambda^* \leq \lambda_D$ are considered to have failed, since they will experience flutter at or below

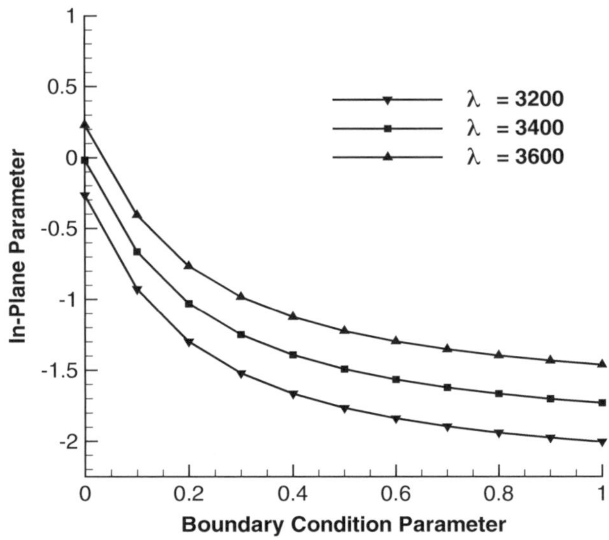


**Fig. 17** Flutter boundaries in the boundary condition parameter ($\alpha$) and the in-plane parameter ($R_x$) for selected values of $\lambda$. For each value of $\lambda$, flutter is encountered below the flutter boundary

the design speed. A probability of failure using Monte-Carlo simulation is then computed as a ratio of the number of failures to the ensemble size, which is varied to achieve a reasonably converged solution. Finally, in Sect. 5.4, these simulation-based results are compared to a probability of failure predicted using the first order reliability method (FORM).

Flutter boundaries in the $\alpha - R_x$ space are found to be nonlinear, primarily reflecting a nonlinear relationship between system damping and $\alpha$. Curves of flutter points are shown in Fig. 17 for three different values of $\lambda$. These paths are computed by modifying the B-method to treat $R_x$ as the flutter parameter. A flutter point is first computed for the specified value of $\lambda$ with $\alpha = 0$; the path is formed by repeating the bifurcation calculation for values of $\alpha$ extending to 1. Examination of the results indicates that flutter occurs below the flutter boundaries, which makes sense, since making the in-plane parameter more negative represents a panel under increasing compressive load and closer to departing from an un-deflected state (the panel in vacuum will buckle at $R_x = -1$; aerodynamic loads stabilize the system, but with diminished effectiveness as $R_x$ becomes increasingly negative), and since decreasing the boundary condition parameter loosens the panel in a rotation sense and renders it more susceptible to flutter. As the value of $\lambda$ is increased, the curves elevate (move in the direction of increasing $R_x$), since less in-plane load is required to trigger flutter.

### 5.3 Computing Flutter Probability of Failure using Monte-Carlo Simulation

As a demonstration of estimating the probability of failure (Pf), a "design" or baseline point is selected at which the panel is known to be stable, and then pseudo-random variations in $\alpha$ and $R_x$ are generated about this point, leading some points to fall within the unstable domain. For these studies, calculations are restricted to $\lambda = \lambda_D = 3{,}400$. In the $\alpha - R_x$ space, the design point is selected to lie at $\alpha_D = 0.3$ and $R_{xD} = -0.75$, which is shown relative to the flutter boundary in Fig. 18a. As the design point lies in the stable domain, any variations in $\alpha$ and $R_x$ must be sufficiently large for the realization to represent an unstable panel. Values of $\alpha$ and $R_x$ are drawn from Gaussian distributions (this ideal choice is made for the sake of creating a simple and effective presentation; there is little justification, particularly for the parameter $\alpha$, which could be indicative of a structural flaw):

$$\alpha = \alpha_D + \sigma_\alpha x_1, \qquad R_x = R_{xD} + \sigma_{R_x} x_2, \tag{59}$$

where $x_1$ and $x_2$ are standard normal random variables (selected in a pseudo-random manner using the MATLAB `randn` function) and $\sigma = [\sigma_\alpha, \sigma_{R_x}] = [0.06, 0.15]$. An ensemble of 10,000 pseudo-random samples is shown in Fig. 18b; 79 points lie in the unstable domain, resulting in Pf $= 0.0079$. Convergence of this estimate with respect to ensemble size is demonstrated by creating ensembles of different quantities of samples (without resetting the pseudo-random number generator). Estimates for ensembles of size 1,000, 5,000, 10,000, 20,000, and 40,000 are 0.001, 0.0044, 0.0079, 0.0078, 0.0077, respectively. While not converged fully in the second digit (Monte-Carlo converges slowly, with error diminishing proportional to $1/\sqrt{N_{\text{MCS}}}$, where $N_{\text{MCS}}$ is the ensemble size), it is clear that there is little sensitivity in Pf beyond 10,000 samples.

More efficient sampling techniques (e.g., the Latin Hypercube scheme) could be easily applied to this problem. Monte-Carlo Simulation is employed here to achieve a simple demonstration. In some cases, these methods may be required to achieve reasonable Pf estimates under a moderate computational budget.

As described above, the flutter speed of each sample is computed with the B-method; the panel is designated as to have failed if $\lambda^* \leq \lambda_D$. The reader should note that applying the B-method in this way is not inexpensive, even if the flutter speed of each sample is easily predicted, since there are so many samples. Furthermore, this approach does not take advantage of techniques to geometrically render the shape of the failure boundary from a much smaller sample set, which allows a sample of an ensemble to be assigned a classification of "fail" or "not-fail" based on its location relative to the failure boundary [6]. The B-method is employed here with Monte-Carlo simulation to provide a straightforward conceptual means of estimating the probability of failure.

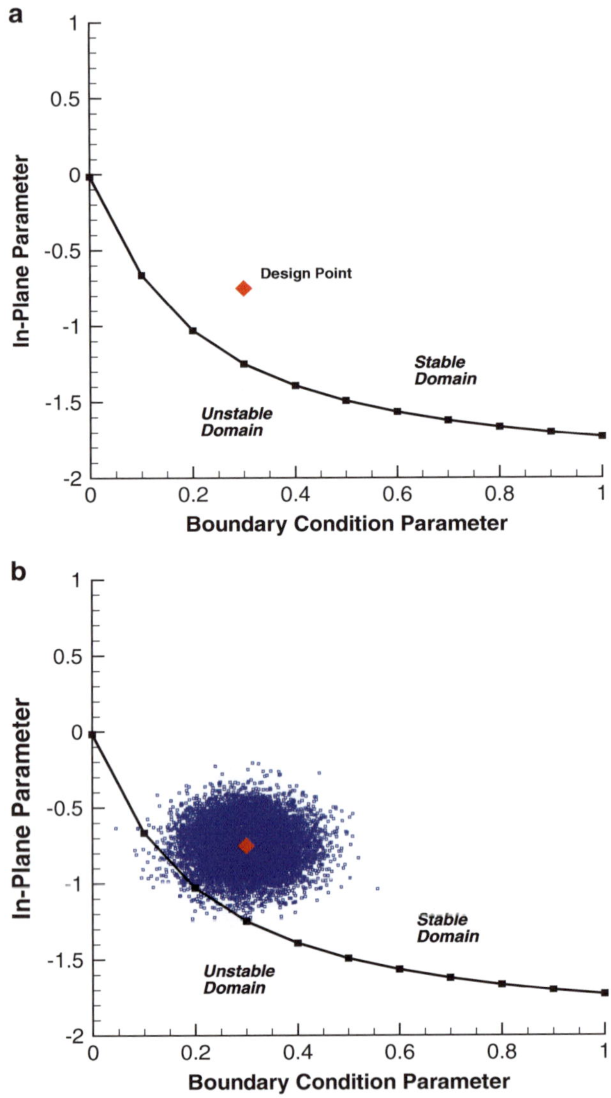


**Fig. 18** (**a**) Location of design point in stable domain associated with $\lambda_D = 3{,}400$; (**b**) locations of 10,000 Monte-Carlo samples, some of which represent unstable panels

## *5.4 Computing Flutter Probability of Failure Using the First-Order Reliability Method*

When the flutter boundary is fairly linear, a First-Order Reliability Method (FORM) can be used with sensitivity information to quickly approximate the probability of failure. Allen and Maute pioneered the application of FORM to aeroelastic models involving CFD [1, 2]. Stanford and Beran demonstrated this process for reliability design of flexible wings susceptible to LCO [42]. Other topical references include [39] and [36]. When applicable, FORM is greatly advantageous, since it can produce high-quality engineering results without the need of Monte-Carlo Simulation. However, when the flutter boundary has a high degree of curvature, the accuracy of FORM degrades, and other methods should be considered: e.g., a Second-Order Reliability Method (SORM) [23, 29], which requires costly second-order derivatives to be obtained, and SORM with Adaptive Approximations, which avoids exact computation of the second-order information [18].

A detailed derivation of FORM is not presented herein; the reader is encouraged to refer to the textbook by Choi et al. [18], the textbook by Melchers [30], or to some other source.

FORM has three basic steps: (1) transformation of the physical variable space into a space defined by standard normal random variables; (2) search over the failure surface (the flutter boundary here) for the most probable point on the surface to occur in this space, designated as MPP, and (3) calculation of Pf based on the distance between the MPP and the design point in the standard normal space. FORM provides an exact computation of Pf if the probabilistic distributions of the physical variables can be mapped into the standard normal space, and if the failure boundary (or hypersurface in higher dimensions) is linear.

In the problem defined above, two standard normal random variables have already been defined, $x_1$ and $x_2$, and the appropriate transformation is given by (59). (For problems involving non-normal distributions, a Rosenblatt transformation is employed to map the problem to standard normal space.) In the transformed $x_1 - x_2$ space, the joint probability density function can be represented as a set of concentric circles about the origin, the transformed mean of the distribution. Higher probabilities of occurrence lie closer to the origin. Thus, the MPP is the point closest to the origin in terms of the Euclidean distance, $d = \sqrt{x_1^2 + x_2^2}$. In terms of the original variables:

$$d = \sqrt{\left(\frac{\alpha^* - \alpha_D}{\sigma_\alpha}\right)^2 + \left(\frac{R_x^* - R_{xD}}{\sigma_{R_x}}\right)^2} \tag{60}$$

where $\alpha^*$ and $R_x^*$ identify a point on the flutter boundary. The MPP is designated to occur at $\alpha^* = \alpha_{\text{MPP}}$ and $R_x^* = R_{x\text{MPP}}$ with $\beta = min(d)$. The third step relates the value of the reliability index $\beta$, the distance to the MPP, to the probability of failure through the equation

$$\text{Pf} = \Phi(-\beta) = \frac{1}{2}\left(1 - \text{erf}\left(\frac{\beta}{\sqrt{2}}\right)\right), \tag{61}$$

where $\Phi(\cdot)$ is the cumulative distribution function of the standard normal distribution.

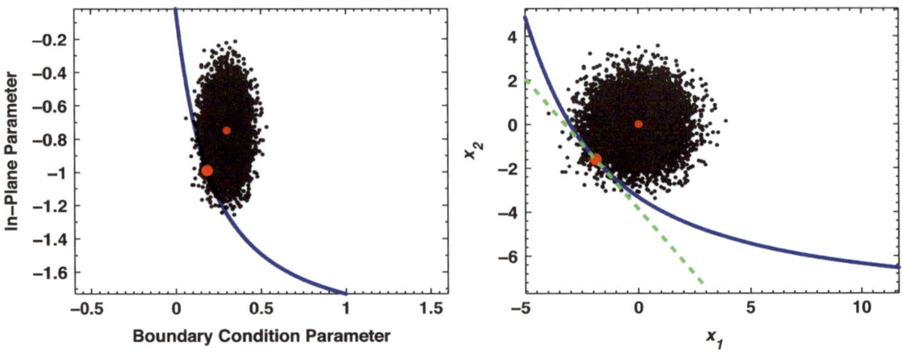


**Fig. 19** Locations of flutter boundary, MPP, design point ($\lambda_D$ = 3,400), and 10,000 samples: (*left*) physical space; (*right*) standard normal space (tangent line at MPP denoted by a *green dashed line*)

The formulation just described is now applied to computing Pf for the panel problem examined in Sect. 5.3 using Monte-Carlo simulation. For this computation, the flutter boundary is formed using a set of closely spaced points computed with the B-method. At each point, $d$ is computed and the minimum value assigned to $\beta$. Evaluation of the simple expression in (61) yields the estimate Pf = 0.0065, which is an approximation since the failure surface is not linear. This is clearly evident in Fig. 19a, b, which shows the flutter boundary, the MPP, the design point, and the samples of the 10,000-sample ensemble in physical and standard normal space, respectively. Figure 19b shows the MPP to be the closest point on the failure surface to the design point, occurring at a point of tangency between the failure curve and the circular probability contour passing through the MPP. It is important to note that the estimate of Pf provided by FORM is *non conservative* in this problem: i.e., $\mathrm{Pf_{FORM}}(0.0065) < \mathrm{Pf_{MCS}}(0.0077)$. This occurs because the curvature of the failure surface enlarges the failure region compared to that produced by a straight line passing through the MPP, and locally tangent to the failure surface (dashed line in Fig. 19b).

Use of computational sensitivities of flutter speed to various parameters has two roles in the FORM process. First, sensitivities as part of a gradient-based and reliability-based optimization strategy can be used to find the MPP on the failure surface. Second, sensitivities of flutter to various key parameters can then be used to re-design a structural component to meet a constraint on its reliability, $R$, where

$$R \equiv 1 - \mathrm{Pf}. \tag{62}$$

These sensitivities are efficiently computed using the techniques described in Sect. 3. The posing of an aeroelastic design problem, and the description of the procedure by which the problem is solved, is outside the scope of this chapter. The interested reader is referred to Refs. [42] and [43], in which certain canonical configurations are studied from deterministic and reliability perspectives.

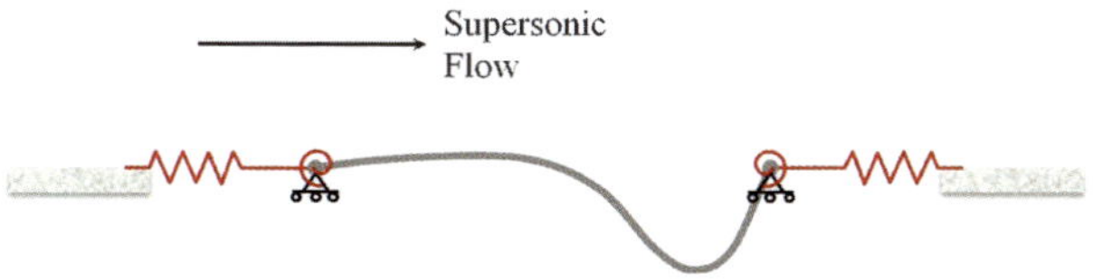


**Fig. 20** Schematic of a flexible panel in supersonic flow (infinite in spanwise direction), with linear axial springs and cubic rotational springs at either end

It is also noted that in extending Sect. 3, this section only addresses reliability from a linearized perspective: i.e., quantifying the risk of the system losing stability to infinitesimal disturbances. However, it may be more important to quantify the risk of the system exhibiting an undesirable nonlinear, aeroelastic behavior. We discuss this quantification in the next section, supported by the material presented above on LCO.

## 5.5 Uncertainty Quantification of LCOs

The previous exercises have all labeled "probability of failure" as the probability that a panel structure will flutter. This final section considers non-deterministic limit cycle behavior, where the character of the LCO is efficiently computed with MMS. Specifically, *a structure is designated to fail if it experiences a subcritical limit cycle oscillation* (i.e., $\beta_{2r}$ is greater than zero). As noted above, the destabilizing nonlinearities inherent to a subcritical LCO can lead to a dangerously large amplitude branch which exists at speeds above and below the flutter point. Contrastingly, supercritical LCOs ($\beta_{2r} < 0$, as seen in Fig. 4) present relatively benign behavior, with small LCOs that monotonically grow with flight speed. For this exercise, stochastic parameters are selected that only affect the nonlinear aspects of the problem, such that the linear flutter point is never altered. This section also considers the case of non-normal random variables.

The panel structure considered here is seen in Fig. 20, an extension of that used in Fig. 3. The sole differences between the two figures are in the boundary conditions at either end of the panel. Rather than pinning these edges, the configuration in Fig. 20 allows either edge to travel in the axial direction, where translational movement is constrained by linear springs. The rotation of the two ends ($w_x$) is also limited by two torsional cubic springs, which may be hardening or softening. These torsional springs are essentially nonlinear, and have no linear rotational resistance. The modifications to the panel equations of motion (Eq. (16)) needed to capture the effect of these end springs are not detailed here, though the interested reader is referred to Refs. [7] and [22].

Similar to Eq. (58), parametric variations in these spring constants are meant to emulate uncertainties in the panel's boundary conditions, specifically in how

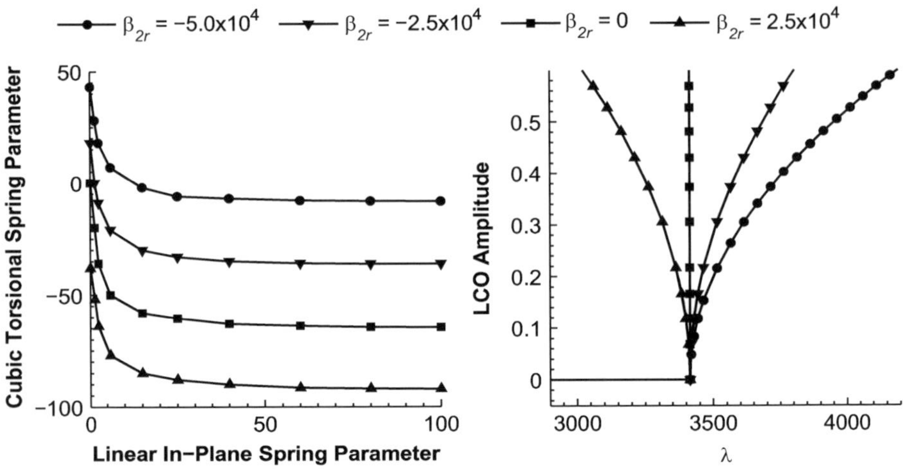


**Fig. 21** LCO behavior as a function of the linear in-plane spring parameter and the cubic torsional spring parameter for selected values of $\beta_{2r}$. For negative values of $\beta_{2r}$, the LCO is supercritical

the panel is connected to a larger supporting substructure. The linear axial spring constant may take any value between 0 and infinity, while the cubic rotational spring constant may have any value at all, where a negative (positive) constant provides softening (hardening) behavior. If the axial spring constants are set to a very large value, and the cubic rotational springs are set to zero, then the setup of Fig. 20 reverts to that of Fig. 3.

As noted above, changing any of these spring constants affects the problem in an essentially nonlinear manner, with no effect at all upon the linear flutter point. That this is true for the cubic rotational springs is obvious, but it is also true for the linear axial springs, as they have no pre-load (i.e., $R_x$), and there is no linear coupling between in-plane and out-of-plane vibrations. Only a nonlinear coupling exists, via Eq. (14). These spring constant parameters then isolate changes in the resultant LCO behavior, as intended. Typical results are given in Fig. 21 in terms of $\beta_{2r}$ ($\beta_1$ is only dependent upon the linear system response about the flutter point, and is therefore always equal to the value found above: $2.22 \times 10^{-4} - 8.23 \times 10^{-6}i$). The in-plane spring parameter is the product of the translational spring constant and $Eh$, while the cubic torsional spring parameter is the product of the rotational spring constant and $ELh^2$. The values of the spring constants are assumed to be symmetric across the panel structure.

For very large values of the in-plane spring constant, and a cubic rotational spring set to zero, the boundary conditions of Fig. 20 revert back to Fig. 3, and $\beta_{2r} = -6.01 \times 10^4$ as found in Eq. (48). The negative sign indicates a supercritical LCO, driven by the hardening nonlinearities of Eq. (14). Decreasing the in-plane spring constant from this baseline point down to zero gradually drops $\beta_{2r}$ to zero as well, which means that the nonlinearities have been completely removed from the system. Because the panel is allowed to axially retract at will, its arc-length doesn't change in response to a transverse deflection, and so $N_x$ of Eq. (14) is always zero. Adding a cubic rotational spring with a positive constant shifts the curves of Fig. 21 up, further hardening the LCOs, and decreasing the resulting $\beta_{2r}$. Alternatively, a negative spring constant softens the LCOs: a sufficiently low value offsets the inherent hardening nonlinearities of the panel structure, and the character of the LCO can become subcritical ($\beta_{2r} > 0$).

Next, both the spring constants (torsional and linear) are given parametric uncertainties. A Gaussian distribution is used for the cubic torsional spring parameter, with a mean value of 0 and a standard deviation of 25: a conversion from the physical space to the standard normal space is as written in Eq. (59). Alternatively, a Weibull distribution is used for the linear in-plane spring parameter, a choice which reflects the fact that this spring constant is bounded between 0 and infinity (a similar argument could have been used for the boundary condition parameter $\alpha$ above). The Rosenblatt transformation between the physical variable space $r$ and the standard normal space $x$ (needed for FORM) is

$$x = \Phi^{-1}(1 - e^{-(r/\alpha_k)^{\gamma_k}}), \tag{63}$$

where $\alpha_k$ is a scale factor, and $\gamma_k$ is a shape factor. The former is set to 100, and the latter to unity. Because of its unit shape factor, the mean axial spring stiffness is also equal to 100, which referencing Fig. 21, is large enough to be considered essentially rigid. In a non-deterministic sense, however, much lower values (and softer LCOs) are possible.

A limit state function is generated such that the panel will fail if $\beta_{2r} > 0$ (subcritical LCO); for the mean design (in-plane spring parameter = 100, cubic torsional spring parameter = 0), $\beta_{2r} = -6.01 \times 10^4$ as found in Eq. (48). Monte-Carlo sampling is then used to estimate the probability of failure, where the MATLAB `randn` function is used to generate samples of the cubic torsional spring parameter, and the `wblrnd` function is used for the in-plane spring parameter. Pf estimates for ensemble sizes of 1,000, 5,000, 10,000, 20,000, and 40,000 are 0.0120, 0.0132, 0.0126, 0.0124, and 0.0122. As above, this process is not fully converged, but the sensitivity above 10,000 samples is weak. The ensemble of 10,000 samples is shown in the left plot of Fig. 22, where the failure boundary corresponds to the curve in Fig. 21 where $\beta_{2r} = 0$.

Finally, FORM can be used to estimate the probability of failure. The process is the same as that used above, except that the transformation from physical space into standard normal space is highly nonlinear, as dictated by Eq. (63). Even if the failure surface is linear in the physical space (which is not the case for this problem), the mapping into the standard normal space introduces curvature at the MPP, and the FORM-based Pf computation will not be correct. This mapping is shown on the right side of Fig. 22, where the nonlinear distortions are particularly strong for low values of the in-plane spring parameter. The MPP on this curve is clearly shown, again resulting in a non-conservative estimate for Pf of 0.0082. FORM is less accurate for this case as compared to the probabilistic flutter case, entirely due

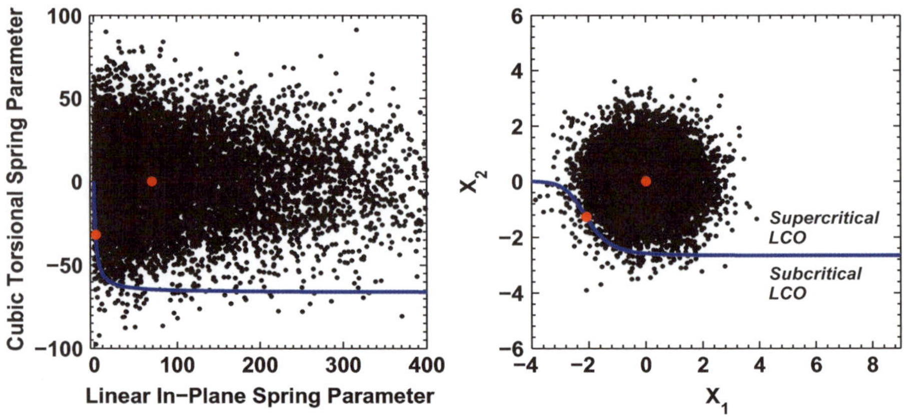


**Fig. 22** Locations of subcritical LCO boundary, MPP, design point, and 10,000 samples: (*left*) physical space; (*right*) standard normal space

to the stronger nonlinearities in the physical space (left side of Fig. 22), and the additional nonlinear distortions from the chosen Weibull distribution (right side of Fig. 22). It is also important to note that the origin in the standard normal space corresponds to the median in the physical space (in-plane spring parameter = 69.3, cubic torsional spring parameter = 0, plotted on the left of Fig. 22, along with the MPP), not the mean. These two points do not coincide for non-normal distributions.

# 6 Summary

A number of different aeroelastic concepts and analysis techniques have been examined with the goal of better understanding how to incorporate reliability concepts in aircraft design. Conventional aeroelastic analysis methods typically offer insufficient precision with which to assess flutter and LCO from a probabilistic standpoint. This manuscript presented two methods for greatly improving precision of flutter prediction, extended their use to the computation of flutter sensitivities, and demonstrated these methods on the problem of a flexible panel in high-speed flow, a simple yet meaningful aeroelastic problem. Sensitivities can be used to gain quantitative understanding of the effects of variation in parameters distributed throughout the aeroelastic system, thus providing a means for identifying the most critical parameters to address during design. The reliability index is an important tool for increasing system safety; a flutter-based First-Order Reliability Method was described to predict reliability of an aeroelastic system.

While this chapter addresses the multidisciplinary interaction of flutter from a probabilistic perspective, emphasis is given to the numerical modeling of the structure and its boundary conditions, and the potential complexity of the aerodynamic physics (e.g., the appearance of a shock or boundary layer) is ignored. Thus, there is a sizable gap between the concepts described here and material presented elsewhere in this text, which emphasizes CFD. This gap has two components. One contribution is fundamental to the aeroelastic interaction: we have said little about how uncertainties in the aerodynamic model and uncertainties in the structural dynamic model interact, particularly in the presence of nonlinear physics. For example, we have not described how time lags introduced in the coupling of the two disciplines influences stability (nor have we contended with the broader issue of numerical convergence).

A second contribution to the gap arises from the methods used to model airflows. Typically, time-domain methods are employed in CFD, especially for aeroelastic problems. As we have seen, time-domain methods inject uncertainty into flutter analysis, obscuring the quantification of other sources of uncertainty. This chapter describes more direct techniques for capturing changes in aeroelastic stability, which necessitate the re-formulation of many CFD techniques (or at least extend them, e.g., in the use of reduced order modeling) used to model the airflow.

Filling the gap between the concepts described herein and state-of-the-art CFD methods is the subject of active research throughout the aeroelasticity community. Progress in this area should enable the construction of more effective aircraft design methods, thereby reducing vehicle development time and cost. Hopefully, when physics are both difficult and costly to observe experimentally and constraining to vehicle operation (e.g., at high speed), these tools will prove beneficial.

**Acknowledgements** The authors wish to thank Prof. Ramana Grandhi (Wright State University) for many discussions that were helpful in the preparation of this manuscript, Dr. Chris Koehler (Universal Technology Corporation) for assistance in developing manuscript graphics, and Dr. Manav Bhatia (Universal Technology Corporation) for reviewing the manuscript. This work was sponsored by the Air Force Office of Scientific Research under Laboratory Task 03VA01COR (monitored by Dr. Fariba Fahroo).

## References

1. Allen, M., Maute, K.: Reliability-Based Design Optimization of Aeroelastic Structures. Structural and Multidisciplinary Optimization **27**(4), 228–242 (2004)
2. Allen, M., Maute, K.: Reliability-Based Shape Optimization of Structures Undergoing Fluid-Structure Interaction Phenomena. Computer Methods in Applied Mechanics and Engineering **194**(30), 3472–3495 (2005)
3. Ashley, H., Zartarian, G.: Piston Theory - A New Aerodynamic Tool for the Aeroelastician. Journal of the Aeronautical Sciences **23**(12), 1109–1118 (1956)
4. Badcock, K., Woodgate, M.: Bifurcation Prediction of Large-Order Aeroelastic Models. AIAA Journal **48**(6), 1037–1046 (2010)
5. Badcock, K., Timme, S., Marques, S,. Khodaparast, H., Prandina, M., Mottershead, J., Swift, A., Da Ronch, A., Woodgate, M.: Transonic Aeroelastic Simulation for Instability Searches and Uncertainty Analysis. Progress in Aerospace Sciences **47**(2), 392–423 (2011)
6. Basudhar, A., Missoum, S.: Update of Explicit Limit State Functions Constructed Using Support Vector Machines. AIAA 2007-1872, April (2007)

7. Beloiu, D., Ibrahim, R., Pettit, C.: Influence of Boundary Conditions Relaxation on Panel Flutter with Compressive In-Plane Loads. Journal of Fluids and Structures **21**(2), 743–767 (2005)
8. Bendiksen, O.O.: Review of Unsteady Transonic Aerodynamics: Theory and Applications. Progress in Aerospace Sciences (47), 136–167 (2011)
9. Bendiksen, O.O.: Unsteady Aerodynamics and Flutter Near Mach 1: Aerodynamic and Stability Reversal Phenomena. IFASD 2011-091, Paris, June (2011)
10. Beran, P.S., Lucia, D.J.: A Reduced Order Cyclic Method for Computation of Limit Cycles. Nonlinear Dynamics **39**(1–2), 143–158 (2005)
11. Beran, P.S., Morton, S.A.: A Continuation Method for the Calculation of Airfoil Flutter Boundaries. Journal of Guidance, Control and Dynamics **20**(6), 1165–1171 (1997)
12. Beran, P.S.: Computation of Limit-Cycle Oscillation Using a Direct Method. AIAA 1999-1462, April (1999)
13. Beran, P.S., Pettit, C.L., Millman, D.R.: Uncertainty Quantification of Limit-Cycle Oscillations. Journal of Computational Physics **217**(1), 217–247 (2006)
14. Beran, P.S.: A Domain-Decomposition Method for Airfoil Flutter Analysis. AIAA 1998-506, January (1998)
15. Bisplinghoff, R., Ashley, H., Halfman, R.: Aeroelasticity. Addison-Wesley, Cambridge, MA (1955)
16. Chandiramani, N., Librescu, L., Plaut, R.: Flutter of Geometrically-Imperfect Shear-Deformable Laminated Flat Panels Using Non-Linear Aerodynamics. Journal of Sound and Vibration **192**(1), 79–100 (1996)
17. Chen, P.C.: Damping Perturbation Method for Flutter Solution: The g-method. AIAA Journal **38**(9), (2000)
18. Choi, S-K., Grandhi, R.V., Canfield, R.A.: Reliability-based Structural Design. Springer, London (2007)
19. Denegri, C.: Limit Cycle Oscillation Flight Test Results of a Fighter with External Stores. Journal of Aircraft **37**(5), 761–769 (2000)
20. Dribusch, C., Missoum, S., Beran, P.: A Multifidelity Approach for the Construction of Explicit Decision Boundaries: Application to Aeroelasticity. Structural Multidisciplinary Optimization **42**(5), 693–705 (2010)
21. Dowell, E., Edwards, J., Strganac, T.: Nonlinear Aeroelasticity. Journal of Aircraft **40**(5), 857–874 (2003)
22. Dowell, E.H.: Nonlinear Oscillations of a Fluttering Plate. AIAA Journal **4**(7), 1267–1275 (1966)
23. Eldred, M., Bichon, B.: Second-Order Reliability Formulations in DAKOTA/UQ. AIAA 2006-1828 (2006)
24. Franklin, J.N.: Matrix Theory. Prentice-Hall, Englewood Cliffs (1968)
25. Ghommem, M., Hajj, M., Nayfeh, A.: Uncertainty Analysis near Bifurcation of an Aeroelastic System. Journal of Sound and Vibration **329**(16), 3335–3347 (2010)
26. Gilliatt, H., Strganac, T., Kurdila, A.: An Investigation of Internal Resonance in Aeroelastic Systems. Nonlinear Dynamics **31**, 1–22 (2003)
27. Griewank, A., Reddien, G.: The Calculation of Hopf Points by a Direct Method. IMA Journal of Numerical Analysis **3**(1), 295–303 (1983)
28. Grigoriu, M.: Stochastic Calculus: Applications in Science and Engineering. Birkhauser, Boston (2002)
29. Koyluoglu, H., Nielsen, S.: New Approximations for SORM Integrals. Structural Safety, **13**(6), 235–246 (1994)
30. Melchers, R.: Structural Reliability: Analysis and Prediction. Wiley, Chichester, UK (1987)
31. Mignolet, M.P., Chen, P.C.: Aeroelastic Analyses with Uncertainty in Structural Properties. Proceedings of the AVT-147 Symposium: Computational Uncertainty in Military Vehicle Design, Athens, Greece, Dec (2007)
32. Morton, S., Beran, P.: Hopf-Bifurcation Analysis of Airfoil Flutter at Transonic Speeds. Journal of Aircraft **36**(2), 421–429 (1999)

33. Murthy, D., Haftka, R.: Derivatives of Eigenvalues and Eigenvectors of a General Complex Matrix. International Journal for Numerical Methods in Engineering **26**(2), 293–311 (1988)
34. Nayfeh, A.H., Balachandran, B.: Applied Nonlinear Dynamics. Wiley, New York (1995)
35. Nayfeh, A., Ghommem, M., Hajj, M.: Normal Form Representation of the Aeroelastic Response of the Goland Wing. Nonlinear Dynamics, DOI: 10.1007/s11071-011-0111-6 (2011).
36. Nikbay, M., Fakkusoglu, N., Kuru, M.: Reliability-Based Aeroelastic Optimization of a Composite Aircraft Wing via Fluid-Structure Interaction of High Fidelity Solvers. Materials Science and Engineering **10**(1), 1–10 (2010)
37. Paolone, A., Vasta, M., Luongo, A.: Flexural-Torsional Bifurcations of a Cantilever Beam Under Potential and Circulatory Forces II. Post-Critical Analysis. International Journal of Non-Linear Mechanics **41**(4), 595–604 (2006)
38. Pettit, C.L., Beran, P.S.: Convergence Studies of Wiener Expansions for Computational Nonlinear Mechanics. AIAA 2006-1993, May (2006)
39. Pettit, C., Grandhi, R.: Optimization of a Wing Structure for Gust Response and Aileron Effectiveness. Journal of Aircraft **40**(6), 1185–1191 (2003)
40. Schijve, J.: Fatigue of Aircraft Materials and Structures. International Journal of Fatigue **16**(1), 21–32 (1994)
41. Stanford, B., Beran, P.: Optimal Structural Topology of a Plate-Like Wing for Subsonic Aeroelastic Stability. Journal of Aircraft **48**(4), 1193–1203 (2011)
42. Stanford, B., Beran, P.: Computational Strategies for Reliability-Based Structural Optimization of Aeroelastic Limit Cycle Oscillations. Structural and Multidisciplinary Optimization **45**(1), 83–99 (2012)
43. Stanford, B., Beran, P.: Direct Flutter and Limit Cycle Computations of Highly-Flexible Wings for Efficient Analysis and Optimization. Journal of Fluids and Structures (in review), (2011)
44. Timme, S., Marques, S., Badcock, K.: Transonic Stability Analysis Using a Kriging-Based Schur Complement Formulation. AIAA 2010-8228, August (2010)

# Robust Uncertainty Propagation in Systems of Conservation Laws with the Entropy Closure Method

**Bruno Després, Gaël Poëtte, and Didier Lucor**

**Abstract** In this paper, we consider hyperbolic systems of conservation laws subject to uncertainties in the initial conditions and model parameters. In order to solve the underlying uncertain systems, we rely on moment theory and the construction of a moment model in the framework of parametric polynomial approximations. We prove the spectral convergence of this approach for the uncertain inviscid Burgers' equation. We also emphasize the difficulties arising when applying the standard moment method in the context of uncertain systems of conservation laws. In particular, we focus on two relevant examples: the shallow water equations and the Euler system. Next, we review the *entropy*-based method inspired by plasma physics and rational extended thermodynamics that we propose in this context. We then study the mathematical structure of the well-posed large systems of discretized partial differential equations arising in this framework. The first aim of this work is the description of some mathematical features of the moment method applied to the modeling of uncertainties in systems of conservation laws. The second objective is to relate theoretical description and understanding to some basic numerical results obtained for the numerical approximation of such uncertain models. All numerical examples come from fluid dynamics inspired problems.

B. Després (✉)
UMR 7598, Laboratoire Jacques-Louis Lions, Université Pierre et Marie Curie, 75252 Paris Cedex 05, France
e-mail: despres@ann.jussieu.fr

G. Poëtte
CEA, Centre DAM, DIF, F-91297 Arpajon, France
e-mail: gael.poette@cea.fr

D. Lucor
UMR 7190, d'Alembert Institute, Université Pierre et Marie Curie, 75252 Paris Cedex 05, France
e-mail: didier.lucor@upmc.fr

H. Bijl et al. (eds.), *Uncertainty Quantification in Computational Fluid Dynamics*, Lecture Notes in Computational Science and Engineering 92,
DOI 10.1007/978-3-319-00885-1_3, 

# 1 Introduction

The study of the uncertain character of fluid dynamics appears as a fundamental problem since the seminal work of Wiener on the modeling of turbulence in the early 1938 [1]: in particular the last section of his paper about the Burgers' equation with stochasticity and the loss of regularity of shock waves equation is inspiring for our purposes. Also fundamental in this direction is the 1947 Cameron-Martin Theorem [2] which justifies the choice a priori of certain expansions with respect to uncertain parameters. Since that period, the focus of the scientific community working on computational fluid dynamics has been on *deterministic* fluid dynamics problem: central theoretical reference is of course the 1973 monograph of Lax [3]. This theoretical accomplishments came along with intense focus on the development of numerical solvers following for example the ideas of [4, 5] and many others. Since the 1990, the question of stochasticity in fluid problems manifests itself again in the context of the extension of numerical incompressible and (more recently) compressible solvers [6–18]: such a constructive approach intends to design algorithms or numerical schemes which can be used in computers for the calculation of stochastic flows of any kind. As this uncertainty may creep from numerous sources: physical and computational domain/geometry (manufacturing process, roughness, domain size, boundary conditions,...), initial/operating conditions, physical/turbulence models, mathematical model assumptions/simplifications (e.g. linearization, adiabaticity, perfect gas,...), discretization and numerical algorithmic errors (round-off or truncation error, numerical dissipation/dispersion, aliasing,...), the application of uncertainty quantification to compressible flows has obvious impact both on physical and mathematical fundamental problems (e.g. Riemann problem), classical aerodynamics problems (e.g. the piston problem [19] or [7], the dual-throat nozzle [7]) as well as more realistic engineering problems [20].

We will not discuss elliptic models for which we refer to [21] and references therein. Modern work on the application of Monte-Carlo methods in the context of fluid solvers can be found in [22].

The point of view that we develop in this work is that these new schemes can also be viewed as special discretized procedure of large systems of balance equations [23, 24]. In this direction a natural question to address is the well posedness of the underlying system of partial differential equations, somehow going back to the old question addressed in Wiener's paper. To address this question and related ones, we will adopt hereafter the entropy moment method [25–27], see also [28] in a different context: this method stresses the interest of the entropy function and the entropy variables to provide insights into the mathematical structure of well posed large systems of partial differential equations [29]. The entropy method is an extremely powerful approach to design and analyze large systems of conservation laws, and it has been used with great success in plasma physics and rational thermodynamics. Our aim is at the description of some mathematical and numerical features of entropy method applied to the propagation of uncertainties in systems of conservation laws [14, 30]. Notice that we will not describe the schemes used for the numerical illustrations: we refer to [14, 30, 31] where all details have been published; in particular it is shown in these references that the extra computational cost generated by the use of the entropy variable is comparable to the cost of the standard fluid solver.

## 2 The Moment Method for Uncertain Systems of Conservation Laws

The model problem that we consider is

$$\begin{cases} \partial_t u + \partial_x f(u) = 0, & x \in \mathbb{R},\ t > 0, \\ u(x, t = 0) = u_0(x),\ x \in \mathbb{R}, \end{cases} \tag{1}$$

where $u \in \mathbb{R}^n$ is the unknown and $f(u) \in \mathbb{R}^n$ is the smooth flux function. For the sake of simplicity we consider only the one-dimensional case and do not introduce any boundary condition. In most realistic problems, the unknown must lie on a certain set $\Omega$ of the whole space, that is

$$u \in \Omega \subset \mathbb{R}^n. \tag{2}$$

The set $\Omega$ is the set of admissible states. A solution of (1) will be called a **deterministic solution**.

Let us assume that the initial solution is uncertain, which means that we would like to solve (1) for many different values of the initial condition, that is

$$u_0 = u_0(x, \xi) \in \Omega \text{ with } \xi \in \Theta \subset \mathbb{R}^P. \tag{3}$$

The variable $\xi \in \Theta$ characterizes what is called the uncertainty, where the set of uncertainties $\Theta = (\Theta, \mathscr{A}, \mathscr{P})$ is a probability space. In some sense if one solves (1) for all different initial solutions that correspond to different $\xi$, then one exactly propagates the uncertainties. To fix the notations the system

$$\begin{cases} \partial_t u(x, t, \xi) + \partial_x f(u(x, t, \xi)) = 0, & x \in \mathbb{R},\ t > 0,\ \xi \in \Theta, \\ u(x, t = 0, \xi) = u_0(x, \xi), & x \in \mathbb{R},\ \xi \in \Theta. \end{cases} \tag{4}$$

will be denoted as an **uncertain initial value problem**. One notices that different values of $\xi$ correspond to different fully decoupled deterministic systems, so in principle there is no difficulty in solving such uncertain problems. The whole problem comes from the fact that exact propagation of uncertainties is very expensive from a computational point of view, and is therefore impossible in practice. In this context a model reduction approach is very attractive, see for instance, [32].

In this direction it is useful to admit that there exists a **given a priori probability law**, such that different $\xi$'s are not equally probable with respect to this probability law. The probability law will be characterized by a function $w : \Theta \to \mathbb{R}^+$ such that

$$\int_\Theta w(\xi)d\xi = 1, \qquad w \geq 0, \tag{5}$$

where $d\xi$ refers to the usual Lebesgue measure. If so it is reasonable to solve in priority for the value of $\xi$'s which have the greatest probability. This is the purpose of Monte-Carlo methods for example. Another possibility, efficient for relatively low stochastic dimensions and smooth integrand consists in maximizing the knowledge on $u$ from the knowledge of a fixed number of moments taken against the polynomial basis $(\phi_q)_{q\in\mathbb{N}}$, orthonormal with respect to $w$. We here introduce equivalent notations.

$$u_q(x,t) = \int_\Theta u(x,t,\xi)\varphi_q(\xi)w(\xi)d\xi = \int u\varphi_q dw, \qquad 0 \leq q \leq Q.$$

This is the goal of what is called Polynomial Chaos approximation. In the following, for simplicity of notations, we assume that $\xi \in \mathbb{R}$, that is $P = 1$, but the following material can be used in any dimension.

The problem (1)–(3) is in fact more general than it looks at first sight. For example uncertainty in the model itself can easily be rewritten as uncertainty in the initial condition. To understand this fact it is sufficient to consider the uncertain problem

$$\begin{cases} \partial_t u + \partial_x f_\gamma(u) = 0, & x \in \mathbb{R},\ t > 0, \\ \gamma = \gamma(x,\xi), \\ u(x,0,\xi) = u_0(x,\xi),\ x \in \mathbb{R}, \end{cases} \tag{6}$$

where the flux which depends on some parameter $\gamma$ is also uncertain. This problem is formally equivalent to the uncertain initial value problem

$$\begin{cases} \partial_t \begin{pmatrix} u \\ \gamma \end{pmatrix} + \partial_x \begin{pmatrix} f(u,\gamma) \\ 0 \end{pmatrix} = 0,\ x \in \mathbb{R},\ t > 0, \\ \text{plus initial conditions.} \end{cases} \tag{7}$$

The structures of (4) and (7) are the same. We will concentrate first on the uncertain initial value problem for expository purposes. Specific features of model uncertainties will be considered in Sect. 6.

The standard method to construct a reduced model is the following. We first define the set of square integrable uncertain functions

$$L^2_w(\Theta) = \left\{ \text{measurable functions } \xi \mapsto f(\xi) \text{ such that } \int_\Theta f^2(\xi)w(\xi)d\xi < \infty \right\}.$$

Under very general conditions there exists a countable family of polynomials

$$\xi \mapsto \varphi_p(\xi), \qquad p \in \mathbb{N}, \tag{8}$$

which are orthonormal

$$\int_\Theta \varphi_p(\xi)\varphi_q(\xi)w(\xi)d\xi = \delta_{pq},$$

and complete in $L^2_w(\Theta)$. Notice that $p \in \mathbb{N}$ is not necessarily the best ordering: in certain cases it is preferable to use a multi-index $p \in \mathbb{N}^P$. However for the simplicity of presentation, we consider that $p$ is here just a single natural number.

At fixed $t$ and $x$, it is natural to look for an approximation of the solution $u$ in the subspace $\text{Span}_{0\le q\le p}\left\{\varphi_q\right\}$ generated by the first $p+1$ polynomials, that is

$$u_p \in \text{Span}_{0\le q\le p}\left\{\varphi_q\right\}.$$

It is immediate to show that

$$u^p = \sum_{0\le q\le p} u_q\varphi_q \qquad u_q(\xi) = \int_\Theta u(\xi)\varphi_q(\xi)w(\xi)d\xi \tag{9}$$

is such that

$$\int_\Theta (u((\xi) - u^p(\xi))^2\, w(\xi)d\xi \le \int_\Theta (u((\xi) - v^p(\xi))^2\, w(\xi)d\xi$$

for all $v^p \in \text{Span}_{0\le q\le p}\left\{\varphi_q\right\}$: that is the expansion (9) is the best one among all possible trials in $\text{Span}_{q\le p}\left\{\varphi_q\right\}$. Since, $\Theta$ being a probability space i.e. $|\Theta| = 1$, $\varphi_0 = \frac{1}{\sqrt{|\Theta|}} = 1$ denotes the normalized constant polynomial, then the mean value of $u$ is $u_0$

$$\text{mean}(u) = u_0.$$

The variance of $u$ is

$$\text{variance}(u) = \int_\Theta (u(\xi) - u_0)^2\, w(\xi)d\xi = u_1^2 + u_2^2 + \dots.$$

The goal of the modeling of uncertainties is to obtain accurate approximation of all $u_q$'s such that, at least, the variance is computed in an accurate way.

In order to compute the $u_q$'s, one can use the fact that $u$ is the solution of a partial differential equation with derivatives with respect to $t$ and $x$. A natural idea is to extend the previous approach under the form

$$u(x,t,\xi) = \sum_{0\le q\le p} u_q(x,t)\varphi_q(\xi). \tag{10}$$

At this stage it is very natural to seek compatibility with the uncertain system of conservation laws by taking the moments of (1)–(3) against each $\varphi_q$, $q \le p$. One finally obtains the **moment model**

$$\partial_t u_q + \partial_x \int_\Theta f\left(\sum_{0\le l\le p} u_l\varphi_l\right)\varphi_q(\xi)w(\xi)d\xi = 0, \quad 0\le q\le p. \tag{11}$$

Since $u$ is a n-dimensional vector, system (11) is a system of $(p+1)\times n$ equations. It is a closed system in the sense that it has exactly $n(p+1)$ equations and $n(p+1)$ unknowns. In the following, system (11) will also be referred as the $P-$**truncated system** of Eq. (1) with standard closure (10).

It is very reasonable to expect that (11) is an accurate approximation of the uncertain initial problem for large $p \gg 1$ (cf. Cameron-Martin's Theorem [2] or some generalization [33, 34]), provided that a solution exists for this system. We will indeed prove spectral accuracy under very general hypotheses in Sect. 3 for the Burgers' equation. After that we will turn to the mathematical structure of the moment model for physically motivated deterministic systems. We will show that the system (11) may not be hyperbolic for data such that $\sum_{0\le l\le p} u_l\varphi_l \in \Omega$. This germ of instability is directly linked to the mathematical structure of the system (11). If (11) is non-hyperbolic, it is not possible to solve it in a stable way. After that we will introduce the entropy method which is hyperbolic by construction. Various estimates on the eigenvalues of the Jacobian matrix of the problem will be developed. The final section is devoted to uncertainties in the model: we will show, as explained in (7), that taking into account uncertainties in the model is equivalent to taking into account uncertainties in the initial condition so that the material developed in the first section can be directly applied to uncertainties in the model parameters. The discussions are illustrated by numerical results.

# 3 Proof of Spectral Accuracy for a Non-linear Scalar Hyperbolic Case

In this section we prove a result of spectral accuracy for the non-linear Burgers' equation. We use a comparison method between a general approximated solution and a smooth exact solution to establish this result. This is based on the weak-strong method for which we refer to the Dafermos' book [29]. To our knowledge it is new in the context of uncertainty. It is also an enhancement of the results published in [11,35] by a different method. It also stresses the importance of the entropy. We refer to [22] for error estimates in the context of advanced Monte-Carlo algorithms.

We start from the Burgers' equation

$$\partial_t u + \partial_x \left( \frac{u^2}{2} \right) = 0 \tag{12}$$

on the periodic domain $x \in [0,1]_{\text{per}}$. Periodic boundary conditions are considered only for convenience without loss of generality. Let us consider the uniform law on $\Theta = ]-1,1[$. We denote by $(L_p)_{p \in \mathbb{N}}$ the Legendre basis, orthonormal with respect to $\xi$'s probability measure.

The initial data is

$$u(x,0,\xi) = u_0(x,\xi). \tag{13}$$

The initial data is supposed to be smooth function for all $\xi$: hence the exact solution is easily constructed using the method of characteristics for all $\xi$: that is the solution is constant, $u(y(X,t,\xi),t,\xi) = u_0(X,\xi)$, along the characteristic

$$y(X,t,\xi) = X + t u_0(X,\xi).$$

We will assume that the time

$$T_\xi = -\frac{1}{\inf_x (\partial_x u_0(x,\xi))}. \tag{14}$$

at which the characteristic construction fail is bounded from below uniformly

$$\exists T, \quad 0 < T < T_\xi \quad \forall \xi.$$

We also assume regularity with respect to the $\xi$ variable. Let $T^\varepsilon = T - \varepsilon < T$: the exact solution is smooth with respect to all variables

$$u \in L^\infty\left((0,1) \times (0,T^\varepsilon) \times (-1,1)\right) \cap L^\infty\left([0,1]_{\text{per}} \times (0,T^\varepsilon) : H^k(-1,1)\right) \tag{15}$$

for all $k \in \mathbb{N}$ where

$$H^k(\Theta) = \left\{ u \in L^2_w(\Theta) | \int \sum_{l=0}^{k} (u^{(l)})^2 d\xi < \infty \right\}.$$

For convenience we define

$$|||u|||_{k,\varepsilon} = \sup_{(t,x) \in [0,T^\varepsilon] \times [0,1]} \|u(t,x,\cdot)\|_{H^k(-1,1)}$$

The uncertain system of conservation laws of size $p+1$ is

$$\begin{cases} \partial_t u_0 + \partial_x \int_\Theta \frac{\left(\sum_{0\le q\le p} u_q\varphi_q(\xi)\right)^2}{2}\varphi_0(\xi)d\xi = 0, \\ \dots \\ \partial_t u_p + \partial_x \int_\Theta \frac{\left(\sum_{0\le q\le p} u_q\varphi_q(\xi)\right)^2}{2}\varphi_p(\xi)d\xi = 0. \end{cases} \tag{16}$$

It is immediate to verify that this system admits an entropy-entropy flux pair. Indeed let us consider a smooth solution of (16). One has

$$\partial_t \frac{\sum_{0\le r\le p} u_r^2}{2} + \sum_{0\le r\le p} u_r \partial_x \int_\Theta \frac{\left(\sum_{0\le q\le p} u_q\varphi_q(\xi)\right)^2}{2}\varphi_r(\xi)d\xi = 0$$

which yields after rearrangement of the flux

$$\partial_t \frac{\sum_{0\le r\le p} u_r^2}{2} + \partial_x \int_\Theta \frac{\left(\sum_{0\le q\le p} u_q\varphi_q(\xi)\right)^3}{3}d\xi = 0. \tag{17}$$

It means that the entropy of the system is the function $S$

$$S(u_0,\dots,u_p) = \frac{\sum_{0\le r\le p} u_r^2}{2} = \int_\Theta \frac{\left(\sum_{0\le q\le p} u_q\varphi_q(\xi)\right)^2}{2}d\xi$$

and the entropy flux is the function $G$

$$G(u_0,\dots,u_p) = \int_\Theta \frac{\left(\sum_{0\le q\le p} u_q\varphi_q(\xi)\right)^3}{3}d\xi.$$

To be fully general we will consider weak entropy solutions to (16). The entropy law becomes an inequality

$$\partial_t \int_\Theta \frac{\left(\sum_{0\le q\le p} u_q\varphi_q\right)^2}{2}d\xi + \partial_x \int_\Theta \frac{\left(\sum_{0\le q\le p} u_q\varphi_q\right)^3}{3}d\xi \le 0 \tag{18}$$

in the sense of distributions. Let us define for convenience

$$u^p(\xi) = \sum_{0\le q\le p} u_q\varphi_q(\xi). \tag{19}$$

Integration of the previous inequality over $\mathscr{I}$ yields that

$$\frac{d}{dt}\int_{\mathscr{I}}\int_\Theta (u^p)^2 dx d\xi \le 0.$$

One obtains the a priori bound

$$\|u^p(t)\|_{L^2(\mathscr{I}\times\Theta)} \le \left\|u_0^p\right\|_{L^2(\mathscr{I}\times\Theta)}.$$

It is therefore natural to seek weak solutions of the uncertain Burgers' equation in the space $L^\infty((0,T^\varepsilon):L^2(\mathscr{I}\times\Theta))$. This is summarized in next definition.

**Definition 1.** A weak solution of the uncertain Burgers' equation is a function $u^p \in L^\infty((0,T^\varepsilon):L^2(\mathscr{I}\times\Theta))$ such that $u^p$ is a polynomial of degree at most $p$ with respect to the $\xi$ variable (a.e. in $(x,t)$), such that

$$\int_{\mathscr{I}\times]0,T[\times\Theta} u_q\partial_t\varphi\, dx\, dt d\xi + \int_{\mathscr{I}\times]0,T[\times\Theta} \frac{\left(\sum_{0\le q\le p} u_q\varphi_q\right)^2}{2}\varphi_q\partial_x\varphi\, dx\, dt d\xi$$

$$+\int_{\mathscr{I}}(u_q)_0\varphi(t=0)\, dx = 0, \quad q\le p,$$

for all smooth test functions $(x,t,\xi)\mapsto\varphi(x,t,\xi)$ such that $\varphi(\cdot,T,\cdot)\equiv 0$.

A weak solution is an entropy weak solution if (18) holds true in the sense of distributions, that is

$$\int_{\mathscr{I}\times]0,T[\times\Theta}\frac{(u^p)^2}{2}\partial_t\varphi\, dx\, dt d\xi + \int_{\mathscr{I}\times]0,T[\times\Theta}\frac{(u^p)^3}{3}\partial_x\varphi\, dx\, dt d\xi \tag{20}$$

$$+\int_{\mathscr{I}\times\Theta}\frac{(u_0^p)^2}{2}\varphi(t=0)\, dx d\xi \ge 0, \quad q\le p,$$

for all non negative smooth test function $\varphi\ge 0$ such that $\varphi(\cdot,T,\cdot)\equiv 0$.

We define $\Pi_p u$ the orthogonal projector of $u$ solution to (12) in $L^2(\Theta)$ onto the space Span $\{\varphi_q\}_{q\le p}$. It is a priori natural to choose the truncated initial condition as the projection of initial condition, that is $u_0^p = \Pi_p u_0$. However our main estimate will be true without this hypothesis. We will only require that

$$\|u_0^p - u_0\|_{L^2(\mathscr{I}\times\Theta)} \quad \text{is small enough} \tag{21}$$

for the result of the theorem to make sense. The point is that $u_0^p$ can admit discontinuities as well. In this case the weak entropy solution also admits discontinuities. These discontinuities are most probably of small amplitude. The theorem of spectral accuracy in fact shows that there remain of small amplitude provided $t\le T^\varepsilon$.

We are now ready to state the main result of the paper.

**Theorem 1 (Convergence of Burgers' approximation).** *Spectral accuracy holds in the following sense: for all k there exists a constant $D_k^\varepsilon$ such that*

$$\|u(t) - u^p(t)\|^2_{L^2(\mathscr{I}\times\Theta)} \le D_k^\varepsilon \left( \|u(0) - u^p(0)\|^2_{L^2(\mathscr{I}\times\Theta)} + \frac{1}{p^k} \right), \qquad t \le T^\varepsilon. \tag{22}$$

The following proof will make use of some useful technical results listed in the Appendix.

*Proof.* **Proposition 1.** *One has the inequality*

$$\begin{aligned} \frac{d}{dt}\|u-u^p\|^2_{L^2(\mathscr{I}\times\Theta)} &\le \left\|\Pi_p u - u^p\right\|^2_{L^2(\mathscr{I}\times\Theta)} \left\|\partial_x \Pi_p u\right\|_{L^\infty(\mathscr{I}\times\Theta)} \\ &+ \|u^p\|_{L^2(\mathscr{I}\times\Theta)} \left\|\Pi_p u - u\right\|_{L^2(\mathscr{I}\times\Theta)} \left\|\partial_x(\Pi_p u + u)\right\|_{L^\infty(\mathscr{I}\times\Theta)} \\ &+ \|u^p\|_{L^2(\mathscr{I}\times\Theta)} \left\|\partial_x(\Pi_p u - u)\right\|_{L^2(\mathscr{I}\times\Theta)} \left\|\Pi_p u + u\right\|_{L^\infty(\mathscr{I}\times\Theta)}. \end{aligned} \tag{23}$$

We based the proof on the following formula which holds true in the sense of distributions i.e.,

$$\partial_t \int_\Theta \frac{(u^p - u)^2}{2} d\xi = \int_\Theta \left( \underbrace{\partial_t \frac{(u^p)^2}{2}}_{\alpha} - \underbrace{u^p \partial_t u}_{\beta} - \underbrace{u \partial_t u^p}_{\gamma} + \underbrace{\partial_t \frac{u^2}{2}}_{\delta} \right) d\xi.$$

Since $u^p$ is a weak entropy solution, one has $\int_\Theta \alpha d\xi + \int_\Theta \partial_x \frac{(u^p)^3}{3} d\xi \le 0$. Since $u$ is a smooth solution one has that $\int_\Theta \beta d\xi + \int_\Theta u^p \partial_x \frac{u^2}{2} d\xi = 0$ and $\int_\Theta \delta d\xi + \int_\Theta \partial_x \frac{u^3}{3} d\xi = 0$. The remaining term is

$$\int_\Theta \gamma d\xi = \int_\Theta \Pi_p u \partial_t u^p d\xi = -\int_\Theta \Pi_p u \frac{\partial_x (u^p)^2}{2} d\xi$$

in the sense of distribution in space. Indeed since $u^p$ is assumed to be just a weak entropy solution of the uncertain system, it is necessary to estimate $\gamma$ in the sense of distributions. So

$$\partial_t \int_\Theta \frac{(u^p - u)^2}{2} d\xi \le \int_\Theta \left( -\partial_x \frac{(u^p)^3}{3} + u^p \partial_x \frac{u^2}{2} + \Pi_p u \frac{\partial_x (u^p)^2}{2} - \partial_x \frac{u^3}{3} \right) d\xi.$$

Integration in space yields

$$\frac{d}{dt} \int_{\mathscr{I}\times\Theta} \frac{(u^p - u)^2}{2} d\xi \le \int_{\mathscr{I}\times\Theta} \left( u^p \partial_x \frac{u^2}{2} + \Pi_p u \frac{\partial_x (u^p)^2}{2} \right) d\xi$$

which we rewrite under the form

$$\frac{d}{dt}\int_{\mathcal{I}\times\Theta}\frac{(u^p-u)^2}{2}d\xi \le \int_{\mathcal{I}\times\Theta}\left(u^p\partial_x\frac{u^2}{2}-\frac{(u^p)^2}{2}\partial_x\Pi_p u\right)d\xi.$$

We rearrange the right hand side using the periodic boundary conditions and the identity $\int_{\mathcal{I}\times\Theta}\partial_x\frac{(\Pi_p u)^3}{6}d\xi=0$. We get

$$\begin{aligned}\frac{d}{dt}\int_{\mathcal{I}\times\Theta}\frac{(u^p-u)^2}{2}d\xi \le &-\int_{\mathcal{I}\times\Theta}\frac{(\Pi_p u-u^p)^2}{2}\partial_x\Pi_p u d\xi\\ &-\int_{\mathcal{I}\times\Theta}u^p\partial_x\left(\frac{(\Pi_p u)^2-u^2}{2}\right)d\xi.\end{aligned}$$

Expansion of this inequality yields

$$\frac{d}{dt}\int_{\mathcal{I}\times\Theta}(u^p-u)^2 d\xi \le -\int_{\mathcal{I}\times\Theta}(\Pi_p u-u^p)^2\partial_x\Pi_p u d\xi$$

$$-\int_{\mathcal{I}\times\Theta}u^p\left(\Pi_p u+u\right)\partial_x\left(\Pi_p u-u\right)d\xi-\int_{\mathcal{I}\times\Theta}u^p\left(\Pi_p u-u\right)\partial_x\left(\Pi_p u+u\right)d\xi,$$

from which (23) follows.

**Proof of Theorem 1.** From the previous proposition, there exists some constants $\alpha^\varepsilon$ and $\beta_k^\varepsilon$ such that

$$\frac{d}{dt}\|u-u^p\|^2_{L^2(\mathcal{I}\times\Theta)} \le \alpha^\varepsilon\left\|\Pi_p u-u^p\right\|^2_{L^2(\mathcal{I}\times\Theta)}+\frac{\beta_k^\varepsilon}{p^k}, \qquad t\le T^\varepsilon.$$

The triangular inequality yields

$$\begin{aligned}\left\|\Pi_p u-u^p\right\|_{L^2(\mathcal{I}\times\Theta)} &\le \left\|\Pi_p u-u\right\|_{L^2(\mathcal{I}\times\Theta)}+\|u-u^p\|_{L^2(\mathcal{I}\times\Theta)}\\ &\le \frac{C_k^\varepsilon}{p^k}+\|u-u^p\|_{L^2(\mathcal{I}\times\Theta)}\end{aligned}$$

therefore

$$\frac{d}{dt}\|u-u^p\|^2_{L^2(\mathcal{I}\times\Theta)} \le \gamma^\varepsilon\|u-u^p\|^2_{L^2(\mathcal{I}\times\Theta)}+\frac{\delta_k^\varepsilon}{p^k}, \qquad t\le T^\varepsilon.$$

To finish the proof we use the Gronwall lemma.

### 3.1 Numerical Application

Let us illustrate the spectral convergence of Theorem 1. We consider Burgers' equation (12) together with zero fluxes boundary conditions. We consider a *smooth* uncertain initial condition. This choice is motivated by the fact that despite this initial smoothness the dynamics of the system stiffen the problem in both the random and the physical space. The initial condition we consider is given by

$$u^0(x,\xi) = K_0\mathbb{I}_{[0,x_0]}(x-\sigma\xi) + K_1\mathbb{I}_{[x_1,L]}(x-\sigma\xi) + Q(x-\sigma\xi)\mathbb{I}_{[x_0,x_1]}(x-\sigma\xi),$$

with coefficients $K_0$, $K_1$ to be defined and

$$Q(x) = ax^3 + bx^2 + cx + d.$$

The coefficients $(a,b,c,d)$ are

$$a = -2\frac{K_0 - K_1}{x_0^3 + 3x_0x_1^2 - x_1^3 - 3x_1x_0^2}, \quad b = \frac{3(K_0-K_1)(x_0+x_1)}{x_0^3 + 3x_0x_1^2 - x_1^3 - 3x_1x_0^2}$$

$$c = -6\frac{(K_0-K_1)x_1x_0}{x_0^3 + 3x_0x_1^2 - x_1^3 - 3x_1x_0^2}, \quad d = \frac{-x_1^3K_0 + 3x_1^2K_0x_0 + K_1x_0^3 - 3K_1x_1x_0^2}{x_0^3 + 3x_0x_1^2 - x_1^3 - 3x_1x_0^2},$$

so that the initial condition and its first derivatives are continuous with respect to the space and stochastic variable, i.e. $u^0(x,\xi)$ verifies the conditions of Theorem 1. The initial condition for several realizations of the random variable $\xi \sim \mathscr{U}([-1,1])$ are presented Fig. 1 (left). The stochastic initial conditions consist in uniformly distributed translations along the $x$-axis of one deterministic curve.

The analytical solution is given by

$$u(x,t,\xi) = \begin{cases} K_0 & \text{if } (x-\sigma\xi) - K_0t < x_0 \\ aX^3(x,t,\xi) + bX^2(x,t,\xi) + cX(x,t,\xi) + d & \text{if } x_0 \le X(x,t,\xi) \le x_1 \\ K_1 & \text{if } X(x,t,\xi) \ge x_1 \end{cases}$$

where

$$X(x,t,\xi) = h(x-\sigma\xi,t) - \frac{a(3ct+3) - b^2}{9a^2th} - \frac{b}{3a}$$

with

$$\begin{aligned} h(x,t) = {} & (((27a^2tx^2 + x(27a^2d^2 + (4b^3 - 18abc)d \\ & + 4ac^3 - b^2c^2)t^3((-54a^2d + 18abc - 4b^3)t^2 + 18abt) \\ & + (-18abd + 12ac^2 - 2b^2c)t^2 + (12ac - b^2)t + 4a)^{\frac{1}{2}})/(6\sqrt{3}a^2t^{\frac{3}{2}}) \\ & - \frac{a^2(27dt - 27x) + ab(-9ct - 9) + 2b^3t}{54a^3t})^{\frac{1}{3}}. \end{aligned} \tag{24}$$

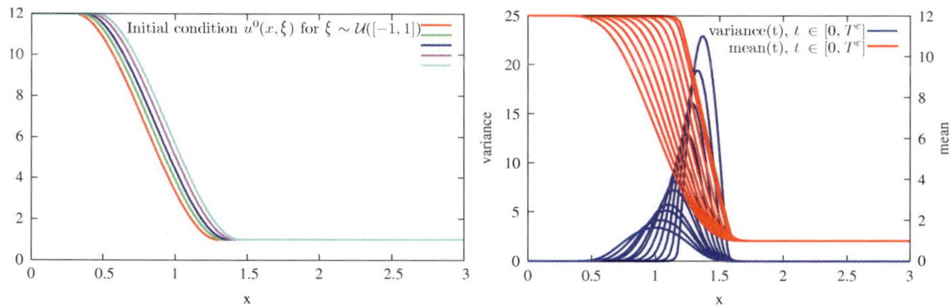


**Fig. 1** *Left*: initial condition $u^0(x,\xi)$ for several realizations of $\xi \sim \mathscr{U}([-1,1])$. *Right*: time evolution of the mean and variance of the solution for $t \in [0, T^\varepsilon]$

In practice, we take $L = 3$, $K_0 = 12$, $K_1 = 1$, $x_0 = 0.5$, $x_1 = 1.5$ and $\sigma = 0.2$ so that $\sigma\xi \in [-0.2, 0.2]$. Besides, the theoretical value of $T^*(\xi)$ is known by formula (14). Here we observe that the critical time

$$T_\xi = \frac{1}{\frac{b^2}{3a} - c}$$

is independent of $\xi$. For numerical tests we take $T^\varepsilon = \frac{1}{\frac{b^2}{3a}-c} - \varepsilon$ with $\varepsilon = 10^{-10}$.

The results are displayed in Figs. 1 and 2. Figure 1 (right) shows the time evolution of the mean and the variance with respect to the space variable. As time passes, the mean gets steeper and the variance increases. The computation is stopped at $T^\varepsilon$ that is just before the appearance of a shock wave in both the stochastic and physical space. In Fig. 2 (left), we display the numerical solution with respect to $\xi$ at point $x = 1.5$ and at different times: so it represents the time evolution at a certain point in space. We observe that the solution also gets steeper with respect to the random parameter as time increases.

Figure 2 presents the numerical results, the relative errors in $L_2(\Theta, \mathscr{I})$ at time $T^\varepsilon$ obtained by the discretization of the $P$−truncated Burgers' system with a Roe solver with, respectively, 500, 1,000 and 2,000 cells: spectral convergence and the result of Theorem 1 are recovered. Note the stagnation in the final portion of the curve which corresponds to spatial discretization limits.

We have demonstrated the spectral convergence of the truncated Burgers' equation with classical closure (cf. Sect. 2) and confirmed the theory on a numerical test case. At this stage, one can wonder whether this kind of approach could lead to similar results when dealing with more complex model, i.e. systems of conservation laws (as opposed to scalar conservation laws) such as shallow water or Euler equations. Next section aims at emphasizing the difficulty to answer the latter question, especially because the previous methodology – application of the moment model (11) with closure (10) to a system (1) (i.e. $n > 1$) – can lead to a loss of hyperbolicity of the $P$−truncated system (11).

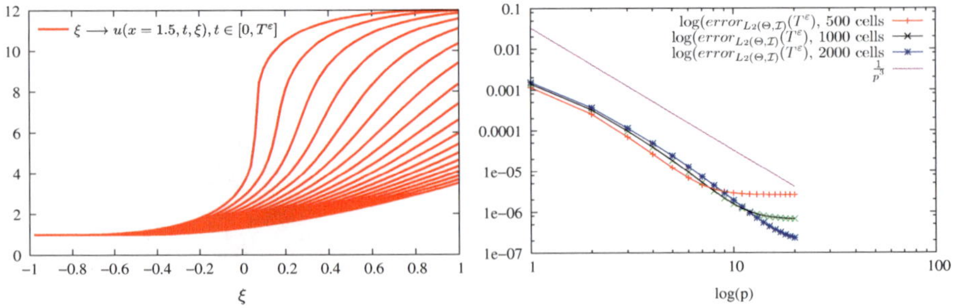

**Fig. 2** Illustration of Theorem 1: Burgers' solution (*left*) spectral convergence (*right*) with respect to polynomial approximation order $p$

# 4 Loss of Hyperbolicity of the Discretized Problem

We introduce hereafter two examples that demonstrate that system (11) is *not* always hyperbolic. Our goal is to explain that the non-linearity of the initial model can generate ill-posedness even for simple reasonable data. This is the motivation for the introduction of the entropy method in next section.

## 4.1 Example 1: Shallow Water Equations

The shallow water or Saint-Venant system may be expressed as

$$\begin{cases} \partial_t h + \partial_x (hv) = 0, \\ \partial_t (hv) + \partial_x \left( hv^2 + g\frac{h^2}{2} \right) = 0, \end{cases} \tag{25}$$

where $h$ is the water height, $v$ the velocity of the water and $g > 0$ is the local gravity constant. In this case $n = 2$, $u = (h, hv)^t$ and $f(u) = \left( hv, hv^2 + g\frac{h^2}{2} \right)$. From now on, we assume that the system bears some level of generic uncertainty represented by a scalar uncertain variable $\xi \in \Theta = ]-1, 1[$ for simplicity. The probability law is the uniform one, that is $w = \frac{1}{2}$. We consider expansions along the first two Legendre polynomials

$$\varphi_0(\xi) = \frac{1}{\sqrt{2}}, \qquad \varphi_1(\xi) = \sqrt{\frac{3}{2}}\xi.$$

System (11) can be recast as

$$\partial_t \begin{pmatrix} u_0 \\ u_1 \end{pmatrix} + \partial_x \begin{pmatrix} \int_{-1}^{1} f\left(u_0\varphi_0(\xi) + u_1\varphi_1(\xi)\right)\, \varphi_0(\xi)d\xi \\ \int_{-1}^{1} f\left(u_0\varphi_0(\xi) + u_1\varphi_1(\xi)\right)\, \varphi_1(\xi)d\xi \end{pmatrix} = 0. \tag{26}$$

The Jacobian matrix $A$ of the total flux with respect to the unknown $u_0, u_1$ is

$$A = \left(\begin{array}{c|c} \int_{-1}^{1} \nabla f\ \varphi_0(\xi)\varphi_0(\xi) & \int_{-1}^{1} \nabla f\ \varphi_1(\xi)\varphi_0(\xi) \\ \hline \int_{-1}^{1} \nabla f\ \varphi_1(\xi)\varphi_0(\xi) & \int_{-1}^{1} \nabla f\ \varphi_1(\xi)\varphi_1(\xi) \end{array}\right) \in \mathbb{R}^{4\times 4}. \tag{27}$$

The Jacobian matrix of the Saint-Venant flux with respect to $u = (h, hv)$ is

$$\nabla f = \begin{pmatrix} 0 & 1 \\ -v^2 + gh & 2v \end{pmatrix} \in \mathbb{R}^{2\times 2}. \tag{28}$$

*Remark 1.* The 2×2 matrix $\nabla f$ is non symmetric in the general case. But if it would be symmetric, then $A$ would also be symmetric and in consequence the system (26) would be hyperbolic.

**Proposition 2.** *Assume that* $u_0 = (\sqrt{2}, 0)$ *and* $u_1 = \left(0, \sqrt{\frac{2}{3}}\right)$. *Then for all* $0 < g < \frac{3}{25}$ *the matrix* $A$ *has complex eigenvalues, so the system (11) is not hyperbolic.*

*Remark 2.* By hypothesis, the height is $h = \sqrt{2}\,\varphi_0(\xi) = 1$. So the chosen height is deterministic and constant. Since $hv = \sqrt{\frac{2}{3}}\,\varphi_1(\xi) = \xi$, it means that the velocity is $v = \xi$. So the velocity is constant in space but with an uncertain level.

We plug this in (28) and compute the moments of this matrix against $\varphi_0$ and $\varphi_1$. Explicit calculations show that

$$A = \left(\begin{array}{cc|cc} 0 & 1 & 0 & 0 \\ -\frac{1}{3} + g & 0 & 0 & \frac{2}{\sqrt{3}} \\ \hline 0 & 0 & 0 & 1 \\ 0 & \frac{2}{\sqrt{3}} & -\frac{3}{5} + g & 0 \end{array}\right).$$

The eigenvectors $Ar = \lambda r$ satisfy

$$\begin{cases} r_2 = \lambda r_1, \\ \left(-\frac{1}{3} + g\right) r_1 + \frac{2}{\sqrt{3}} r_4 = \lambda r_2, \\ r_4 = \lambda r_3, \\ \frac{2}{\sqrt{3}} r_2 + \left(-\frac{3}{5} + g\right) r_3 = \lambda r_4, \end{cases} \Longrightarrow \begin{cases} \left(-\frac{1}{3} + g\right) r_1 + \frac{2}{\sqrt{3}}\lambda r_3 = \lambda^2 r_1, \\ \frac{2}{\sqrt{3}}\lambda r_1 + \left(-\frac{3}{5} + g\right) r_3 = \lambda^2 r_3. \end{cases}$$

Therefore the eigenvalues $\lambda$ are roots of the characteristic polynomial

$$\left(-\frac{1}{3}+g-\lambda^2\right)\left(-\frac{3}{5}+g-\lambda^2\right)-\frac{4}{3}\lambda^2=0. \tag{29}$$

That is $\mu=\lambda^2$ is solution of

$$\mu^2-\left(\frac{2}{5}+2g\right)\mu+\frac{1}{5}-\frac{14}{15}g+g^2=0. \tag{30}$$

The determinant of this second order polynomial is

$$\Delta=\left(\frac{2}{5}+2g\right)^2-4\left(\frac{1}{5}-\frac{14}{15}g+g^2\right)=16\left(-\frac{1}{25}+\frac{g}{3}\right)$$

So $\Delta<0$ for all $0<g<\frac{3}{25}$. Therefore there exists a root of (30) which belongs to $\mathbb{C}-\mathbb{R}^+$. It turns into two non real different $\lambda$ solution of the characteristic polynomial (29). It proves the claim.

In order to numerically illustrate the difficulties encountered when dealing with non-hyperbolic systems, let us solve the latter truncated shallow water system ($p=1$) with a numerical method and set the state $(h_0,h_1,(hv)_0,(hv)_1)=(1,0,0,\frac{1}{\sqrt{3}})$ with $0<g=0.1<\frac{3}{25}$ as initial condition. Note that the analytical solution is stationary and homogeneous (i.e. constant with respect to $\forall(x,t)\in[0,1]\times\mathbb{R}^+$). Suppose we are interested in the solution a time $t=0.2$. The truncation order $p=1$ should allow recovering the analytical stochastic solution $\forall(x,t,\xi)\in[0,1]\times\mathbb{R}^+\times\Theta$.

The numerical results are displayed in Fig. 3 in several configurations for 100 cells: on Fig. 3 top-left, when solving our problem, numerical instabilities appear in the center of the domain and make the solution non physical. This is due to the non-hyperbolicity of the model solved: instabilities are growing exponentially fast with time. On Fig. 3 bottom-left and top-right, we revisit the same problem but we apply more and more diffusive numerical schemes. We denote by $D_n$ the numerical diffusion coefficient of our scheme. The increase in numerical diffusion artificially smoothes the solution and even makes it look physical at the same time $t=0.2$ on Fig. 3 top-right, whereas it only consists in a numerical trick. Indeed, if we consider the same resolution scheme as before but are interested in the solution at time $t=0.4$, the small oscillations occurring at time $t=0.2$ keep growing exponentially with time leading to Fig. 3 bottom-right.

## 4.2 *Example 2: Euler Equations*

The Euler equations of compressible gas dynamics is of fundamental interest for applications. The number of equations being larger than for the shallow water equations, the algebra is a little more involved. This is why we will develop a simplified approach. The result is nevertheless very similar, that is there exists reasonable and physical states such that the moment model is not hyperbolic.

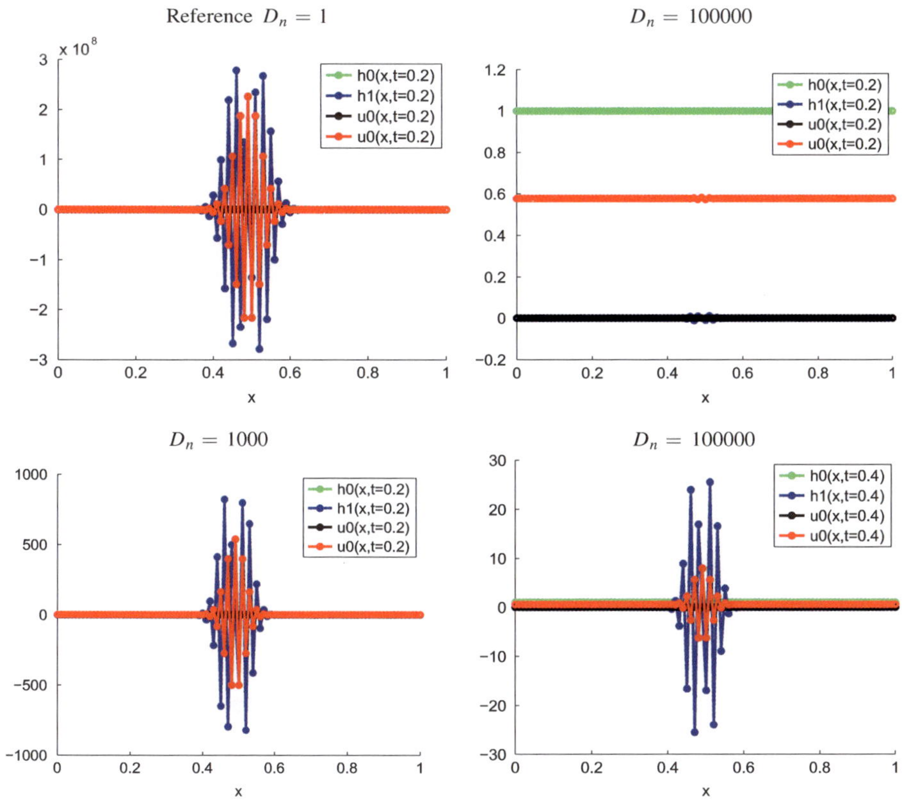


**Fig. 3** Illustration of the non-hyperbolicity of the shallow water truncated system (Data are specified in Proposition 2). A very small germ of oscillations increases exponentially fast. The numerical diffusion coefficient is $D_n \in \{1, 10^3, 10^5\}$. Even artificially large numerical diffusion $D_n = 10^5$ is not able to control it for sufficiently large time

The Euler equations of non viscous compressible gas dynamics are

$$\begin{cases} \partial_t \rho + \partial_x (\rho v) = 0, \\ \partial_t (\rho v) + \partial_x \left(\rho v^2 + p\right) = 0, \\ \partial_t (\rho e) + \partial_x (\rho v e + p v) = 0, \end{cases} \tag{31}$$

where $\rho > 0$ is the density, $v$ the velocity, $e$ the total energy and $p$ the pressure. We assume a perfect gas law $p = (\gamma - 1)\rho\varepsilon$ where $\varepsilon = e - \frac{1}{2}v^2$ is the internal energy. With the general notations (1), this system corresponds $u = (\rho, \rho v, \rho e)$ and $f(u) = (\rho v, \rho v^2 + p, \rho v e + p v)$. We consider once again expansion using $\varphi_0$ and $\varphi_1$.

The Jacobian matrix (27) of the uncertain system (26) may be calculated using the well known formula

$$\nabla f = \begin{pmatrix} 0 & 1 & 0 \\ -\frac{3-\gamma}{2}\frac{b^2}{a^2} & (3-\gamma)\frac{b}{a} & (\gamma-1) \\ -\gamma\frac{bc}{a^2} + (\gamma-1)\frac{b^3}{a^3} & \gamma\frac{c}{a} - \frac{3\gamma-3}{2}\frac{b^2}{a^2} & \gamma\frac{b}{a} \end{pmatrix}.$$

with $a = \rho$, $b = \rho v$ and $c = \rho e$.

**Proposition 3.** *There exist states such that the uncertain system (11) is not hyperbolic when applied to compressible gas dynamics (31).*

In the following, we assume that $\gamma = 3$. This numerical value is not physically pertinent but simplifies a lot the analysis. Similar conclusions are reached for the case of a perfect gas with $\gamma = \frac{5}{3}$ at the expense of more tedious algebra, see results in Remark 3. We then have

$$\nabla f = \begin{pmatrix} 0 & 1 & 0 \\ 0 & 0 & 2 \\ -3ve + 2v^3 & 3e - 3v^2 & 3v \end{pmatrix}.$$

Assume that $\mathbf{u}_0 = (1, 0, \alpha)$ and $\mathbf{u}_1 = (0, \sqrt{\frac{1}{3}}, 0)$. In this case $\rho = 1$, $v = \xi$ and $e = \alpha$. So

$$\nabla f = \begin{pmatrix} 0 & 1 & 0 \\ 0 & 0 & 2 \\ -3\alpha\xi + 2\xi^3 & 3\alpha - 3\xi^2 & 3\xi \end{pmatrix}.$$

Therefore

$$A = \left(\begin{array}{ccc|ccc} 0 & 1 & 0 & 0 & 0 & 0 \\ 0 & 0 & 2 & 0 & 0 & 0 \\ 0 & 3\alpha - 1 & 0 & \sqrt{3}(-\alpha + \frac{2}{5}) & 0 & \sqrt{3} \\ \hline 0 & 0 & 0 & 0 & 1 & 0 \\ 0 & 0 & 0 & 0 & 0 & 2 \\ \sqrt{3}(-\alpha + \frac{2}{5}) & 0 & \sqrt{3} & 0 & 3\alpha - \frac{9}{5} & 0 \end{array}\right).$$

The eigenvectors of this matrix are continuous with respect to the parameter $\alpha$. Therefore they are close to the eigenvalues of the matrix in case $\alpha = \frac{1}{2}$: these eigenvalues are solution to

$$\begin{cases} r_2 = \lambda r_1, \\ 2r_3 = \lambda r_2, \\ \frac{1}{2}r_2 - \frac{\sqrt{3}}{10}r_4 + \sqrt{3}r_6 = \lambda r_3, \\ r_5 = \lambda r_4, \\ 2r_6 = \lambda r_5, \\ -\frac{\sqrt{3}}{10}r_1 + \sqrt{3}r_3 - \frac{3}{10}r_5 = \lambda r_6. \end{cases}$$

After elimination of $r_2$, $r_3$, $r_5$ and $r_6$ it yields

$$\begin{cases} \frac{\lambda}{2}r_1 - \frac{\sqrt{3}}{10}r_4 + \frac{\sqrt{3}}{2}\lambda^2 r_4 = \frac{1}{2}\lambda^3 r_1, \\ -\frac{\sqrt{3}}{10}r_1 + \frac{\sqrt{3}}{2}\lambda^2 r_1 - \frac{3}{10}\lambda r_4 = \frac{1}{2}\lambda^3 r_4. \end{cases}$$

The characteristic polynomial is

$$p(\lambda) = \left(\frac{\lambda}{2} - \frac{1}{2}\lambda^3\right)\left(-\frac{3}{10}\lambda - \frac{1}{2}\lambda^3\right) - \left(-\frac{\sqrt{3}}{10} + \frac{\sqrt{3}}{2}\lambda^2\right)^2.$$

Setting $\mu = \lambda^2$ one gets $p(\lambda) = q(\mu)$ with

$$\begin{aligned} q(\mu) &= \mu\left(\frac{1}{2} - \frac{1}{2}\mu\right)\left(-\frac{3}{10} - \frac{1}{2}\mu\right) - \left(-\frac{\sqrt{3}}{10} + \frac{\sqrt{3}}{2}\mu\right)^2 \\ &= \mu\left(\frac{1}{4}\mu^2 + \frac{1}{20}\mu - \frac{3}{20}\right) - 3\left(\frac{\mu^2}{4} - \frac{1}{10}\mu + \frac{1}{100}\right) \\ &= \frac{1}{4}\mu^3 - \frac{7}{10}\mu^2 + \frac{3}{20}\mu - \frac{3}{100} \end{aligned}$$

Let us proof by contradiction that $q$ cannot admit three real roots: so we assume that there exists three real roots $\mu_1 \leq \mu_2 \leq \mu_3$ with

$$q(\mu) = \frac{1}{2}(\mu - \mu_1)(\mu - \mu_2)(\mu - \mu_3).$$

Between two roots there exists roots of $q'(\mu)$

$$q(z_1) = q(z_2), \quad \mu_1 \leq z_1 \leq \mu_2 \leq z_2 \leq \mu_3.$$

Since $q(-\infty) < 0$ then $q(z_1) \geq 0$. On the other hand

$$3q(\mu) - \mu q'(\mu) = -70\mu^2 + 30\mu - 9 < 0 \quad \forall \mu$$

since

$$\Delta = 30^2 - 4 \times 70 \times 9 = -1{,}620 < 0.$$

Therefore the polynomial $q(\mu)$ has just one real root, which yields the fact that $p(\lambda)$ has non real roots. That is $A$ is not hyperbolic for $e = \frac{1}{2}$. By continuity it proves the claim for all energies $e$ in a neighboring of $\frac{1}{2}$.

*Remark 3.* We have checked the same previous state (i.e. $\mathbf{u}_0 = (1, 0, \alpha = \frac{1}{2})$ and $\mathbf{u}_1 = (0, \sqrt{\frac{1}{3}}, 0))$ also leads to a loss of hyperbolicity in the case of a perfect gas with $\gamma = \frac{5}{3}$. The results in this case show that some eigenvalues of the Jacobian matrix of the flux

$$\nabla f = \begin{pmatrix} 0 & 1 & 0 \\ 0 & 0 & \gamma - 1 \\ 0 & \frac{-\gamma+1}{2} + \gamma\alpha & 0 \\ 0 & 0 & 0 \\ -\frac{1}{6}(3-\gamma)\sqrt{3} & \frac{1}{3}(3-\gamma)\sqrt{3} & 0 \\ \frac{1}{5}(\gamma-1)\sqrt{3} - \frac{1}{3}\alpha\gamma\sqrt{3} & 0 & \frac{1}{3}\gamma\sqrt{3} \\ 0 & 0 & 0 \\ -\frac{1}{6}(3-\gamma)\sqrt{3} & \frac{1}{3}(3-\gamma)\sqrt{3} & 0 \\ \frac{1}{5}(\gamma-1)\sqrt{3} - \frac{1}{3}\alpha\gamma\sqrt{3} & 0 & \frac{1}{3}\gamma\sqrt{3} \\ 0 & 1 & 0 \\ 0 & 0 & \gamma - 1 \\ 0 & \frac{-9\gamma+9}{10} + \gamma\alpha & 0 \end{pmatrix}.$$

have imaginary eigenvalues.

In this last section, we have put forward the fact that the truncation procedure (moment model (11) with closure (10)) presented in Sect. 2 can lead to systems which can not be solved in a stable way when applied to systems (i.e. $n > 1$). In the next section, we show that an alternative exists in order to build a hyperbolic moment model (11) from a native hyperbolic system of conservation laws. The method is based on the use of an entropy in order to close the truncated system rather than on (10).

## 5 Ensuring Hyperbolicity via Entropy Closure

The entropy closure method is very convenient to construct systems of moment which are hyperbolic. We assume that there exists two scalar functions, the entropy $u \mapsto s(u)$ which is a twice differentiable strictly convex and the entropy flux $u \mapsto g(u)$ which is differentiable, such that

$$\forall u \in \Omega, \quad \underbrace{\nabla s}_{\in \mathbb{R}^n} \underbrace{\nabla f}_{\in \mathbb{R}^{n \times n}} = \underbrace{\nabla g}_{\in \mathbb{R}^n} . \tag{32}$$

It is convenient to define the entropy variable

$$v = \nabla s \in \mathbb{R}^n. \tag{33}$$

Of note, there is no ambiguity between the vector $v$ in (32) and the scalar velocity variable used in (25) of (31). Assuming that $\Omega$ is a non empty open convex set, the transformation $u \mapsto v$ is a diffeomorphism from $\Omega$ to $\tilde{\Omega} = \nabla s(\Omega)$. This is easily proved thanks to the strict convexity of the entropy function $s$. The inverse function $h : \Omega \to \tilde{\Omega}$ satisfies

$$h(\nabla s(u)) = u \quad \forall u \in \Omega. \tag{34}$$

We will commonly use the notation $u = h(v)$. Smooth solutions of the initial system may be rewritten as

$$\nabla h \partial_t v + \nabla(f \circ h)\partial_x v = 0, \tag{35}$$

where $\nabla(f \circ h)$ is a symmetric matrix. Applying standard results of system of conservation laws with an entropy, one can prove the hyperbolicity of (1). We readily obtain a result for the spectral radius $\rho(v)$

$$\rho(v) = \max_{\lambda \in \mathrm{Sp}(\nabla f(u))} |\lambda| = \max_{\lambda \in \mathrm{Sp}(\nabla h(v)^{-1} \nabla f \circ h(v))} |\lambda|, \qquad u = h(v). \tag{36}$$

The spectral radius $\rho(v)$ is the maximum of the modulus of the eigenvalues of the Jacobian matrix.

Physically motivated problems such as the shallow water system (25) or the Euler system (31) are such that $\tilde{\Omega}$ is also a convex set. For example the adjoint variable of the compressible Euler system is $v = \left(\frac{\mu}{T}, -\frac{u}{T}, \frac{1}{T}\right)$ where $T > 0$ is the temperature and $\mu$ is the Gibbs potential. On this form is reasonable to admit, for example, the solutions we are interested in are such that $\frac{1}{T}$ takes values in a segment $a \le \frac{1}{T} \le b$. In this case this assumption is very natural and is not a restriction. For the simplicity of the mathematical theory we will assume, in the general case, that

$$\tilde{\Omega} \text{ is a non empty open convex set.} \tag{37}$$

Let us consider a smooth function $\xi \mapsto u(\xi) \in \Omega$. We associate $\xi \mapsto v(\xi) = \nabla s(u(\xi)) \in \tilde{\Omega}$. Under usual convergence assumptions, an infinite expansion of the function $u$

$$u(\xi) = \sum_{p=0}^{\infty} u_p \varphi_p(\xi)$$

is equivalent to infinite expansion of the function $v$

$$v(\xi) = \sum_{p=0}^{\infty} v_p \varphi_p(\xi).$$

The idea of the entropy closure method is based on the remark that it is much better to truncate the series for $v$ than for $u$. In other words we consider the moment model

$$v(x,t,\xi) = \sum_{0 \le q \le p} v_q(x,t)\varphi_q(\xi) \tag{38}$$

together with

$$\partial_t u_q + \partial_x \int_\Theta f\left(h\left(\sum_{0 \le l \le p} v_l \varphi_l\right)\right) \varphi_q(\xi) w(\xi) d\xi = 0, \quad \forall q \le p, \tag{39}$$

where the correspondence between $(u_q)$ and $(v_q)$ is obtained through

$$u_q = \int_\Theta h\left(\sum_{0 \le r \le p} v_r \varphi_r(\xi)\right) \varphi_q(\xi) w(\xi) d\xi, \qquad \forall q \le p. \tag{40}$$

**Definition 2.** We define the uncertain entropy closure of (1) as the system (38)–(40).

## 5.1 Reformulation

It is convenient to reformulate (38)–(40) as a new enlarged system of conservation law

$$\partial_t U + \partial_x F(U) = 0 \tag{41}$$

with $U = (u_q)_{0 \le q \le p} \in \mathbb{R}^{n \times (p+1)}$ and

$$F(U) = \left(\int_\Theta f\left(h\left(\sum_{0 \le l \le p} v_l \varphi_l\right)\right) \varphi_q(\xi) w(\xi) d\xi\right)_{0 \le q \le p} \in \mathbb{R}^{n \times (p+1)}. \tag{42}$$

**Definition 3.** We define the set $\tilde{\Omega}_p \subset \mathbb{R}^{n\times(p+1)}$ such that $V = (v_q)_{0\leq q\leq p} \in \tilde{\Omega}_p$ if and only if

$$\sum_{0\leq q\leq p} v_q\varphi_q(\xi) \in \tilde{\Omega} \qquad \forall \xi \in \Theta. \tag{43}$$

We also define

$$\Omega_p = \left\{U = (u_q)_{q\leq p};\ \exists V \in \tilde{\Omega}_p,\ \text{such that (40) holds}\right\}.$$

*Remark 4.* It should be noticed that the polynomial $\sum_{q\leq p} v_q\varphi_q(\xi)$ takes a priori infinite values when $|\xi| \to \infty$. In most physically driven hyperbolic problems (1), the physical domains, $\Omega$ as well as $\tilde{\Omega}$, are unbounded: for example $\Omega = [0,\infty]\times\mathbb{R}$ for the shallow water system (25) is unbounded. A similar remark holds for the Euler system (31). As a consequence it is almost necessary that the domain of uncertainties $\Theta$ is bounded in order that non trivial polynomial exists in $\tilde{\Omega}_p$, and that uncertain propagations make sense. For the simplicity of the presentation, we will assume that

$$\Theta \text{ is a bounded set in } \in \mathbb{R}^P. \tag{44}$$

**Proposition 4.** *Assuming (37), $\tilde{\Omega}_p$ is a non-empty open convex set.*

By definition $\varphi_0(\xi) = 1$ is a constant non-zero polynomial. Let us take $v \in \tilde{\Omega}$ which is non empty (37) such that $v = v_0 = \int \varphi_0 v dw$, in other words, consider a deterministic admissible state, then in this case

$$V = (v_0, 0, \dots) \in \tilde{\Omega}_p,$$

since the condition (43) is trivially satisfied. Therefore $\tilde{\Omega}_p$ is non empty.

Take $V_1 = (v_q^1)_{0\leq q\leq p} \in \tilde{\Omega}_p$, $V_2 = (v_q^2)_{0\leq q\leq p} \in \tilde{\Omega}_p$ and $\alpha \in [0,1]$. Set $V_3 = \alpha V_1 + (1-\alpha)V_2 = (v_q^3)_{0\leq q\leq p}$. So

$$\sum_q v_q^3\varphi_q(\xi) = \alpha \underbrace{\sum_q v_q^1\varphi_q(\xi)}_{\in\tilde{\Omega}} + (1-\alpha)\underbrace{\sum_q v_q^2\varphi_q(\xi)}_{\in\tilde{\Omega}} \in \tilde{\Omega}, \quad \forall \xi \in \Theta.$$

Therefore $V_3 \in \tilde{\Omega}_p$ which is a convex set. It ends the claim.

An important preliminary result is the following.

**Proposition 5.** *For all $U \in \Omega_p$ the function $F(U)$ is well defined from $\Omega_p$ into $\mathbb{R}^{n\times(p+1)}$.*

The main point consists in showing that $V = (v_q)_{0\leq q\leq p}$ can be determined in an unique manner from $U \in \Omega_p$. Let us define $B(V) = (b_{qr})_{0\leq q,r\leq p}$ the Jacobian matrix of the function (40) which associates $U \in \Omega_p$ to $V \in \tilde{\Omega}$

$$b_{qr} = \int_\Theta \nabla h \left( \sum_{0 \le r \le p} v_r \varphi_r(\xi) \right) \varphi_q(\xi) \varphi_r(\xi) w(\xi) d\xi.$$

Let $Z = (z_q)_{q \le p} \in \mathbb{R}^{n \times (p+1)}$ be an arbitrary vector. Therefore

$$\begin{aligned}(Z, B(V)Z) &= \sum_{0 \le q,r \le p} \int_\Theta \nabla h \left( \sum_{0 \le r \le p} v_r \varphi_r(\xi) \right) z_q \varphi_q(\xi) \cdot z_r \varphi_r(\xi) w(\xi) d\xi \\ &= \int_\Theta \left( z(\xi) \cdot \nabla h \left( \sum_{0 \le r \le p} v_r \varphi_r(\xi) \right) z(\xi) \right) w(\xi) d\xi\end{aligned}$$

where the polynomial function $z(\xi)$ is given by

$$z(\xi) = \sum_{0 \le q \le p} z_q \varphi_q(\xi).$$

Since $\nabla h > 0$ and $z(\xi)$ is a non zero function, then $(Z, BZ) > 0$. It shows that the transformation $V \mapsto U$ is strictly convex. It proves the uniqueness of $V$ for any given $U$. So the flux (42) is indeed uniquely defined with respect to $U$. The claim is proved.

**Proposition 6.** *Assume the system (1) has an entropy-entropy flux pair. Then the system (41) is hyperbolic at all $U \in \Omega_p$.*

We need to show that the matrix $\nabla F(U) \in \mathbb{R}^{[n(p+1)] \times [n(p+1)]}$ has a complete set of real eigenvectors and eigenvalues. It is a standard matter for symmetrizable system. Let us define the matrix $C(V) = \nabla_V F(U) = (c_{qr})_{q,r \le p}$ with

$$c_{qr} = \int_\Theta \nabla (f \circ h) \left( \sum_{l \le p} v_l \varphi_l(\xi) \right) \varphi_q(\xi) \varphi_r(\xi) w(\xi) d\xi \in \mathbb{R}^{n \times n}$$

A standard result for systems of conservation laws with an entropy is that since $\nabla(f \circ h)$ is a symmetric matrix, the matrix $C$ in the quasi-linear reformulation of (41)

$$B(V) \partial_t V + C(V) \partial_x V = 0,$$

is itself a symmetric matrix. Since $B = B^T > 0$ and $C = C^T$, the matrix $B^{-1}C$ admits a complete set of real eigenvectors and eigenvalues. It ends the proof.

In the next section, we study the mathematical structure of the built truncated system. We especially study the behavior of the characteristic waves of the uncertain entropy closure (38)–(40) of (1) with respect to the ones of (1).

## 5.2 Wave Velocities

The wave velocities are the eigenvalues of $\nabla F(U)$. They are equal to the eigenvalues of the matrix $B^{-1}C$. In this section we relate these eigenvalues to the eigenvalues of $\nabla f(u)$.

We start from the eigenvalue problem

$$C(V)X_l = \mu(V)_l B(V)X_l, \quad l = 1, \dots n \times (p+1). \tag{45}$$

The spectral radius is

$$\rho(V) = \max_l |\mu_l| = \max_{X \neq 0} \frac{|(X, C(V)X)|}{(X, B(V)X)}.$$

For any given $X \in \mathbb{R}^{n \times (p+1)}$ we define the polynomial function

$$x^p(\xi) = \sum_{0 \le q \le p} X_q \varphi_q(\xi).$$

**Proposition 7.** *Assume $\Theta$ is bounded. One has the formula for all $X \neq 0$*

$$\frac{(X, C(V)X)}{(X, B(V)X)} = \frac{\int_\Theta \left( \nabla (f \circ h) \left( \sum_{0 \le r \le p} v_r \varphi_r(\xi) \right) x^p(\xi), x^p(\xi) \right) w(\xi) d\xi}{\int_\Theta \left( \nabla h \left( \sum_{0 \le r \le p} v_r \varphi_r(\xi) \right) x^p(\xi), x^p(\xi) \right) w(\xi) d\xi}. \tag{46}$$

This is evident from the definitions of $B$ and $C$.

Let us define for convenience the symmetric positive matrix

$$D_V(\xi) = \nabla h \left( \sum_{0 \le r \le p} v_r \varphi_r(\xi) \right) = D_V(\xi)^t > 0$$

and the symmetric matrix

$$E_V(\xi) = \nabla (f \circ h) \left( \sum_{0 \le r \le p} v_r \varphi_r(\xi) \right).$$

For a given polynomial $x^p(\xi)$ we define $y^p(\xi) = D_V(\xi)^{\frac{1}{2}} x^p(\xi)$ which is well defined since $\nabla h = \nabla h^T > 0$. Using the convenient normalization $\int_\Theta y^p(\xi)^2 w(\xi) d\xi = 1$, we obtain the formula

$$\frac{(X, C(V)X)}{(X, B(V)X)} = \int_\Theta (M_V(\xi) y^p(\xi), y^p(\xi)) \, w(\xi) d\xi \tag{47}$$

where the matrix is $M_V(\xi) = D_V(\xi)^{-\frac{1}{2}} E_V(\xi) D_V(\xi)^{-\frac{1}{2}}$.

**Theorem 2.** *Let $V \in \tilde{\Omega}_p$. One has the bound*

$$\rho(V) \le \sup_{\xi \in \Theta} \rho(v(\xi)) \tag{48}$$

This is immediate from (47).

A more precise characterization of the eigenvalues stems from the min-max theorem.

**Proposition 8.** *One has*

$$\mu_k = \min_{S_k} \max_{X \in S_k, \|X\|=1} \frac{\int_\Theta \left(\nabla(f \circ h)\left(\sum_{r \le p} v_r \varphi_r(\xi)\right) x^p(\xi), x^p(\xi)\right) w(\xi) d\xi}{\int_\Theta \left(\nabla h \left(\sum_{r \le p} v_r \varphi_r(\xi)\right) x^p(\xi), x^p(\xi)\right) w(\xi) d\xi}. \tag{49}$$

*where $S_k \subset \mathbb{R}^{n \times (p+1)}$ denotes any subspace of dimension equal to $k$.*

This is the min-max principle applied to the problem (45).

Note that the result of Theorem 2 is very useful in practice for the numerical resolution of the truncated system. Indeed, explicit numerical schemes needs an a priori estimate of the highest eigenvalue of the solved system in order to satisfy a stability criterion (CFL number). Theorem 2 gives this estimate. Furthermore, Proposition 8 is a first step toward building new characteristic-based numerical schemes for the resolution of the uncertain entropy closure (38)–(40) of (1).

### 5.2.1 Case of Euler System in Lagrangian Coordinates

In the case of Euler system in Lagrangian coordinates, the wave structure can be investigated in more details. This is because the min-max principle (49) gives in certain cases accurate bounds for the eigenvalues. In the following we detail the case of uncertain Lagrangian gas dynamic system.

The deterministic system is

$$\begin{cases} \partial_t \tau - \partial_m v = 0, \\ \partial_t v + \partial_m p = 0, \\ \partial_t e + \partial_m p v = 0, \end{cases} \tag{50}$$

where $\tau = \frac{1}{\rho}$ is the specific volume, $v$ the velocity and $e$ the total energy. The mass variable is denoted as $m$ as usual.

The entropy variable is $v = (v_1, v_2, v_3)$ with

$$v_1 = -\frac{p}{T}, \quad v_2 = \frac{v}{T} \text{ and } v_3 = -\frac{1}{T}$$

where $p$ is the pressure and $T$ the temperature. There is no ambiguity between the pressure and the index $p$ which gives the size of the uncertain system, and also no ambiguity between the adjoint variable (a vector) and the velocity of the fluid (a scalar in dimension one). With these notations

$$\nabla f \circ h = \nabla_v \begin{pmatrix} \frac{v_2}{v_3} \\ \frac{v_1}{v_3} \\ -\frac{v_1 v_2}{v_3^2} \end{pmatrix} = \begin{pmatrix} 0 & \frac{1}{v_3} & -\frac{v_2}{v_3^2} \\ \frac{1}{v_3} & 0 & -\frac{v_1}{v_3^2} \\ -\frac{v_2}{v_3^2} & -\frac{v_1}{v_3^2} & 2\frac{v_1 v_2}{v_3^3} \end{pmatrix}. \tag{51}$$

**Proposition 9.** *The eigenvalues of the uncertain Lagrangian system based on (50) have the following sign*

$$\begin{cases} \mu_k < 0, \; k \le p+1, \\ \mu_k > 0, \; 2p+2 < k \le 3(p+1). \end{cases}$$

*Moreover there exists* $k \in ]p+1, 2p+2]$ *such that* $\mu_k = 0$.

Let us use the min-max formula (49) with

$$S_{p+1} = \left\{ X = (X_0, \ldots, X_p) \in \mathbb{R}^{n \times (p+1)}, \; X_q = \alpha_q \begin{pmatrix} 1 \\ 1 \\ 0 \end{pmatrix} \text{ with } \alpha_q \in \mathbb{R} \; \forall q \right\}.$$

The polynomial in (49) is

$$x(\xi) = \left( \sum_{q \le p} \alpha_q \varphi_q(\xi) \right) \begin{pmatrix} 1 \\ 1 \\ 0 \end{pmatrix}.$$

So

$$\left( \nabla(f \circ h) \left( \sum_{r \le p} v_r \varphi_r(\xi) \right) x(\xi), x(\xi) \right) = \frac{2}{v_3(\xi)} \left( \sum_{q \le p} \alpha_q \varphi_q(\xi) \right)^2 < 0$$

since $x(\xi) \neq 0$ and $v_3 < 0$ by definition. That is $\mu_{p+1} < 0$. To prove that $\mu_{2p+3} > 0$ it is sufficient to apply the same method to the eigenvalue problem

$$C'(V) X_l = \mu'(V)_l B(V) X_l, \quad l = 1, \ldots n \times (p+1).$$

with $C' = -C$ and $\mu'_l = -\mu_{n(p+1)-l}$.

Finally we notice that the flux $(-v, p, pv)$ is homogeneous of degree 0 with respect to the entropy variable. Therefore the Euler relation implies

$$(\nabla f \circ h)\begin{pmatrix} v_1 \\ v_2 \\ v_3 \end{pmatrix} = 0 \tag{52}$$

which is easily checked from (51). So $C(V)V = \left(\sum_{r \le p} c_{qr} v_r\right)_{q \le p}$ with

$$\sum_{r \le p} c_{qr} v_r = \int_\Theta \nabla(f \circ h)\left(\sum_{r \le p} v_r \varphi_r(\xi)\right) \varphi_q(\xi) \left(\sum_{r \le p} v_r \varphi_r(\xi)\right) w(\xi) d\xi = 0$$

thanks to (52). Therefore $C(V)V = 0$ which shows that $C(V)$ has at least one vanishing eigenvalue. It ends the proof.

## 5.3 Entropy Choice

As usual with entropy closure of hyperbolic systems, the entropy of the initial system is also an entropy for the final system: in our case the entropy of the deterministic system (1) yields an entropy for the uncertain system (41).

**Proposition 10.** *The system (41) is endowed with an entropy*

$$S(U) = \int_\Theta s\left(h\left(\sum_{q \le p} v_q \varphi_q(\xi)\right)\right) w(\xi) d\xi$$

*and entropy flux*

$$G(U) = \int_\Theta g\left(h\left(\sum_{q \le p} v_q \varphi_q(\xi)\right)\right) w(\xi) d\xi.$$

By definition of the entropy function $s$, one has that

$$dS(U) = \int_\Theta \underbrace{\left(\sum_{q \le p} v_q \varphi_q(\xi)\right)}_{=v^p} \cdot \underbrace{dh\left(\sum_{q \le p} v_q \varphi_q(\xi)\right)}_{=\partial_t u^p} w(\xi) d\xi$$

$$= \sum_{q \le p} v_q \cdot d\left(\int_\Theta \varphi_q(\xi) h(\sum_{q \le p} v_q \varphi_q(\xi)) w(\xi) d\xi\right) = \sum_{q \le p} v_q \cdot d u_q.$$

That is

$$dS(U) = V \cdot U.$$

Similarly

$$dG(U) = \int_{\Theta} \underbrace{\left(\sum_{q\leq p} v_q\varphi_q(\xi)\right)}_{=v^p} \cdot \underbrace{df \circ h \left(\sum_{q\leq p} v_q\varphi_q(\xi)\right)}_{=\partial_x f(u^p)} w(\xi)d\xi$$

$$= \sum_{q\leq p} v_q \cdot \left(\int_{\Theta} df \circ h \left(\sum_{q\leq p} v_q\varphi_q(\xi)\right) v_q\varphi_q(\xi)w(\xi)d\xi\right)$$

That is

$$dG(U) = V \cdot dF(U).$$

It has two consequences.

The first consequence is that smooth solutions of (41) are such that

$$\partial_t S(U) + \partial_x G(U) = 0, \tag{53}$$

since $\partial_t S(U) + \partial_x G(U) = V \cdot (\partial_t U + \partial_x F(U))$. It shows that a non trivial additional conservation law is satisfied by smooth solutions.

The second consequence is that $S(U)$ is a strictly convex functional with respect to $U$. Indeed $\nabla_U S = V$. So $\nabla_U^2 S = \nabla_U V$ is a symmetric matrix. This matrix, $\nabla_U V$, is positive since it is by definition the inverse matrix of $B(V)$. The proof is ended.

In the next section, we consider two hydrodynamic test cases emphasizing the stability of the uncertain entropy closure (38)–(40) of (1) (see Proposition 10) together with the behaviour of its wave velocities (see Theorem 2 and Proposition 8).

## 5.4 *Numerical Applications*

### 5.4.1 Stochastic Riemann Problem for Euler Equations

In order to illustrate the above material, we consider Euler system (31) together with a stochastic Riemann problem with initial conditions are given by

$$\begin{cases} \rho(x,0,\Xi) &= \begin{cases} 1 \text{ if } x \leq x_{interface}(\Xi), \\ 0.125 \text{ elsewhere,} \end{cases} \\ \rho u(x,0,\Xi) &= 0, \\ \rho e(x,0,\Xi) &= \begin{cases} 2.5 \text{ if } x \leq x_{interface}(\Xi), \\ 0.25 \text{ elsewhere,} \end{cases} \end{cases} \tag{54}$$

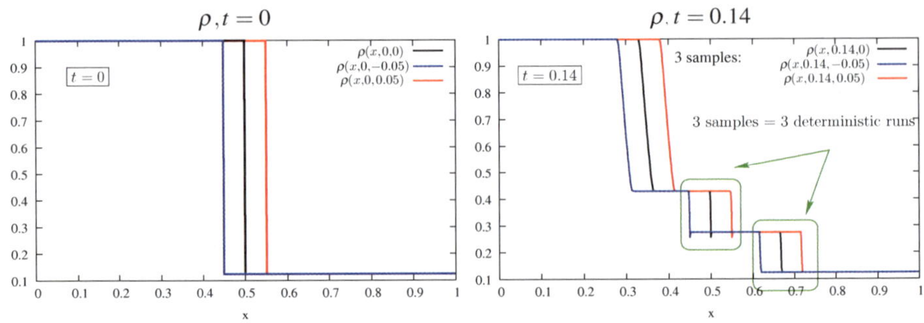


**Fig. 4** Stochastic Riemann Problem whose initial conditions are given by (54). The *left column* shows the initial conditions for $\rho$ (three realizations). The *right column* shows the solution $\rho$ at final time $t = 0.14$ (three realizations)

so that the uncertainty is initially carried by the position of the interface between the two fluids at rest. This problem is inspired from the Sod shock tube problem [36, 37]. Figure 4 presents the initial conditions (left) and three deterministic runs for three realizations of the random variable (right). For one realization of the random variable, the solution consists in three waves: a rarefaction wave in the heavy fluid, smooth part of the curves in Fig. 4 (right), the interface position and a shock wave in the light fluid. Let us now solve the $p$–truncated Euler system and study the behavior of these three waves under uncertainties for different stochastic discretizations ($p = 4$ and $p = 20$). To do so, we rely on the numerical scheme described in [31]: it is a third-order Lagrange+Remap scheme with 1,000 cells, stable under cfl condition taken to 0.9.

Figure 5 presents the mean and the variance of the mass density $\rho$ at time $t = 0.14$ with respect to $x$ for two polynomial orders $p = 4$ and $p = 20$. The wave velocities (5 for $p = 4$ and 21 for $p = 20$) are identifiable in the variability zones of the shock for example. More precisely we identify five small oscillations reminiscent of shocks in the interval $x \in [0.68, 0.83]$ in Fig. 5. On the right part the number of shocks is greater, but the amplitude is also much smaller. This is why the solution behave like a smooth curve, even if we still distinguish the shocks. Figure 5 also allows emphasizing that the waves in the vicinity of the interface or the shock behave differently. The study of the nature (linearly degenerate, genuinely nonlinear) of the waves of the $p$–truncated system is complicated in general as the size of the system makes the analytical expressions of the eigenvectors of the Jacobian of the flux hard to obtain. At least we have not obtained any convincing result in this direction.

### 5.4.2 Shock Hitting an Uncertain Interface Between Two Fluids

In order to illustrate the well-posedness and stability of the moment built $p$-truncated system, we consider an hydrodynamic problem, described by Euler

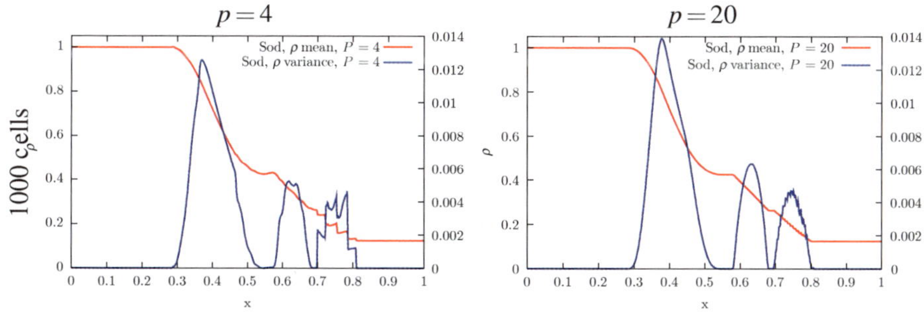


**Fig. 5** Illustration of the existence of multiple waves, see Proposition 8. The phenomenon is especially emphasized for the mean in the vicinity of the probable shock region ($x \in [0.7, 0.8]$) where $P + 1 = 5$ discontinuities are visible on the *left picture* and $P + 1 = 21$ discontinuities are visible on the *right* one

equation (31), in which a shock hits an interface between two states which initial position is uncertain:

$$\left\{\begin{array}{l} \rho(x,0,\Xi) = \begin{cases} 4, & \text{if } x \leq x_{interface}(\Xi). \\ 1, & \text{if } x_{interface}(\Xi) \leq x \leq x_{shock}. \\ \frac{2\gamma-\frac{\gamma}{2}+\frac{1}{2}}{2\gamma-\frac{\gamma}{2}-\frac{1}{2}}, & \text{if } x \geq x_{shock}. \end{cases} \\ u(x,0,\Xi) = \begin{cases} 0, & \text{if } x \leq x_{interface}(\Xi). \\ 0, & \text{if } x_{interface}(\Xi) \leq x \leq x_{shock}. \\ -\sqrt{\frac{(\rho-1)}{\rho}}, & \text{if } x \geq x_{shock}. \end{cases} \\ p(x,0,\Xi) = \begin{cases} 1, & \text{if } x \leq x_{interface}(\Xi). \\ 1, & \text{if } x_{interface}(\Xi) \leq x \leq x_{shock}. \\ 2, & \text{if } x \geq x_{shock}. \end{cases} \end{array}\right. \tag{55}$$

The initial conditions in mean and variance are presented in Fig. 6 left column.

To the left of the interface, the heavy fluid is at rest, see Fig. 6 (left). Note that the uncertainty at time $t = 0$ is only carried by the mass density in the vicinity of the interface, pressure and velocity are completely deterministic (zero variance for $u$ and $p$ in Fig. 6 left column).

On the right of the interface, a shock is initialized at $x_{shock} = 0.7$ in the light fluid. For $t > 0$, for every realization of the random variable, the shock propagates in the direction of the interface. In practice, we take $\gamma = 1.4$. The initial uncertain interface position is modeled by a random variable $x_{interface}(\Xi) = 0.5 + 0.05\Xi$ where $\Xi \sim \mathscr{U}([-1, 1])$. For every realization of the uncertain parameter, the shock hits the interface and reflects/refracts in the light/heavy fluid. In order to solve this problem, we rely on the numerical scheme described in [31]: it is a third-order Lagrange+Remap scheme with 1,000 cells, stable under cfl condition taken to 0.9. Figure 6 presents the means and variances of the mass density, velocity and pressure at times $t = 0$ and $t = 0.34$ for $P = 20$. The result illustrates the result of Proposition 10: the computation is stable and physical.

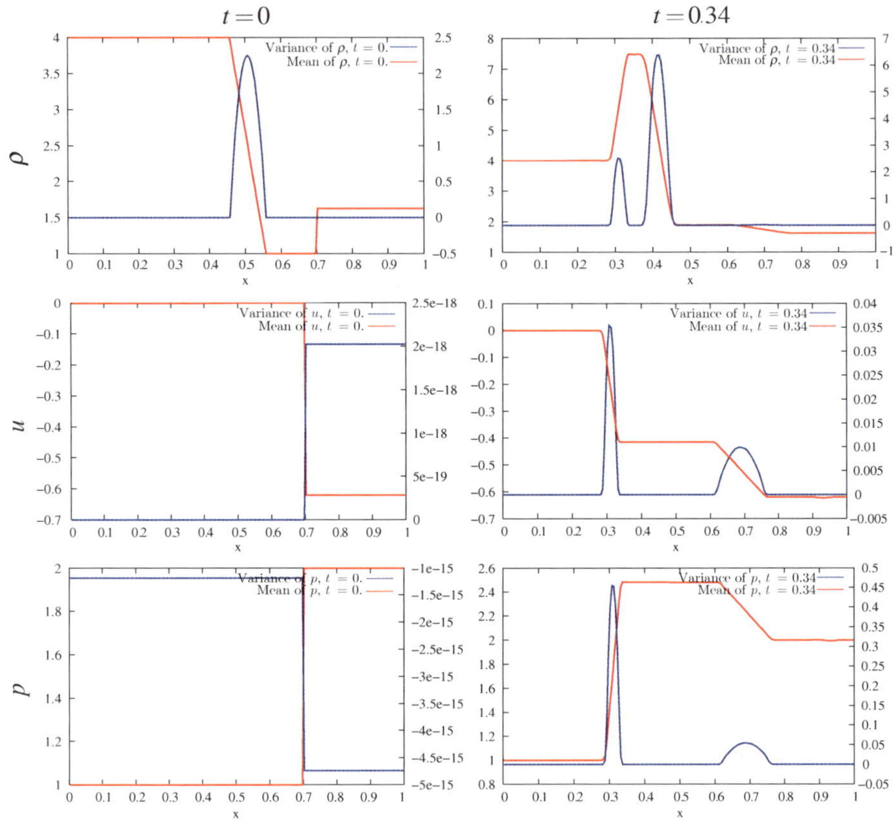


**Fig. 6** Illustration of the result of Proposition 10: well-posedness of the $p$—truncated system on a hydrodynamic problem, $P = 20$ with 200 cells

Let us now describe more precisely the test problem. Initially, only the mass density is uncertain. After the shock hitting the interface, the uncertainty is distributed to the different waves, interface, reflected and refracted shocks and the different physical quantities, velocity $u$ and pressure $p$: on Fig. 6 (right column), the variance in the vicinity of the refracted and reflected is important for every variable $\rho, u, p$. Note that on this same problem, the classical Polynomial Chaos closure approach leads to a crash of the code as the shock passes the interface: the amplitude of the discontinuity in the random space is such that the positivity of the mass density is not ensured.

Figure 7 shows the same quantities as Fig. 6 but with 1,000 cells instead of 200 cells. With 1,000 cells on Fig. 7, the stochastic errors (depending on $P$) becomes preponderant with respect to the spatial errors and the $P+1$ waves of Proposition 8 are not anymore "masked" by the numerical diffusion of the spatial scheme. This is especially visible in the vicinity of the probable shock location.

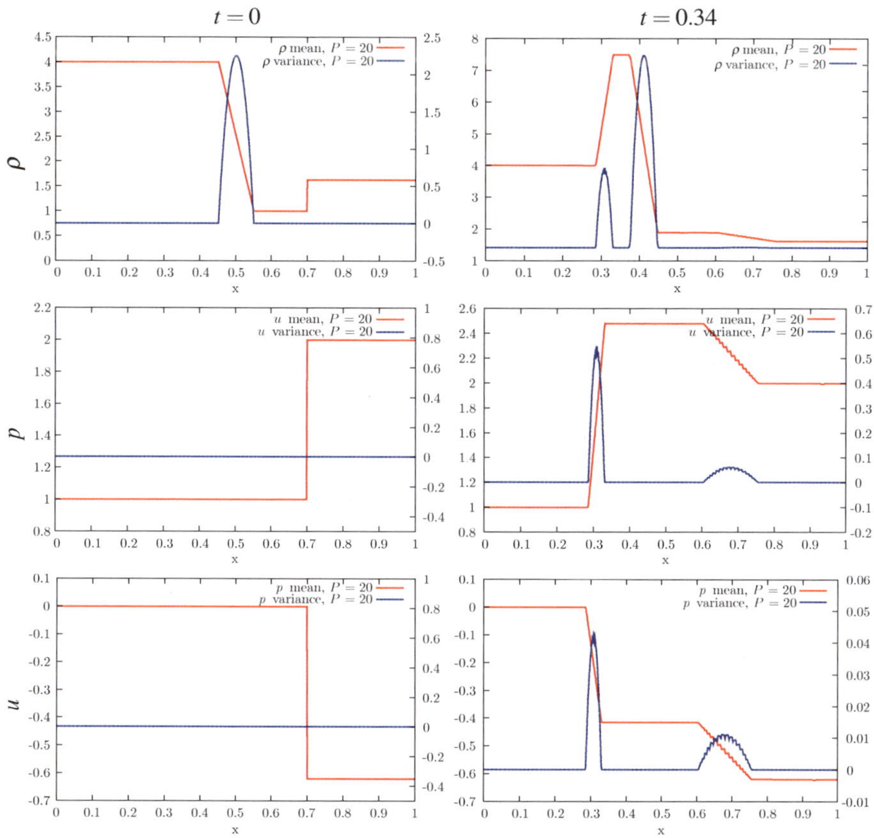


**Fig. 7** Illustration of the result of Proposition 10: well-posedness of the $p-$truncated system on a hydrodynamic problem, $P = 20$ with 1,000 cells

In the precedent sections, we suggested an entropy based closure approach to the moment problem of Sect. 2, stated a stability result and other technical properties for the resulting system. For the sake of simplicity, the base of the study was to take into account uncertainties in the initial conditions. In the next section, we show and illustrate that the latter results can be applied in the context of uncertainties in the model parameters.

## 6 Parametric Uncertainty of the Model

As pointed out in the introduction, uncertainty in the model may be addressed by considering the enlarged system of conservation laws (7). It will appear that a specific compatibility condition is attached to (7).

The main point is the following. If the enlarged system (7) is itself a hyperbolic system of conservation laws with an entropy, then it is sufficient to apply the previous strategy to construct a hyperbolic uncertain system where the model parameter $\gamma$ is uncertain.

Let us consider the system

$$\partial_t u + \partial_x f_\gamma(u) = 0 \tag{56}$$

which admits by hypothesis an entropy-entropy flux pair $\left(s_\gamma(u), g_\gamma(u)\right)$, that is: (a) $s_\gamma(u)$ is strictly convex with respect to $u$; and (b) we define $v_\gamma$ as

$$ds_\gamma(u) = v_\gamma \cdot du \text{ together with } dg_\gamma(u) = v_\gamma \cdot df_\gamma(u).$$

These differential forms hold for constant $\gamma$. We also notice that the entropy variable shows a dependence with respect to $\gamma$ since it is formerly defined as the gradient (33) of the entropy. The domain in which the parameter $\gamma$ lives is

$$\gamma \in \Gamma \subset \mathbb{R}^\lambda, \qquad \lambda \in \mathbb{N}^*.$$

Let us define $f(u,\gamma) = f_\gamma(u)$ and rewrite (56) as a system

$$\begin{cases} \partial_t u + \partial_x f(u,\gamma) = 0, \\ \partial_t \gamma \qquad\qquad\quad\; = 0. \end{cases} \tag{57}$$

A tentative entropy-entropy flux pair for this system is

$$S(u,\gamma) = s_\gamma(u) + \frac{|\gamma|^2}{\varepsilon} \text{ and } G(u,\gamma) = g_\gamma(u).$$

Here $\varepsilon > 0$ is a parameter that is chosen sufficiently small as explained in the following proposition.

**Proposition 11.** *Assuming that $\varepsilon > 0$ is sufficiently small, the entropy $S(u,\gamma)$ is strictly convex with respect to $(u,\gamma)$.*

One has $dS(u,\gamma) = v_\gamma \cdot du + \left(\partial_\gamma s_\gamma(u) + \frac{\gamma}{\varepsilon}\right) \cdot d\gamma$. So the Hessian of $S$ is

$$\nabla^2 S(u,\gamma) = \begin{pmatrix} \nabla_u v_\gamma & \partial^2_{u,\gamma} s_\gamma(u)^T \\ \partial^2_{u,\gamma} s_\gamma(u) & \frac{1}{\varepsilon} \end{pmatrix}.$$

For $z \in \mathbb{R}^n$ and $w \in \mathbb{R}^\lambda$, we set $y = (z,w)$ and compute

$$\left(y, \nabla^2 S(u,\gamma) y\right) = \left(z, \nabla_u v_\gamma z\right) + \frac{|w|^2}{\varepsilon} + 2\left(y, \partial^2_{u,\gamma} s_\gamma(u) w\right)$$

Since

$$2\left|\left(y,\partial^2_{u,\gamma}s_\gamma(u)w\right)\right|\leq 2\left|\partial^2_{u,\gamma}s_\gamma(u)\right|\,|z|\,|w|$$
$$\leq\alpha\left|\partial^2_{u,\gamma}s_\gamma(u)\right||z|^2+\frac{1}{\alpha}\left|\partial^2_{u,\gamma}s_\gamma(u)\right||w|^2,\qquad\forall\alpha>0.$$

Therefore

$$\left(y,\nabla^2S(u,\gamma)y\right)\geq\left[(z,\nabla_u v_\gamma z)-\alpha\left|\partial^2_{u,\gamma}s_\gamma(u)\right||z|^2\right]+\left[\frac{1}{\varepsilon}-\frac{1}{\alpha}\left|\partial^2_{u,\gamma}s_\gamma(u)\right|\right]|w|^2.$$

Next we choose $\alpha>0$ sufficiently small such that $\alpha\left|\partial^2_{u,\gamma}s_\gamma(u)\right|<\lambda_{\min}\left(\nabla_u v_\gamma\right)$ where $\lambda_{\min}\left(\nabla_u v_\gamma\right)>0$ is the smallest positive eigenvalue of the symmetric positive matrix $\nabla_u v_\gamma$. Finally we choose $\varepsilon>0$ sufficiently small such that $\frac{1}{\varepsilon}-\frac{1}{\alpha}\left|\partial^2_{u,\gamma}s_\gamma(u)\right|>0$. In this case there exists a constant $C(u,\gamma)>0$ such that

$$\left(y,\nabla^2S(u,\gamma)y\right)\geq C(u,\gamma)|y|^2.$$

It shows that the Hessian is a positive matrix. The claim is proved.

**Proposition 12.** *Assuming one of these two conditions is fulfilled*

$$\partial_x\gamma=0\text{ or }\partial_\gamma g_\gamma(u)-v_\gamma\partial_\gamma f_\gamma(u)=0,\tag{58}$$

*then $(S(u,\gamma),G(u,\gamma))$ is an entropy-entropy flux pair in the sense that smooth solutions of (57) satisfy the additional conservation law*

$$\partial_t S(u,\gamma)+\partial_x G(u,\gamma)=0.\tag{59}$$

By definition one has

$$\partial_t S(u,\gamma)=v_\gamma\cdot\partial_t u+\left(\partial_\gamma s_\gamma(u)+\frac{\gamma}{\varepsilon}\right)\cdot\partial_t\gamma=v_\gamma\cdot\partial_t u$$

and

$$\partial_x G(u,\gamma)=\nabla_u g_\gamma(u)\cdot\partial_x u+\partial_\gamma g_\gamma(u)\cdot\partial_x\gamma=v_\gamma\nabla_u f_\gamma(u)\cdot\partial_x u+\partial_\gamma g_\gamma(u)\cdot\partial_x\gamma$$
$$=v_\gamma\cdot\partial_x f_\gamma(u)-v_\gamma\partial_\gamma f_\gamma(u)\cdot\partial_x\gamma+\partial_\gamma g_\gamma(u)\cdot\partial_x\gamma.$$

So

$$\partial_t S(u,\gamma)+\partial_x G(u,\gamma)=\left(\partial_\gamma g_\gamma(u)-v_\gamma\partial_\gamma f_\gamma(u)\right)\cdot\partial_x\gamma.$$

So if one of the two terms on the right hand side vanishes, then any smooth solutions of (41) satisfies the additional entropy law. The claim is proved.

The compatibility law (58) expresses that the deterministic system admits an entropy law. Since the entropy is strictly convex, it is immediate to apply the entropy closure method to define a uncertain hyperbolic system, where both the unknown and the parameters are uncertain. In the sequel we detail the two cases (58).

## 6.1 First Case: The Model Parameter Is a Random Variable

In this case, the random parameter $\gamma$ is constant in space for instance.

The entropy variable derived from $S(\hat{u})$ (with $\hat{u} = (u, \gamma)$) is

$$\hat{v} = (v_\gamma, \frac{\gamma}{\varepsilon}).$$

Expansion of $\hat{v}$ over the polynomials $\varphi_q(\xi)$ yields

$$v_\gamma = \sum_{q\leq} v_q \varphi_q(\xi)$$

and

$$\frac{\gamma}{\varepsilon} = \sum_{q\leq p} \tilde{\gamma}_q \varphi_q(\xi).$$

The coefficients $v_q$ have already been discussed. Considering the moment method applied to the equation $\partial_t \gamma = 0$, and the fact that $\partial_x \gamma = 0$, it implies that the coefficients $\gamma_q$ are constant in time **and** space. We notice that one can set $\tilde{\gamma}_q = \frac{\gamma_q}{\varepsilon}$ so that the parameter $\gamma$ expands as

$$\gamma = \sum_{q\leq p} \gamma_q \varphi_q(\xi).$$

The parameter $\varepsilon$ does not show up in this formula. It means that it is possible to write directly the limit system as $\varepsilon \to 0^+$. We find the uncertain system with constant in time and space uncertain parameters

$$\begin{cases} \partial_t u_q + \partial_x \int_\Theta f\left(h\left(\sum_{l\leq p}(v_\gamma)_q\varphi_q\right), \gamma\right)\varphi_q(\xi)w(\xi)d\xi = 0, \ \forall q \leq p, \\ u_q = \int_\Theta h\left(\sum_{r\leq p}(v_\gamma)_r\varphi_r(\xi)\right)\varphi_q(\xi)w(\xi)d\xi, \qquad \forall q \leq p, \\ \gamma = \sum_{q\leq p}\gamma_q\varphi_q(\xi) \text{ with } \gamma_q \text{ constant in time and space.} \end{cases} \tag{60}$$

This system is hyperbolic in the sense that (59) holds for all $\varepsilon$. So we can pass to the limit and identify the resulting equation. The specific form of $S(u,\gamma)$ and $G(u,\gamma)$ shows that smooth solutions of (60) actually satisfy the additional classical entropy relation

$$\partial_t \int_\Theta s_\gamma(h_\gamma(\sum_{q\le p}(v_\gamma)_q\varphi_q(\xi)))w(\xi)d\xi$$
$$+\partial_x \int_\Theta g_\gamma(h_\gamma(\sum_{q\le p}(v_\gamma)_q\varphi_q(\xi)))w(\xi)d\xi = 0.$$

This is simply the $\gamma$ extension of (53).

In the next section, we illustrate numerically the stability of the uncertain entropy closure (38)–(40) of (1) in the context of uncertainties carried by the model parameter: we will consider an uncertain adiabatic coefficient of a perfect gas for the Euler system in similar configurations to problems of Sects. 5.4.1 and 5.4.2.

### 6.1.1 Application to Uncertain $\gamma$−Law Gas Value

In this section we revisit the two precedent problems (Sect. 5.4.1 "The Stochastic Riemann Problem" and Sect. 5.4.2 "The Shock Hitting Interface Problem") considering uncertainties in the $\gamma$ parameter of the perfect gas closure rather than on the interface position.

This leads to the initial condition (61) for the stochastic Riemann problem

$$\begin{cases} \rho(x,0) &= \begin{cases} 1 \text{ if } x \le x_{interface} = 0.5 \\ 0.125 \text{ elsewhere,} \end{cases} \\ \rho u(x,0) &= 0, \\ \rho e(x,0) &= \begin{cases} 2.5 \text{ if } x \le x_{interface} = 0.5 \\ 0.25 \text{ elsewhere.} \end{cases} \end{cases} \tag{61}$$

In other words, the initial condition is deterministic, the uncertainty only affects $\gamma(\Xi) = 1.4 + 0.25\Xi$ where $\Xi \sim \mathscr{U}([-1,1])$ so that every realizations of the uncertain parameter are physically relevant (i.e. $\gamma(\xi) > 1, \forall\xi \in [-1,1]$). Figure 8 presents the mean and variance profiles of the mass density, velocity and pressure at time $t = 0.14$. The stability and coherence of the calculation illustrate Proposition 12. At final time, the most sensitive parts are localized in the vicinities of the discontinuities, the interface and the shock. Contrary to the computations of Sect. 5.4.1, the waves are not the only uncertain regions even if concentrating the variability. The mass density and the velocity are importantly affected in between the foot of the rarefaction fan and the shock. The pressure, on the contrary, is not.

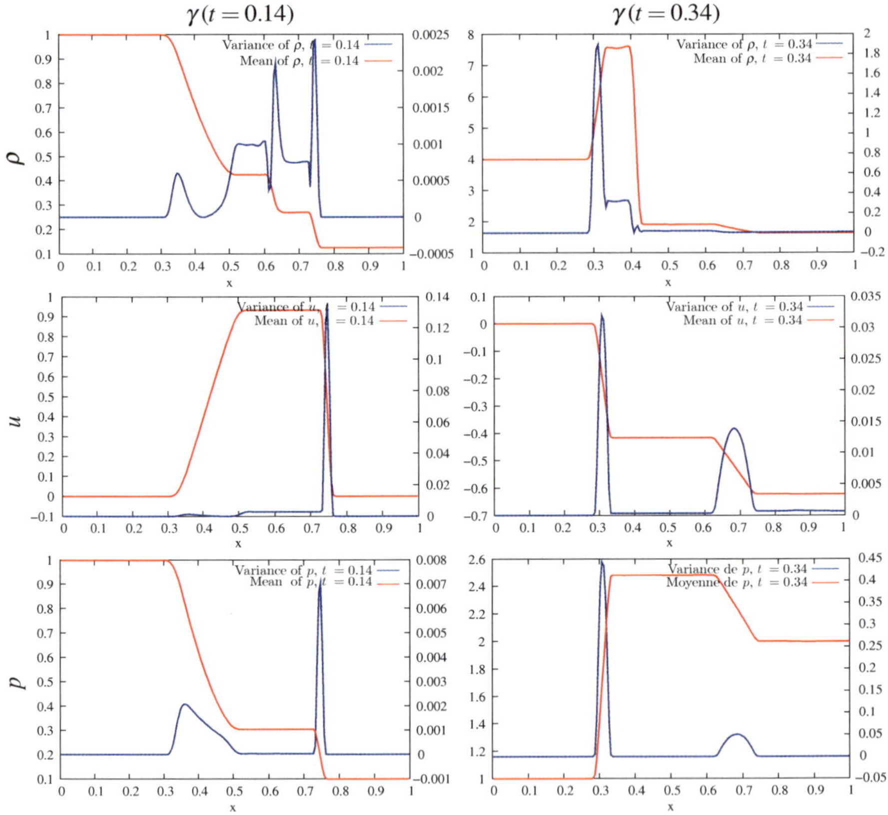


**Fig. 8** Illustration of Proposition 12: uncertain $\gamma$ parameter in two configurations, the computations are stable and coherent. The computations were made using $P = 20$ and 200 cells

For the second case we have initial condition (62):

$$
\begin{cases}
\rho(x,0,\Xi) = \begin{cases} 4, & \text{if } x \le x_{interface} = 0.5. \\ 1, & \text{if } 0.5 = x_{interface} \le x \le x_{shock} = 0.7. \\ \frac{2\gamma(\Xi) - \frac{\gamma(\Xi)}{2} + \frac{1}{2}}{2\gamma(\Xi) - k\frac{\gamma(\Xi)}{2} - \frac{1}{2}}, & \text{if } x \ge x_{shock} = 0.7. \end{cases} \\
u(x,0,\Xi) = \begin{cases} 0, & \text{if } x \le x_{interface} = 0.5. \\ 0, & \text{if } 0.5 = x_{interface} \le x \le x_{shock} = 0.7. \\ -\sqrt{\frac{(\rho(x,0,\Xi)-1)}{\rho(x,0,\Xi)}}, & \text{if } x \ge x_{shock} = 0.7. \end{cases} \\
p(x,0) \quad = \begin{cases} 1, & \text{if } x \le x_{interface} = 0.5. \\ 1, & \text{if } x_{interface} = 0.5 \le x \le x_{shock} = 0.7. \\ 2, & \text{if } x \ge x_{shock} = 0.7. \end{cases}
\end{cases}
\tag{62}
$$

Once again, we consider $\gamma(\Xi) = 1.4 + 0.25\Xi$ where $\Xi \sim \mathscr{U}([-1,1])$.

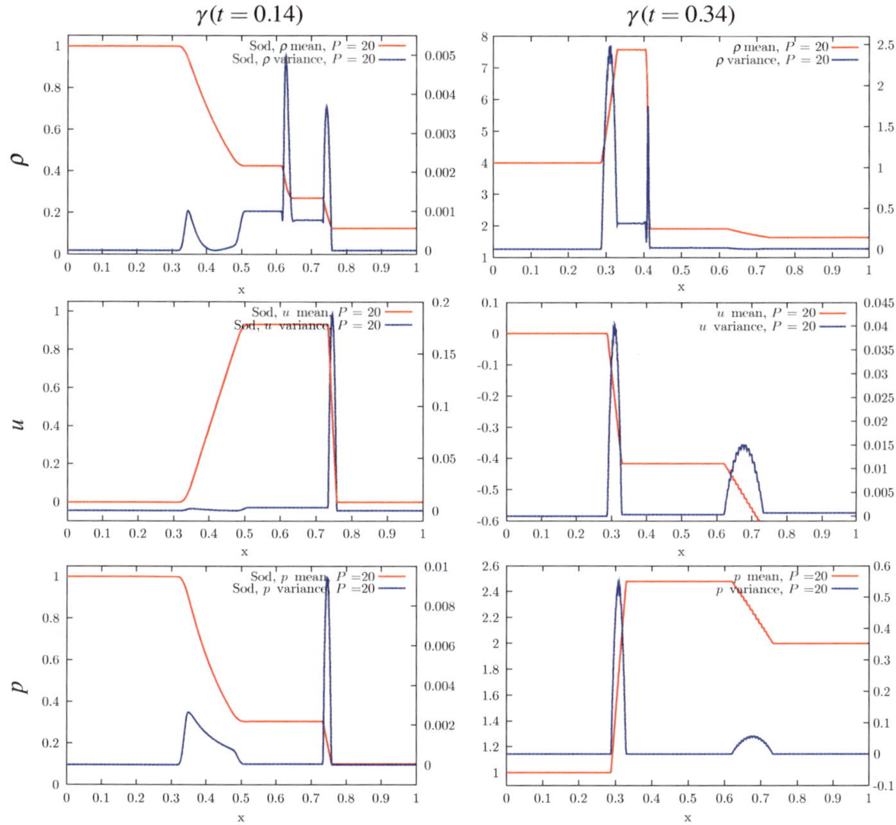


**Fig. 9** Illustration of Proposition 12: uncertain $\gamma$ parameter in two configurations, the computations are stable and coherent. The computations were made using $P = 20$ and 1,000 cells

Both problems are solved in the same condition, we again rely on the numerical scheme described in [31] with 200 cells (Fig. 8) and 1,000 cells (Fig. 9), stable under cfl condition taken to 0.9. Figure 8 (right) shows the means and variances of the mass density, velocity and pressure at time $t = 0.34$ for problem (62) for $P = 20$. Once again, in agreement with Proposition 12, the computation is stable and coherent. The results are also particular in the sense for this problem, the vicinity of the interface does not concentrate the uncertainty. The most sensitive region of the calculation corresponds to the vicinity of the refracted shock for the mass density, the velocity and the pressure. Figure 9 also illustrates well the results of Proposition 8 concerning the appearance of $P + 1$ waves when the spatial error is way below the stochastic error. When this is not the case, e.g. when the numerical diffusion of the deterministic scheme is of the same order, the oscillations fade out, see Fig. 9.

## 6.2 Second Case: The Model Parameter Is a Random Process

In this case, the random parameter $\gamma$ is not constant in space.

The first case is interesting but of limited use if one desires to address uncertainties depending on the region of the space. In this case $\partial_x \gamma \neq 0$ so it is necessary to rely on the second compatibility condition

$$\partial_\gamma g_\gamma(u) - v_\gamma \partial_\gamma f_\gamma(u) = 0. \tag{63}$$

Instead of developing a general theory for such systems, we prefer to show that the physical underground of the Lagrangian system (50) makes him a good candidate to study this second compatibility condition.

So let us consider the system

$$\begin{cases} \partial_t \tau - \partial_m v = 0, \\ \partial_t v + \partial_m p_\gamma = 0, \\ \partial_t e + \partial_m p_\gamma v = 0, \end{cases} \tag{64}$$

where the pressure is given by

$$p_\gamma = (\gamma - 1)\rho \left( e - \frac{1}{2} v^2 \right).$$

**Proposition 13.** *The compatibility condition (63) holds for the system (64).*

The classical entropy flux is zero, namely $g_\gamma = 0$. One has

$$\partial_\gamma f_\gamma(u) = \partial_\gamma \begin{pmatrix} -v \\ p_\gamma \\ p\gamma v \end{pmatrix} = \begin{pmatrix} 0 \\ \partial_\gamma p_\gamma \\ \partial_\gamma p \gamma v \end{pmatrix}$$

Therefore

$$v_\gamma \cdot \partial_\gamma f_\gamma(u) = \frac{1}{T} \begin{pmatrix} -p \\ v \\ -1 \end{pmatrix} \cdot \begin{pmatrix} 0 \\ \partial_\gamma p_\gamma \\ \partial_\gamma p \gamma v \end{pmatrix} = 0.$$

It ends the proof.

We find the uncertain system with constant in time uncertain parameters

$$\begin{cases} \partial_t u_q + \partial_x \int_\Theta f\left( h\left( \sum_{l \le p} (v_\gamma)_q \varphi_q \right), \gamma \right) \varphi_q(\xi) w(\xi) d\xi = 0, \ \forall q \le p, \\ u_q = \int_\Theta h\left( \sum_{r \le p} (v_\gamma)_r \varphi_r(\xi) \right) \varphi_q(\xi) w(\xi) d\xi, \qquad \forall q \le p, \\ \gamma = \sum_{q \le p} \gamma_q \varphi_q(\xi) \text{ with } \gamma_q \text{ constant in time and but not in space.} \end{cases} \tag{65}$$

This system is hyperbolic.

### 6.3 Modeling Parameter Uncertainties in Eulerian Systems

Once again we quote that the compatibility condition is not true for all deterministic systems of conservation laws. Our observation is that (63) is convenient for the modeling of uncertainty in Lagrangian systems of conservation laws. If the deterministic system is Eulerian, a modification must be done. In what follows we describe this modification for the Eulerian system of compressible gas dynamics (31).

Let us start from

$$\begin{cases} \partial_t \rho + \partial_x (\rho v) = 0, \\ \partial_t (\rho v) + \partial_x \left(\rho v^2 + p_\gamma\right) = 0, \\ \partial_t (\rho e) + \partial_x \left(\rho v e + p_\gamma v\right) = 0, \end{cases} \tag{66}$$

As before the first task is to model the propagation of parameter uncertainty in order that the enlarged system is hyperbolic with a similar entropy. We propose to consider

$$\partial_t(\rho\gamma) + \partial_x(\rho v\gamma) = 0. \tag{67}$$

The system (66)–(67) is the Eulerian version of the Lagrangian system (64) with $\gamma$ constant in time. Therefore it is hyperbolic with entropy $\tilde{S} = \rho s$ gnd entropy flux $v\rho S$. It must be noted that the entropy variable is

$$\tilde{v} = \frac{1}{T}(-\mu, v, 1, -\gamma)$$

where $\mu$ is the Gibbs potential. The method presented above allows to design a hyperbolic model with uncertainties both in the initial condition and in the parameter $\gamma$.

## 7 Conclusion and Open Problems

In this paper, we considered hyperbolic systems of conservation laws subject to uncertainties in the initial conditions and model parameters. In order to solve the underlying uncertain system of conservation laws, we have relied on moment theory and the construction of a moment model. We first proved spectral convergence of the moment model for a non-linear *scalar* equation: the uncertain inviscid Burgers' equation. We then emphasized the difficulties arising when applying the moment method in the context of uncertain *systems* of conservation laws. In particular, we have shown that the moment model for the shallow water equations and Euler system may not be always well-posedness.

We have then suggested a new *entropy*-based spectral discretisation inspired by plasma physics and rational extended thermodynamics, constructed in such a way that it always preserves the hyperbolicity of the original stochastic system. We have also investigated the mathematical structure (wave velocities) of the resulting well-posed large truncated systems of partial differential equations. Finally, we have presented the natural extension of the proposed numerical framework to the case of model parametric uncertainty.

Among all the questions raised in this paper, we distinguish some of which seems fundamental even for very different reasons.

The first problem is of theoretical nature. It consists in the development of a general method for the study of the difference between the deterministic solution and the one issued from the moment method. The major difficulty is weak solutions.

The second problem is related to the curse of dimensionality. Indeed the uncertain variable can live in a space of high dimension. Special quadrature procedures have to be invented. Our preliminary tests show that positivity of the solutions (that is the respect of the natural constraint of the problem, example a density must be non negative) is difficult to ensure.

A third problem is the realizability of the moment problem. It is well known since [38] that moment models with large number of moments and bounded physical space $\Omega$ have tendency to be singular near $\partial\Omega$.

All these questions have to be addressed together with the development of efficient innovative numerical and computational methods.

## Appendix

The following propositions are useful for the proof of Theorem 1.

**Proposition 14.** *There exists a constant $C > 0$ such that for all $p \geq 0$*

$$\left\| \Pi_p u \right\|_{L^\infty(\mathscr{I}\times]0,T^\varepsilon[\times\Theta)} \leq C |||u|||_{\varepsilon,k+1}, \qquad \forall k \geq 1. \tag{68}$$

It comes from the infinite expansion $u(x,t,\xi) = \sum_p \langle u(x,t,\cdot), L_p \rangle L_p(\xi)$ where the usual $L^2$ scalar product with respect to the variable $\xi$ is denoted as $< f, g >= \int fg d\xi$. Since $\frac{d}{d\xi}\left((1-\xi^2)\frac{d}{d\xi}L_p(\xi)\right) + p(p+1)L_p(\xi) = 0$, one has that

$$\frac{\partial}{\partial \xi}\left((1-\xi^2)\frac{\partial}{\partial \xi}u(x,t,\xi)\right) = \sum \langle u(x,t,\cdot), L_p \rangle\, p(p+1)L_p(\xi).$$

The first and second derivatives of $u$ with respect to $\xi$ being bounded in $L^2_\xi$ by hypothesis (15), one has that $\sum \langle u(x,t,\cdot), L_p \rangle^2 p^2(p+1)^2 < \infty$. This is true at any order and everywhere, that is

$$\sum_p \langle u(x,t,\cdot), L_p \rangle^2 (p^{2k}+1) < \infty, \qquad \forall k \in \mathbb{N}, \quad \forall t > 0,\ x \in \mathscr{I}.$$

On the other hand one has the bound for all $t$ and $x$

$$\|\Pi_p u(x,t,\cdot)\|_{L^\infty(\Theta)} \le \sum_{n\le p} |\langle u(x,t,\cdot), L_n\rangle| \, \|L_n\|_{L^\infty(\Theta)}$$

$$\le \left(\sum_{n\le p} \langle u(x,t,\cdot), L_n\rangle^2 (n^{2k}+1)\right)^{\frac{1}{2}} \left(\sum_{n\le p} \frac{\|L_n\|^2_{L^\infty(\Theta)}}{n^{2k}+1}\right)^{\frac{1}{2}}. \tag{69}$$

The first sum is bounded by $|||u|||_{\varepsilon,k+1}$. One the other hand one has the bound [39]

$$\|L_n\|_{L^\infty(\Theta)} \le C(n+1)^{\frac{1}{2}}. \tag{70}$$

Choosing $k \ge 1$, the second sum is also bounded uniformly with respect to $k, p \ge 1$. The proof is ended.

**Proposition 15.** *There exist constants $C^\varepsilon > 0$ which depend on $T^\varepsilon = T - \varepsilon$ and the solution $u$ such that*

$$\left\|u - \Pi_p u\right\|_{L^\infty(\mathscr{I}\times]0,T^\varepsilon[\times\Theta)} \le \frac{C^\varepsilon}{p^k}, \qquad p,k \ge 1. \tag{71}$$

One has the formula $u(x,t,\xi) - \Pi_p u(x,t,\xi) = \sum_{n\ge p+1} \langle u(x,t,\cdot), L_n\rangle L_n(\xi)$. Using the same trick as in (69) we obtain

$$\|u(x,t,\cdot) - \Pi_p u(x,t,\cdot)\|_{L^\infty(\Theta)} \le \left(\sum_{p<n} \langle u(x,t,\cdot), L_n\rangle^2 (n^{2k'}+1)\right)^{\frac{1}{2}} \left(\sum_{p<n} \frac{\|L_n\|^2_{L^\infty(\Theta)}}{n^{2k'}+1}\right)^{\frac{1}{2}}.$$

The first sum is bounded. Taking $k' = k+1$ and using the estimate (70) one gets $\sum_{p<n} \frac{\|L_n\|^2_{L^\infty(\Theta)}}{n^{2k'}+1} \le \frac{c_p}{p^k}$ for some constant $c_p$. It ends the proof.

**Proposition 16.** *There exist constants $C^\varepsilon > 0$ which depend on $T^\varepsilon = T - \varepsilon$ and the solution $u$ such that*

$$\left\|\partial_x \Pi_p u\right\|_{L^\infty(\mathscr{I}\times]0,T^\varepsilon[\times\Theta)} \le C^\varepsilon, \qquad p \ge 1$$

*and*

$$\left\|\partial_x (u - \Pi_p u)\right\|_{L^\infty(\mathscr{I}\times]0,T^\varepsilon[\times\Theta)} \le \frac{C^\varepsilon_p}{p^k}, \qquad p,k \ge 1.$$

One has $\partial_x \Pi_p u = \Pi_p \partial_x u$ since the operators commute. Next we use (68) and (71) with $v = \partial_x u$. The proof is ended.

## References

1. N. Wiener. The Homogeneous Chaos. *Amer. J. Math.*, 60:897–936, 1938.
2. R.H. Cameron and W.T. Martin. The Orthogonal Development of Non-Linear Functionals in Series of Fourier-Hermite Functionals. *Annals of Math.*, 48:385–392, 1947.
3. P. D. Lax. Hyperbolic systems of conservation laws and the theory of shock waves. *SIAM*, 1973. Philadelphia.
4. S. K. Godunov. A Difference Scheme for Numerical Solution of Discontinuous Solution of Hydrodynamic Equations. *Math. Sbornik*, 47:271–306, 1959. translated US Joint Publ. Res. Service, JPRS 7226, 1969.
5. P. L. Roe. Approximate Riemann solvers, parameter vectors and difference schemes. *Journal of Computational Physics*, 43:357–372, 1981.
6. L. Mathelin, M.Y. Hussaini, and T.A. Zang. Stochastic approaches to uncertainty quantification in CFD simulations. *Numer. Algo.*, 38:209–236, 2005.
7. Qian-Yong Chen, David Gottlieb, and Jan S. Hesthaven. Uncertainty analysis for the steady-state flows in a dual throat nozzle. *Journal of Computational Physics*, 204(1):378–398, 2005.
8. G. Lin, S.-H. Su, and G.E. Karniadakis. Predicting shock dynamics in the presence of uncertainties. *Journal of Computational Physics*, 217:260–276, 2006.
9. G. Lin, C.-H. Su, and G. E. Karniadakis. Random roughness enhances lift in supersonic flow. *Phys. Rev. Lett.*, 99(10):104501, 2007.
10. D. Lucor, C. Enaux, H. Jourdren, and P. Sagaut. Multi-Physics Stochastic Design Optimization: Application to Reacting Flows and Detonation. *Comp. Meth. Appl. Mech. Eng.*, 196:5047–5062, 2007.
11. D. Gottlieb and D. Xiu. Galerkin Method for Wave Equations with Uncertain Coefficients. *Commun. Comp. Phys.*, 3:505–518, 2008.
12. T. Chantrasmi, A. Doostan, and G. Iaccarino. Padé-Legendre approximants for uncertainty analysis with discontinuous response surfaces. *Journal of Computational Physics*, 228(19):7159–7180, 2009.
13. Pettersson P., Iaccarino G., and Nordstrom J. Numerical analysis of the Burgers equation in the presence of uncertainty. *Journal of Computational Physics*, 228:8394–8412, 2009.
14. G. Poëtte, B. Després, and D. Lucor. Uncertainty Quantification for Systems of Conservation Laws. *J. Comp. Phys.*, 228(7):2443–2467, 2009.
15. F. Simon, P. Guillen, P. Sagaut, and D. Lucor. A gPC-based approach to uncertain transonic aerodynamics. *Compu. Meth. Appl. Mech. Eng.*, 199:1091–1099, 2010.
16. J.-C. Chassaing and D. Lucor. Stochastic investigation of flows about airfoils at transonic speeds. *AIAA J.*, 48(5):938–950, 2010.
17. T.J. Barth. On the propagation of statistical model parameter uncertainty in CFD calculations. *Theor. Comput. Dyn.*, pages 1–28, 2011.
18. G. Poëtte, L. Lucor, and H. Jourdren. A stochastic surrogate model approach applied to calibration of unstable fluid flow experiments. *C.R. Acad. Sci. paris, Ser. I*, 350(5-6342):319–324, 2012.
19. G. Lin, C.-H. Su, and G. E. Karniadakis. The Stochastic Piston Problem. *PNAS*, 101(45):15840–15845, 2004.
20. P.M. Congedo, P. Colonna, C. Corre, J.A.S. Witteveen, and G. Iaccarino. Backward uncertainty propagation method in flow problems: Application to the prediction of rarefaction shock waves. *Computer Methods in Applied Mechanics and Engineering*, 213-216(0):314–326, 2012.
21. Ch. Schwab A. Barth and N. Zollinger. Multilevel Monte-Carlo Method for Elliptic PDEs with Stochastic Coefficients. *Num. Math.*, 2011.
22. Ch. Schwab S. Mishra and J. Sukys. Multi-level Monte Carlo finite volume methods for non-linear systems of conservation laws in multi-dimensions. Technical report, ETHZ, 2011.
23. R. Abgrall. A Simple, Flexible and Generic Deterministic Approach to Uncertainty Quantifications in Non Linear Problems: Application to Fluid Flow Problems. *Rapport de Recherche INRIA*, 2007.

24. J. Tryoen, O. Le Maître, M. Ndjinga, and A. Ern. Intrusive Galerkin methods with upwinding for uncertain nonlinear hyperbolic systems. *Journal of Computational Physics*, 229:6485–6511, 1 September 2010. Original Research Article.
25. G. Boillat and T. Ruggeri. Hyperbolic principal subsystems: entropy convexity and subcharacteristic conditions. *Arch. Ration. Mech. Anal.*, 137(4):305–320, 1997.
26. G. Chen, C. Levermore, and T. Liu. Hyperbolic Conservation Laws with Stiff Relaxation Terms and Entropy. *Comm. Pure Appl. Math.*, 47:787–830, 1994.
27. I. Müller and T. Ruggeri. *Rational Extended Thermodynamics, 2nd ed.* Springer. Tracts in Natural Philosophy, Volume 37, 1998. Springer-Verlag, New York.
28. Bruno Després. A geometrical approach to nonconservative shocks and elastoplastic shocks. *Archive for Rational Mechanics and Analysis*, 186(2):275–308(34), 2007.
29. C. Dafermos. *Hyperbolic conservation laws in continuum physics.* Springer Verlag 325, Berlin, 2000.
30. G. Poëtte and D. Lucor. Non Intrusive Iterative Stochastic Spectral Representation with Application to Compressible Gas Dynamics. *J. of Comput. Phys.*, 231:3587–3609, 2012.
31. G. Poëtte, B. Després, and D. Lucor. Treatment of Uncertain Interfaces in Compressible Flows. *Comp. Meth. Appl. Math. Engrg.*, 200:284–308, 2010.
32. K. Veroy, C. Prud'homme, and A. T. Patera. Reduced-basis approximation of the viscous burgers equation: rigorous a posteriori error bounds. *Comptes Rendus Mathématique*, 337(9):619–624, 2003.
33. D. Xiu and G. E. Karniadakis. The Wiener-Askey Polynomial Chaos for Stochastic Differential Equations. *SIAM J. Sci. Comp.*, 24(2):619–644, 2002.
34. X. Wan and G.E. Karniadakis. Beyond Wiener-Askey Expansions: Handling Arbitrary PDFs. *SIAM J. Sci. Comp.*, 27(1–3), 2006.
35. D. Xiu and G.E. Karniadakis. Modeling Uncertainty in Steady State Diffusion Problems via generalized Polynomial Chaos. *Comp. Meth. Appl. Mech. Engrg.*, 191:4927–4948, 2002.
36. D. Serre. *Systèmes Hyperboliques de Lois de Conservation, partie I.* Diderot, 1996. Paris.
37. E.F. Toro. *Riemann solver and numerical methods for fluid dynamics.* Springer-Verlag, 1997.
38. Michael Junk. Maximum Entropy for Reduced Moment Problems. *Math. Mod. Meth. Appl. Sci.*
39. Milton Abramowitz and Irene A. Stegun. *Handbook of Mathematical Functions with Formulas, Graphs, and Mathematical Tables.* Dover, New York, ninth dover printing, tenth gpo printing edition, 1964.

# Adaptive Uncertainty Quantification for Computational Fluid Dynamics

**Richard P. Dwight, Jeroen A.S. Witteveen, and Hester Bijl**

**Abstract** Two different approaches to propagating uncertainty are considered, both with application to CFD models, both adaptive in the stochastic space. The first is "*Adaptive Stochastic Finite Elements*", which approximates the model response in the stochastic space on a triangular grid with quadratic reconstruction on the elements. Limiting reduces the reconstruction to linear in the presence of discontinuities, and the mesh is refined using the Hessian of the response as an indicator. This construction allows for UQ in the presence of strong shocks in the response, for example in case of a transonic aerofoil with uncertain Mach number, and where variability in surface pressure is of interest. The second approach is "*Adaptive Gradient-Enhanced Kriging for UQ*", which uses a Gaussian process model as a surrogate for the CFD model. To deal with the high cost of interpolating in high-dimensional spaces we make use of the adjoint of the CFD code, which provides derivatives in all stochastic directions at a cost independent of dimension. The Gaussian process framework allows this information to be incorporated into the surrogate, as well as providing regression of both CFD output values and derivatives according to error estimates. It also provides an interpolation error estimate on which an adaptivity indicator is based, weighted with the input uncertainty. A transonic aerofoil with four uncertain shape parameters is given as an example case.

R.P. Dwight (✉) · H. Bijl
Faculty of Aerospace, TU Delft, Delft, The Netherlands
e-mail: r.p.dwight@tudelft.nl; h.bijl@tudelft.nl

J.A.S. Witteveen
Center for Turbulence Research, Stanford University, Stanford, CA, USA
e-mail: jasw@stanford.edu

H. Bijl et al. (eds.), *Uncertainty Quantification in Computational Fluid Dynamics*, Lecture Notes in Computational Science and Engineering 92,
DOI 10.1007/978-3-319-00885-1_4, 

# 1 Introduction

Many sources of uncertainty are present when attempting to model fluids in real engineering situations with CFD. Operational uncertainties, geometric uncertainties, and model inadequacy are commonly encountered by engineers. In the design of aeroplanes examples of operational uncertainties include free stream conditions like pressure, velocity, temperature or angle of incidence. Geometrical uncertainties arise from production imprecision and deterioration under use.

In this chapter we discuss two adaptive methods for uncertainty propagation designed for application to expensive models. It is assumed that probability density functions (pdfs) of the uncertain input parameters are known, and we do not consider the problem of eliciting these distributions [16]. The only issue is then to obtain the corresponding output pdfs. Since our main interest is fluid dynamics and CFD, the efficiency of the uncertainty propagation methods are of primary importance. A deterministic CFD simulation can take days. If an engineer has to employ 20 times this computational effort for uncertainty propagation compared to a deterministic simulation, the information gained should be significant. Much more than 20 solves is impractical in many applications. Adaptivity provides a means by which the cost can be minimized for a given level of error.

Two distinct adaptive approaches are discussed: "*Adaptive Stochastic Finite Elements*" in Sect. 2, and "*Adaptive Gradient-Enhanced Kriging for UQ*" in Sect. 3. In common with most UQ methods they both attempt to approximate the response of the simulation code – the mapping from the space spanned by the uncertain input parameters to the output of interest. For this purpose the former uses piecewise quadratic reconstruction, and the latter a Gaussian Process Model. In both cases an error estimate is devised, leading to an adaptivity indicator. Both methods are then applied to testcases in aerodynamics.

# 2 Method 1: Adaptive Stochastic Finite Elements with Newton-Cotes Quadrature

Discontinuous solutions (shocks) and bifurcation phenomena in aeroelastic systems can lead to a high output sensitivity to small input variations. Global polynomial approximations of Polynomial Chaos methods can for these cases result in unreliable predictions of unphysical realizations due to oscillatory overshoots and undershoots at the singularity. More robust multi-element Polynomial Chaos methods based on Gauss-quadrature in hyperrectangular elements can still result locally in unphysical oscillations in the elements and they require completely recomputing the solution in adaptively refined domains. In this chapter an alternative non-intrusive Adaptive Stochastic Finite Elements (ASFE) method based on Newton-Cotes quadrature in simplex elements is developed. The method does not result in unphysical predictions since it preserves the extrema of the samples.

The required number of deterministic solves is relatively low, since the samples are both used in approximating the response in multiple elements and reused after refinements. Fourth-order convergence results for a piston problem, and transonic flow over a NACA0012 airfoil illustrate that the method resolves the amplification of the input randomness in these problems with discontinuous solutions reliably.[1]

## 2.1 Background

In the proposed Adaptive Stochastic Finite Elements approach based on Newton-Cotes quadrature in simplex elements the response is represented by a piecewise polynomial approximation by subdividing probability space into multiple elements. In the elements the response is approximated by collocating the problem in Newton-Cotes quadrature points. Simplex elements are employed, since they are the natural elements for Newton-Cotes quadrature in multiple dimensions. The quadrature approximation in the elements leads to a non-intrusive approach, in which uncoupled deterministic problems are solved for varying parameter values. The required accuracy is obtained by adaptively refining the elements using a refinement measure based on the curvature of the approximation of the response surface weighted by the probability represented by the elements. As measure for the curvature the largest absolute eigenvalue of the Hessian in the elements is used.

The required number of deterministic solves is relatively low compared to a Gauss quadrature adaptive multi-element method based on hypercube elements with respect to the following three points:

1. The tensor grid of Gauss quadrature points for constructing a polynomial approximation of order $p$ results for $n$ uncertain parameters in $(p + 1)^n$ samples per element. For Newton-Cotes quadrature in simplex elements the number of samples per element of $(n + p)!/n!p!$ increases less rapidly with $p$ and $n$.
2. In the Gauss quadrature discretization the decoupled elements all contain $(p+1)^n$ samples. In contrast, especially at low orders, many Newton-Cotes quadrature points are located on the boundaries of the elements. The samples are, therefore, used in approximating the response in multiple elements. In the examples it is illustrated that this reduces the average number of samples per element to approximately 2 instead of $(n + p)!/n!p!$.
3. Refining an element using Gauss quadrature points, requires the computation of $(p + 1)^n$ new samples in every new element. The deterministic solves on intermediate refinement levels are, therefore, not directly used in the final

[1] Based on: Jeroen A.S. Witteveen, *"Efficient and Robust Uncertainty Quantification for Computational Fluid Dynamics and Fluid-Structure Interaction"*, Ph.D. thesis, Delft University of Technology, 2009 (Chap. 4).

approximation. The Newton-Cotes quadrature points in the new elements include those of the refined element, such that all samples are reused after successive refinements.

These advantages of Newton-Cotes quadrature typically come forward in a multi-element discretization. For a single-element approximation Gauss quadrature can be more favorable. The number of $3^n$ samples for the initial grid of $n!$ simplex elements can become large for higher dimensional probability spaces.

Since lower-order expansions are efficient for complex problems [13], the degree of the Newton-Cotes quadrature is limited in this chapter to two, which is known in one-dimension as Simpson's rule. This results in a piecewise quadratic approximation of the response. To preserve extrema of the samples in the piecewise polynomial approximation, the elements are subdivided in subelements with a linear trapezoidal rule approximation of the response where necessary.

The Stochastic Finite Elements formulation considered in this chapter is developed in Sect. 2.2. The adaptive Stochastic Finite Elements approach with Newton-Cotes quadrature in simplex elements is in Sect. 2.3 applied to typical flow problems involving shock waves and bifurcations, which result in singularities in probability space. The properties of the adaptive Stochastic Finite Elements approach are first studied using a one-dimensional piston problem. Separate cases with a discontinuity, a discontinuous derivative, and a smooth response are considered. The input uncertainty is given by up to three independent uncertain parameters. The extension to more than three random dimensions is a geometrical exercise. The effect of the adaptive grid refinement and the degree of the Newton-Cotes quadrature is investigated. The results are compared to those of a global polynomial Stochastic Collocation method and Monte Carlo simulation. The comparison with the global polynomial approximation of the Stochastic Collocation method should be interpreted as mainly a qualitative comparison. Finally, the adaptive Stochastic Finite Elements approach is applied to transonic flow over a NACA0012 airfoil. The transonic flow field proves to be sensitive to the free stream conditions.

## 2.2 *Adaptive Stochastic Finite Elements*

In this subsection the adaptive Stochastic Finite Elements approach with Newton-Cotes quadrature in simplex elements is developed. The stochastic adaptive grid refinement strategy is considered in Sect. 2.2.2.

### 2.2.1 Newton-Cotes Quadrature in Simplex Elements

In a probabilistic description of uncertainty, one is typically interested in the statistical moments of the response. The $m^{\text{th}}$ moment $\mathbb{E}(u(\mathbf{x}, t, \omega)^m)$ of the probability distribution of the response $u(\mathbf{x}, t, \omega)$ given by

$$\mathbb{E}(u(\mathbf{x},t,\omega)^m) = \int_{\Omega} u(\mathbf{x},t,\omega)^m \mathrm{d}\omega, \tag{1}$$

is an integral quantity over probability space. The space of functions mapping probability space $\Omega$ onto parameter space $A$ is denoted by $\Theta$. The mapping $\Theta$ gives the parameter values which correspond to a realization in probability space $\Omega$. This mapping is defined by the probability distribution of the uncertain parameters.

Stochastic Finite Elements methods divide the integral (1) over probability space $\Omega$ in a summation of integrals over $N_\Omega$ non-overlapping elements $\Omega_i$ for $i = 1, \ldots, N_\Omega$

$$\mathbb{E}(u(\mathbf{x},t,\omega)^m) = \sum_{i=1}^{N_\Omega} \int_{\Omega_i} u(\mathbf{x},t,\omega)^m \mathrm{d}\omega. \tag{2}$$

To obtain an uncoupled non-intrusive sampling based approach the integrals over the elements $\Omega_i$ are approximated by a quadrature integration rule based on $N_\mathrm{s}$ deterministic samples in each element

$$\mathbb{E}(u(\mathbf{x},t,\omega)^m) \approx \sum_{i=1}^{N_\Omega} \sum_{j=1}^{N_\mathrm{s}} c_{i,j} u_{i,j}(\mathbf{x},t)^m, \tag{3}$$

where $c_{i,j}$ are the quadrature weights and $u_{i,j}(\mathbf{x},t)$ are the realizations of the response $u(\mathbf{x},t,\omega)$ for the parameter values $\{a_1(\omega), \ldots, a_n(\omega)\}_{i,j}$ in the $N_\mathrm{s}$ quadrature points in element $i$.

Here Newton-Cotes quadrature points are employed, since many of these points are located on the boundaries of the elements. In this way the samples at the quadrature points are used to construct the approximation in multiple elements. The choice of Newton-Cotes quadrature points also implies that some of the samples are located on the outer boundary of parameter domain. If the probability approaches zero at the parameter domain boundary as for a unimodal beta distribution, the sampling points on the domain boundary contribute nonetheless to the construction of the response approximation in the interior of the parameter domain.

The quadrature weights $c_{i,j}$ are defined by the mapping $\Theta^{-1}$ of the $n$-dimensional Newton-Cotes formula of degree $d$ from parameter space $A$ to probability space $\Omega$. The Newton-Cotes quadrature weights are normally given by the integrals of the Lagrange basis polynomials through the quadrature points. Here the weights $c_{i,j}$ are given by the integrals weighted by the probability density of the uncertain input parameters

$$c_{i,j} = \int_{A_i} L_{i,j}(a_1, \ldots, a_n) f_\mathrm{a}(a_1, \ldots, a_n) \mathrm{d}\mathbf{a}, \tag{4}$$

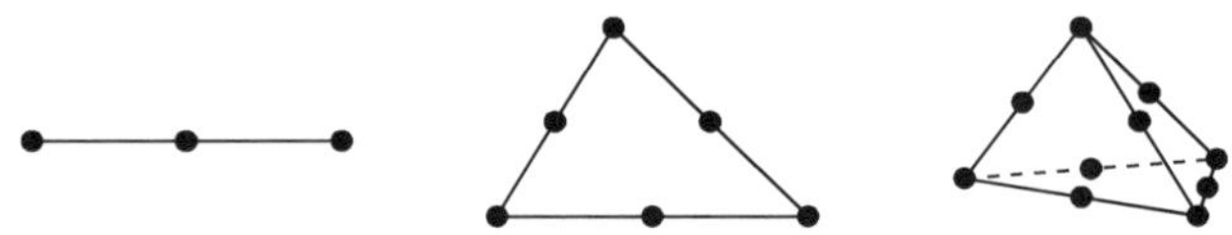


**Fig. 1** The $n$-simplex elements in parameter space $A$ with the $\binom{n+2}{2}$ samples of the second-degree Newton-Cotes quadrature approximation

for $i = 1, \ldots, N_\Omega$, where $\mathbf{a} = \{a_1, \ldots, a_n\}$ is the vector of uncertain input parameters and $A_i$ is the mapping $\Theta$ of the element $\Omega_i$ to parameter space. This results in a "Polynomial Chaos" formulation of the Newton-Cotes formulas. The values for the weights $c_{i,j}$ are computed numerically for each element using "normal" Newton-Cotes integration in the element $A_i$ on a fine subgrid with $N_{A_{\mathrm{sub}}}$ $n$-simplex subelements with $N_{s_{\mathrm{sub}}}$ quadrature points

$$c_{i,j} \approx \sum_{k=1}^{N_{A_{\mathrm{sub}}}} \sum_{l=1}^{N_{s_{\mathrm{sub}}}} e_l L_{i,j,k,l} f_{\mathrm{a}}(a_1, \ldots, a_n)_{i,k,l}, \tag{5}$$

for $i = 1, \ldots, N_\Omega$ and $j = 1, \ldots, N_{\mathrm{s}}$, where $e_l$ are the "normal" Newton-Cotes quadrature weights, and $L_{i,j,k,l}$ and $f_{\mathrm{a}}(a_1, \ldots, a_n)_{i,k,l}$ are the values of the Lagrange polynomial $L_{i,j}(a_1, \ldots, a_n)$ and the probability density of the uncertain parameters $f_{\mathrm{a}}(a_1, \ldots, a_n)$ in the quadrature points on the fine subgrid in element $A_i$. The weights $c_{i,j}$ can be different for every element, since the mapping $\Theta^{-1}$ between $A$ and $\Omega$ is in general different for different elements.

The elements $\Omega_i$ in probability space are defined by discretizing parameter space $A$ by $N_\Omega$ $n$-simplex elements $A_i$, with $i = 1, \ldots, N_\Omega$. An $n$-simplex is the $n$-dimensional analogue of a triangle, which results for one-dimensional parameter space in a line segment, for two-dimensions in a triangle, for three-dimensions in a tetrahedron, etc., see Fig. 1. For more random parameters ($n > 3$) $n$-dimensional simplex elements can be used. The $n$-simplex elements are the natural elements for Newton-Cotes quadrature in $n$ dimensions. The volume $V_i$ of the $n$-simplex elements in parameter space $A$ is given by

$$V_i = \frac{1}{n!} |\det(a_{i,0} - a_{i,1} \quad a_{i,1} - a_{i,2} \quad \ldots \quad a_{i,n-1} - a_{i,n})|. \tag{6}$$

where $a_{i,j}$ are the $n+1$ vertices of the $n$-simplex $A_i$ in parameter space $A$.

The choice of the degree of the Newton-Cotes quadrature is a balance between high-order accuracy in smooth regions and the effectiveness of lower-order approximations near singularities. For complex problems lower-order representations are therefore more effective than higher-order ones. In most of the paper second-degree ($d = 2$) Newton-Cotes quadrature is used, which is known as Simpson's rule. This corresponds to a piecewise quadratic approximation of the response. Extensions to

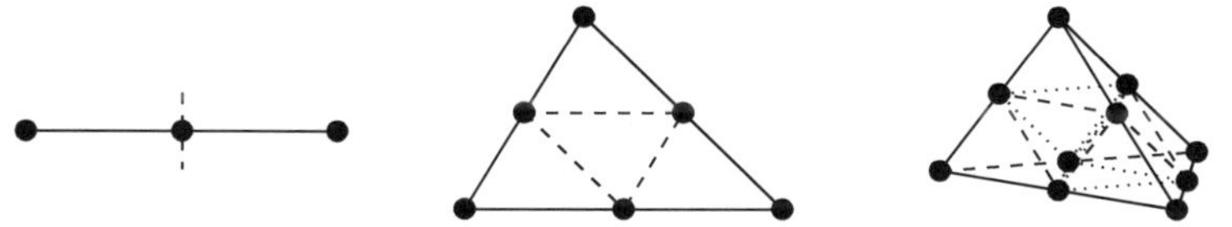


**Fig. 2** The $2^n$ subelements given by the *dashed lines* with $n+1$ samples for the linear trapezoidal rule approximation of the response for preserving the extrema of the samples

higher orders are possible by using Newton-Cotes quadrature rules of higher degree such as Boole's rule. Second-degree Newton-Cotes quadrature results in $N_s = \binom{n+2}{2}$ samples per element $A_i$, see Fig. 1. Here the notation $\binom{n}{k} = \frac{n!}{k!(n-k)!}$ refers to the binomial coefficients. Most the quadrature points are located on the boundary of the elements $\partial A_i$ such that they can be used for approximating the response in multiple elements.

Near singularities the piecewise quadratic approximation of the second-degree Newton-Cotes quadrature can result in unphysical oscillations. To preserve the extrema of the samples in the piecewise polynomial approximation, where necessary the elements are subdivided in $n$-simplex subelements with a linear trapezoidal rule approximation of the response. The $2^n$ subelements each contain $n+1$ of the $\binom{n+2}{2}$ samples of the original element, see Fig. 2. An element is split into subelements when the polynomial approximation of the response has an extremum in the element other than in a quadrature point.

### 2.2.2 Stochastic Adaptive Grid Refinement

For complex, high-dimensional problems multi-element approaches may result in large computational costs. An adaptive refinement strategy can lead to more efficient approximations of complex situations. Here the refinement measure is based on the curvature of the approximation of the response surface in the elements weighted by the probability represented by the elements. As measure for the curvature of the approximation of the response surface the largest absolute eigenvalue of the Hessian in the elements is used, which is common in adaptive refinement of deterministic finite element methods. The Hessian of the polynomial approximation in the element $A_i$ in parameter space is given by

$$H_i(u(\mathbf{x},t,\omega)) = \begin{bmatrix} \frac{\partial^2 u}{\partial a_1^2} & \frac{\partial^2 u}{\partial a_1 \partial a_2} & \cdots & \frac{\partial^2 u}{\partial a_1 \partial a_n} \\ \frac{\partial^2 u}{\partial a_2 \partial a_1} & \frac{\partial^2 u}{\partial a_2^2} & \cdots & \frac{\partial^2 u}{\partial a_2 \partial a_n} \\ \vdots & \vdots & \ddots & \vdots \\ \frac{\partial^2 u}{\partial a_n \partial a_1} & \frac{\partial^2 u}{\partial a_n \partial a_2} & \cdots & \frac{\partial^2 u}{\partial a_n^2} \end{bmatrix}_i, \tag{7}$$

which is constant in the element for the piecewise quadratic approximation of the second-degree Newton-Cotes quadrature. The second-order derivatives are derived from the quadratic approximation of the response through the sampled quadrature points in the elements. In one-dimension the Hessian reduces to the absolute value of the second-order derivative of the response. This refinement measure can be extended to higher-order approximations ($d > 2$) by using the maximum of the Hessian in the elements or by employing higher-order derivatives. In the step of determining the Hessian, no elements are subdivided into subelements with a linear trapezium rule approximation.

The refinement measure is weighted by the probability $P_i$ represented by the elements. For the refinement measure this probability is approximated by a similar relation as (6), since the probability of an element is equivalent to its volume in probability space

$$P_i = \frac{1}{n!}|\det(\omega_{i,0} - \omega_{i,1} \quad \omega_{i,1} - \omega_{i,2} \quad \ldots \quad \omega_{i,n-1} - \omega_{i,n})|, \tag{8}$$

where $\omega_{i,j}$ is given by the mapping $\Theta^{-1}$ of the vertices $a_{i,j}$ of element $A_i$ to probability space $\Omega$. The refinement measure $r_i$ in the $i$th element is then defined as

$$r_i = P_i V_i^{N_V} \max(|\mathrm{eig}_1(H_i(u))|, \ldots, |\mathrm{eig}_n(H_i(u))|), \tag{9}$$

where, the factor $V_i^{N_V}$ compensates for the, in general, increase of the second-order derivatives in smaller elements if they contain a singularity.

The value of $N_V$ is chosen here based on the following theoretical argument. Consider an element $\Delta a$ in a one-dimensional parameter space $a \in A$, which contains the discontinuity of a step function response surface $u(a)$, see Fig. 3a. The Hessian (7) then reduces to the second derivative in finite difference approximation

$$\frac{\partial^2 u}{\partial a^2} \approx \frac{u_{-1} - 2u_0 + u_{+1}}{\Delta a^2}. \tag{10}$$

Assume that the element is refined to a smaller element $\Delta\hat{a}$ for which holds $\hat{u}_{-1} = u_{-1}$, $\hat{u}_0 = u_0$, and $\hat{u}_{+1} = u_{+1}$, see Fig. 3b. The Hessian in the new element is

$$\frac{\partial^2 \hat{u}}{\partial a^2} \approx \frac{\hat{u}_{-1} - 2\hat{u}_0 + \hat{u}_{+1}}{\Delta\hat{a}^2}. \tag{11}$$

Factor $N_V$ is then chosen such that the contribution of the Hessian to refinement measure (9) in the element with the discontinuity is independent of the size of the element

$$\Delta a^{N_V} \frac{u_{-1} - 2u_0 + u_{+1}}{\Delta a^2} = \Delta\hat{a}^{N_V} \frac{\hat{u}_{-1} - 2\hat{u}_0 + \hat{u}_{+1}}{\Delta\hat{a}^2}, \tag{12}$$

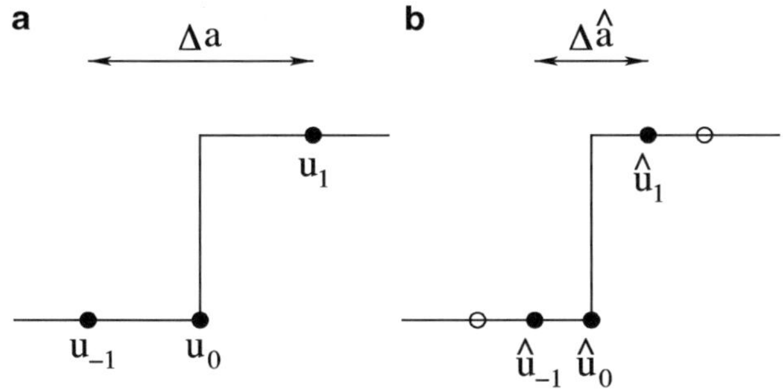


**Fig. 3** Sketch of 2 elements of size $\Delta a$ and $\Delta \hat{a}$ containing the discontinuity of a step function for deriving the general value $N_V = 2$. (**a**) Element $\Delta a$. (**b**) Element $\Delta \hat{a}$

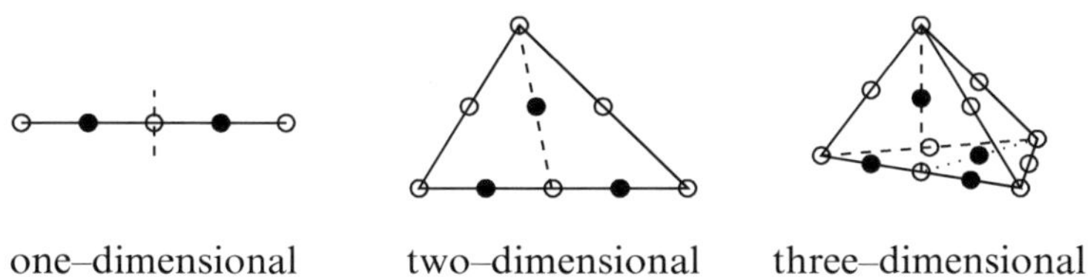


**Fig. 4** Refinement of the $n$-simplex elements into 2 elements with the $n+1$ new samples and the $\binom{n+2}{2}$ samples of the original element given by the *dots* and the *open circles*, respectively, for second-degree Newton-Cotes quadrature

$$\left(\frac{\Delta a}{\Delta \hat{a}}\right)^{N_V} = \frac{\hat{u}_{-1} - 2\hat{u}_0 + \hat{u}_{+1}}{u_{-1} - 2u_0 + u_{+1}} \frac{\Delta a^2}{\Delta \hat{a}^2} = \left(\frac{\Delta a}{\Delta \hat{a}}\right)^2, \tag{13}$$

which results in $N_V = 2$. The rigorous derivation for this abstract example results in a general value for $N_V$, since every discontinuous response surface can locally be approximated by a step function.

The element with the highest value of the refinement measure is then refined into two $n$-simplex elements, see Fig. 4. The longest edge of the element is split into two halves of equal length. An alternative is to use the eigenvector corresponding to the highest absolute eigenvalue to determine which edge to split.

Due to the Newton-Cotes quadrature points there is no need to completely recompute the solution in the refined elements. In fact, only a maximum of $n+1$ new samples has to be computed for the refinement of second-degree Newton-Cotes quadrature, even though both new elements contain $\binom{n+2}{2}$ quadrature points each. First, all samples in the original element are reused in the refined elements. Second, part of the new samples is used by both refined elements, since most Newton-Cotes quadrature points are located on the boundaries of the elements. Furthermore, the new samples are located on the boundary of the original element can already have been computed while refining neighboring elements.

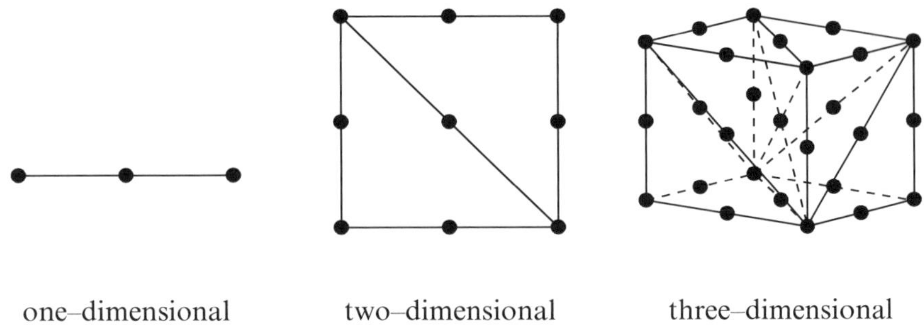


**Fig. 5** Initial discretization of the $n$-dimensional cuboid describing the ranges of parameter space $A$ with $n!$ $n$-simplex elements and $3^n$ second-degree Newton-Cotes quadrature points

For comparison, a standard multi-element, piecewise quadratic Stochastic Collocation approach based on the tensor product of Gauss quadrature points in hexahedral elements would require $2 \cdot 3^n$ new deterministic solves for refining an element into two new elements instead of a maximum of $n + 1$ for Newton-Cotes quadrature. This would result for three uncertain input parameters in 54 deterministic solves instead of a maximum of 4 for Newton-Cotes quadrature. A sparse grid approach can reduce the number of deterministic solves required in Stochastic Collocation [23]. Due to the adaptive stochastic grid refinement the number of required Newton-Cotes quadrature samples for discretizing an $n$ dimensional parameter space of $n$ random parameters scales with less than the $n$-dimensional tensor product.

After the refinement the new refinement measure is computed in the refined elements and the element with the largest refinement measure is again refined, etc. The refinement is stopped when a threshold value for the maximum refinement measure or the maximum number of samples is reached.

The initial grid is given by the coarsest discretization of the $n$-dimensional rectangle describing the ranges of parameter space using $n!$ $n$-simplex elements, see Fig. 5. For more than $n = 3$ random parameters the $n$-dimensional hyperrectangle describing parameter space is divided in $n!$ $n$-dimensional simplex elements. Finding the initial grid discretization for more than $n = 3$ random parameters is a geometrical exercise in the $n$-dimensional parameter space. Finite ranges of parameter space can reasonably be obtained by truncating an infinite domain at a threshold value for the distribution without affecting the accuracy significantly in practical applications. The number of samples in the initial grid is given by $3^n$.

The algorithm for adaptive Stochastic Finite Elements with Newton-Cotes quadrature in simplex elements can be summarized as follows:

1. Solve the $3^n$ deterministic problems for the parameter values corresponding to the collocation points in the initial grid of Fig. 5;
2. Determine the refinement measure (9) in the elements of the initial grid;

3. Refine the element with the highest value of the refinement measure according to Fig. 4;
4. Solve the maximal $n + 1$ deterministic problems for the parameter values corresponding to the new collocation points in the refined element, if they have not been computed before;
5. Determine the refinement measure in the two new elements;
6. Return to step 3 if the threshold value of the maximum refinement measure or the maximum number of samples has not been reached;
7. Split the elements into subelements as in Fig. 2 with a linear approximation of the response if the quadratic approximation has a maximum in the element other than in a quadrature point;
8. Determine the $N_s$ quadrature weights $c_{i,j}$ in the $N_\Omega$ elements using (5), with $i = 1, \dots, N_\Omega$ and $j = 1, \dots, N_s$;
9. Determine the statistical moments of the output $\mathbb{E}(u(\mathbf{x}, t, \omega)^m)$ using (3).

The probability distribution function can be found by sorting the response $u(\mathbf{x}, t, \omega)$ to a monotonically increasing function in $\omega$, with $\omega \in [0, 1]$. The algorithm can be parallelized by solving the maximal $n + 1$ deterministic problems in step 4 in parallel and by refining multiple elements with the highest value of the refinement measure simultaneously instead of a single element in step 3.

## 2.3 *Numerical Results*

In this section numerical results of the adaptive Stochastic Finite Elements approach with Newton-Cotes quadrature in simplex elements are presented for a piston problem, and transonic flow over a NACA0012 airfoil with uniformly and lognormally distributed random parameters. These test problems are typical examples of practical flow applications involving shock waves and bifurcations, which result in singularities in probability space. The effect of the adaptive grid refinement on the Newton-Cotes quadrature rule is investigated. Throughout quadratic elements (i.e. with 3 support points per element in 1d, 6 support points in 2d) are used. The results are compared to those of a Stochastic Collocation approach based on a global polynomial approximation of the response through Gauss quadrature points. The comparison with the Stochastic Collocation approach has mainly to be considered as a qualitative assessment. Reference solutions are obtained by Monte Carlo simulation.

### 2.3.1 Piston Problem

The properties of the method are studied for a one-dimensional piston problem chosen such that discontinuities exist in the response. The effect of a 1–2 uncertain input parameters on the instantaneous and total mass flow is considered.

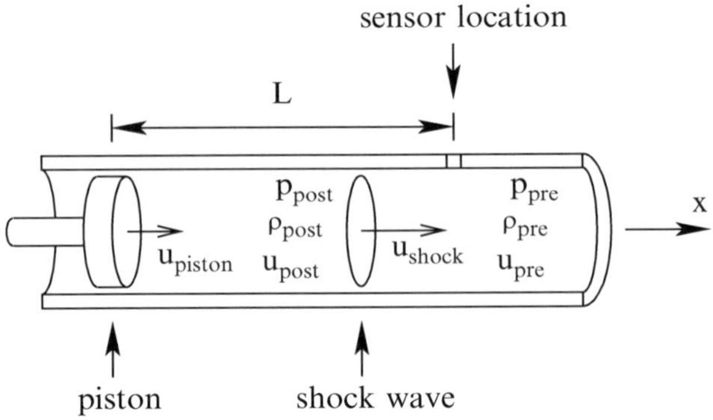


**Fig. 6** The piston problem

The piston problem, see Fig. 6, consists of a one-dimensional flow domain filled with air enclosed by a piston at its left. The piston starts to move to the right at $t = 0$ with a velocity $u_{\text{piston}} > 0$. A shock wave runs with velocity $u_{\text{shock}}$ into the ideal gas with constant initial conditions for the pressure $p_{\text{pre}}$, density $\rho_{\text{pre}}$ and velocity $u_{\text{pre}} = 0$. Neglecting the effects of viscosity the uniform pressure $p_{\text{post}}$, density $\rho_{\text{post}}$ and velocity $u_{\text{post}} = u_{\text{piston}}$ behind the shock wave are governed by the Euler equations. The pressure $p_{\text{post}}$ is given by the Rankine-Hugoniot relation

$$p_{\text{post}} - p_{\text{pre}} = \rho_{\text{pre}} c_{\text{pre}} (u_{\text{post}} - u_{\text{pre}}) \sqrt{1 + \frac{\gamma + 1}{2\gamma} \frac{p_{\text{post}} - p_{\text{pre}}}{p_{\text{pre}}}}, \tag{14}$$

with initial speed of sound $c_{\text{pre}} = \sqrt{\frac{\gamma p_{\text{pre}}}{\rho_{\text{pre}}}}$ and the ratio of specific heats $\gamma = 1.4$.

The other flow conditions can be determined from the one-dimensional shock wave relations [2] using the Mach number of the shock wave $\text{Ma}_{\text{shock}}$ given by

$$\text{Ma}_{\text{shock}} = \sqrt{1 + \frac{\gamma + 1}{2\gamma} \left( \frac{p_{\text{post}}}{p_{\text{pre}}} - 1 \right)}. \tag{15}$$

The instantaneous mass flow, $m(t)$ is the output of interest. Its behavior is considered at a sensor location at a distance $L$ to the right of the initial position of the piston. The response surfaces for $m(t)$ and $M(t)$ contain a discontinuity and a discontinuous derivative, respectively. In terms of the pre- and post-shock conditions the instantaneous mass flow $m(t)$ can be written as

$$m(t) = \begin{cases} \rho_{\text{pre}} u_{\text{pre}}, & t < \frac{L}{u_{\text{shock}}}, \\ \rho_{\text{post}} u_{\text{post}}, & t > \frac{L}{u_{\text{shock}}}. \end{cases}, \tag{16}$$

and the total mass flow as:

$$M(t) = \int_0^t m(\tau)\mathrm{d}\tau. \tag{17}$$

First the piston velocity $u_{\text{piston}}$ is assumed to be uncertain with a lognormal distribution and coefficient of variation $CV_{u_{\text{piston}}} = 10\,\%$. The infinite domain of the probability distribution is truncated at a threshold value $\varepsilon$ for the distribution. A convergence study demonstrated that the effect of the threshold value of $\varepsilon = 10^{-4}$ on the results is negligible.

In Fig. 7 the response surface $u_{\text{piston}}$–$m$ and the resulting probability distribution function of $m$ are given for time $t = 0.5$. The adaptive Stochastic Finite Elements approximation with second-degree Newton-Cotes quadrature is compared with the exact solution for 2, 4 and 8 elements. The samples are given by the dots and the boundaries of the elements are denoted by the bars. The exact solution shows a discontinuity in probability space for the $u_{\text{piston}}$ value at which $u_{\text{shock}} = \frac{L}{t}$, see (16). In the approximation with 2 elements, the second element is split into 2 subelements with a linear approximation of the response, see Fig. 7a, b. Even for this coarse approximation the solution preserves the monotonicity of the samples, and no artificial oscillations or unphysical values are predicted. The error due to the coarse approximation of the discontinuity is restricted to the subelement containing the discontinuity.

The results of using stochastic adaptive grid refinement are given in Fig. 7c–f for an approximation with 4 and 8 elements. The region around the discontinuity is refined with smaller elements while the elements in the rest of the domain remain relatively large. This results in an efficient discretization of probability space, which gives for 8 elements a sharp resolution of the discontinuity in both the response surface and the probability distribution function, see Fig. 7e, f. The approximation with 8 elements requires 17 deterministic samples.

Now we consider a case in which two independent input parameters are assumed to be uncertain. Next to the uncertain piston velocity $u_{\text{piston}}$, also the initial pressure $p_{\text{pre}}$ is described by a lognormal distribution with mean $\mu_{p_{\text{pre}}} = 1$ and coefficient of variation $CV_{p_{\text{pre}}} = 10\,\%$. This results in a two-dimensional probability space $\Omega$, to which two-dimensional Stochastic Finite Elements are applied for resolving the effect on the instantaneous mass flow $m$.

The adaptated grid and solution is shown for a discretization with 2, 10, and 50 elements in Fig. 8. The results for the initial grid with 2 elements are given in Fig. 8a, b. One of the two elements with a quadratic approximation of the solution is split into four subelements with a linear approximation of the response. In Fig. 8c, d the refinement of the grid to 10 elements is shown. The elements are mainly refined to better capture the discontinuity. For the discretization with 50 elements also the domain where the instantaneous mass flow $m$ changes continuously is refined, see Fig. 8e, f.

In Fig. 9 the grid in probability space is given for a discretization with 50 elements. The grid in Fig. 9 is the result of the mapping $\Theta^{-1}$ of the grid in

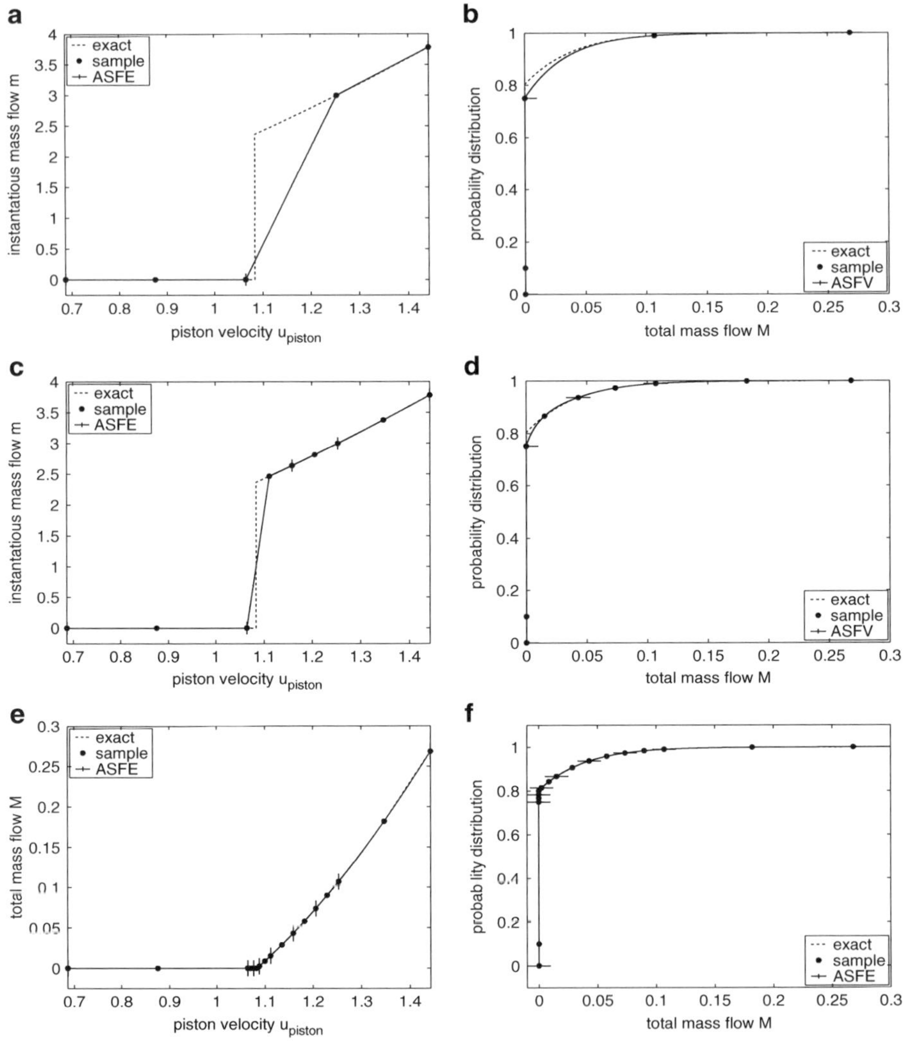


**Fig. 7** Response surface and probability distribution of the instantaneous mass flow $m$ at $t = 0.5$ with an uncertain piston velocity $u_{\text{piston}}$ by adaptive Stochastic Finite Elements (ASFE) for the piston problem. (**a**) Response, 2 elements. (**b**) distribution, 2 elements. (**c**) Response, 4 elements. (**d**) Distribution, 4 elements. (**e**) Response, 8 elements. (**f**) Distribution, 8 elements

parameter space $A$ of Fig. 8f to probability space $\Omega$. The grid in probability space is considerably different from the grid in parameter space. For example, the elements that capture the discontinuity are approximately of the same size, since in parameter space the elements are more refined in the region where the uncertain parameters $u_{\text{piston}}$ and $p_{\text{pre}}$ have a higher probability density, near $u_{\text{piston}}, p_{\text{pre}} \approx 1$. The topology of both grids is the same.

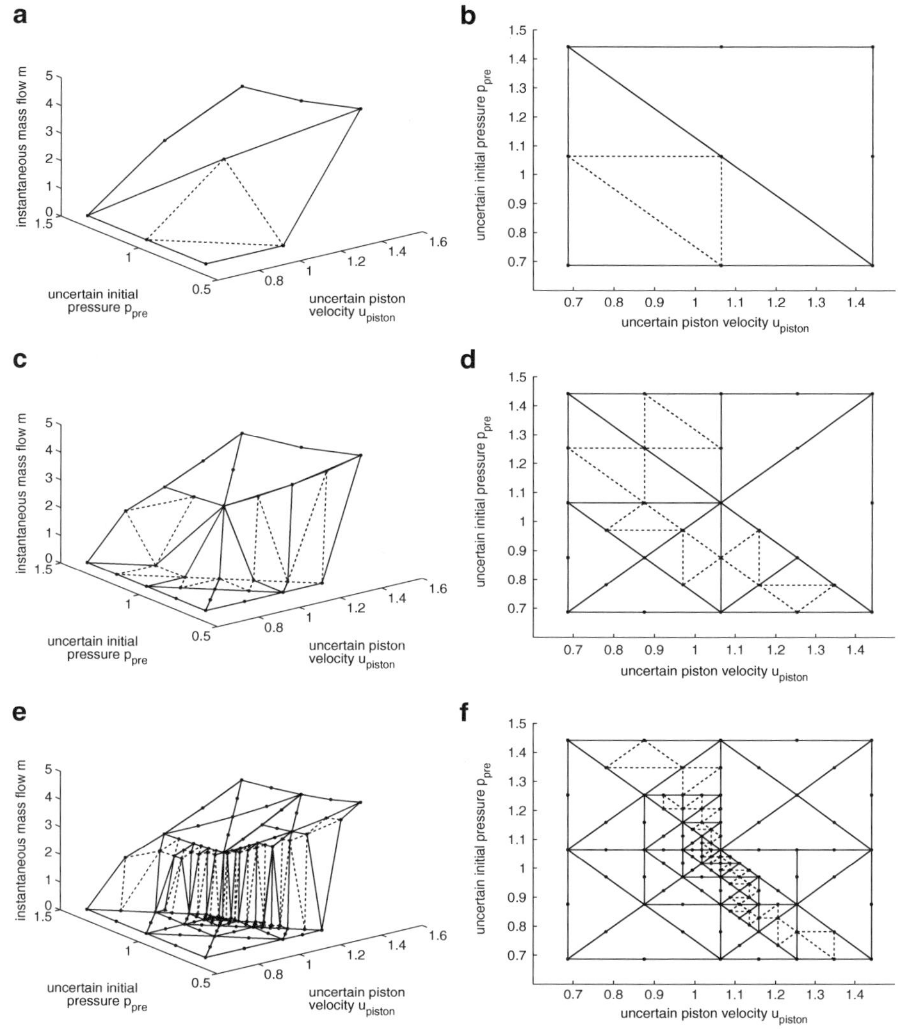


**Fig. 8** Response surface and the grid for the instantaneous mass flow $m$ at $t = 0.5$ as function of the uncertain piston velocity $u_{\text{piston}}$ and initial pressure $p_{\text{pre}}$ by adaptive Stochastic Finite Elements (ASFE) with 2, 10, and 50 elements for the piston problem. *Dotted lines* show elements in which the discretization is reduced to linear. (**a**) 2 elements, $m$. (**b**) 2 elements, grid. (**c**) 10 elements, $m$. (**d**) 10 elements, grid. (**e**) 50 elements, $m$. (**f**) 50 elements, grid

The approximation of the mean and the variance of the instantaneous mass flow $m$ of adaptive Stochastic Finite Elements is compared in Tables 1 and 2 to the result of Monte Carlo simulations. In Table 1 the results are given for a smooth response surface with $t = 1$ and $\mu_{u_{\text{piston}}} = 0.5$. Adaptive Stochastic Finite Elements are converged already for a discretization with 10 elements to a mean of $\mu_{\text{m}} = 0.750$

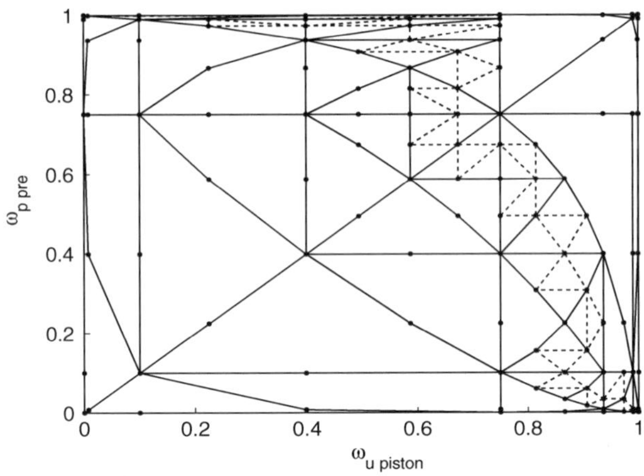


**Fig. 9** Grid in probability space for the instantaneous mass flow $m$ at $t = 0.5$ as function of the uncertain piston velocity $u_{\text{piston}}$ and initial pressure $p_{\text{pre}}$ by adaptive Stochastic Finite Elements (ASFE) with 50 elements for the piston problem

**Table 1** Mean and the variance of the instantaneous mass flow $m$ at $t = 1$ with an uncertain piston velocity $u_{\text{piston}}$ with $\mu_{u_{\text{piston}}} = 0.5$ and initial pressure $p_{\text{pre}}$ by adaptive Stochastic Finite Elements (ASFE) and Monte Carlo (MC) simulations

| Adaptive Stochastic Finite Elements (ASFE) | | | | Monte Carlo (MC) | | |
|---|---|---|---|---|---|---|
| Elements | Samples | Mean | Variance | Samples | Mean | Variance |
| 10 | 31 | 0.750 | 0.011 | 10 | 0.808 | 0.009 |
| $10^2$ | 230 | 0.750 | 0.011 | $10^2$ | 0.741 | 0.009 |
| $10^3$ | 2,084 | 0.750 | 0.011 | $10^3$ | 0.754 | 0.012 |
| | | | | $10^4$ | 0.749 | 0.011 |
| | | | | $10^5$ | 0.749 | 0.011 |
| | | | | $10^6$ | 0.749 | 0.011 |

and a variance of $\sigma_m^2 = 0.011$. The Monte Carlo results converge to the same values for the mean and variance at a much higher number of samples. Minor differences with the Monte Carlo results can be explained by the inherent variability in Monte Carlo simulations.

The results for the case with $t = 0.5$ are given in Table 2. Both methods converge less fast than for the case with $t = 1$ due the high output variance and the presence of the discontinuity in the response surface. Adaptive Stochastic Finite Elements predict $\mu_m = 0.566$ and $\sigma_m^2 = 1.087$ using $10^3$ second-degree elements. This result is obtained using 2,217 deterministic solves. It can be seen in Table 2 that standard Monte Carlo simulation for approximately the same number of deterministic solves results in a significantly less accurate approximation. Monte Carlo simulation requires $10^5$–$10^6$ samples to obtain a comparable accuracy.

Even though every second-order element contains 6 quadrature points, for this case the average number of deterministic solves per element seems to approach 2.

**Table 2** Mean and the variance of the instantaneous mass flow $m$ at $t = 0.5$ with $\mu_{u_{piston}} = 0.5$ and initial pressure $p_{pre}$ by adaptive Stochastic Finite Elements (ASFE) and Monte Carlo (MC) simulations

| Adaptive Stochastic Finite Elements (ASFE) | | | | GEK (Sect. 3) | | | Monte Carlo (MC) | | |
|---|---|---|---|---|---|---|---|---|---|
| Elements | Samples | Mean | Variance | Samples | Mean | Variance | Samples | Mean | Variance |
| 10 | 30 | 0.863 | 0.879 | 36 | 0.565 | 0.875 | 10 | 0.793 | 1.654 |
| $10^2$ | 232 | 0.562 | 1.044 | 256 | 0.562 | 0.998 | $10^2$ | 0.693 | 1.261 |
| $10^3$ | 2,217 | 0.566 | 1.087 | 2,209 | 0.561 | 1.041 | $10^3$ | 0.549 | 1.081 |
| | | | | | | | $10^4$ | 0.573 | 1.100 |
| | | | | | | | $10^5$ | 0.561 | 1.085 |
| | | | | | | | $10^6$ | 0.563 | 1.088 |

This number of samples is relatively low, since the samples which are located on the boundaries of the elements are used for the polynomial approximation in multiple elements. Furthermore, all samples are reused in the successive refinement steps. For complex computational problems in the order of $10^3$ deterministic solves may result in high computational costs, however, already an approximation with a realistic 5 % error for practical applications is obtained with between 10 and 100 elements.

The relatively high coefficient of variation of the instantaneous mass flow $CV_m = 1.842$ compared to the input coefficients of variation $CV_{u_{piston}}, CV_{p_{pre}} = 0.1$ demonstrates that singularities in probability space can result in a high output sensitivity on the input uncertainty.

### 2.3.2 Transonic Flow over a NACA0012 Airfoil

In this section transonic Euler flow over a NACA0012 airfoil subject to an uncertain free stream Mach number $Ma_\infty$ is considered. This is an example of uncertainty in a practical flow problem with a shock wave, which results in a discontinuity in probability space. A transonic flow problem is also of interest since it is known that a transonic flow field can be sensitive to small input variations. The distribution of $Ma_\infty$ is assumed to be lognormal with a mean Mach number of $\mu_{Ma_\infty} = 0.8$ and a coefficient of variation of $CV_{Ma_\infty} = 1\,\%$. Due to the small input coefficient of variation, the free stream Mach numbers $Ma_\infty$ with a significant probability are restricted to the transonic range. The angle of attack is equal to 1.25° and the airfoil has a chord with length $c$. The two-dimensional flow problem is discretized using a second-order upwind spatial finite volume scheme on a unstructured hexahedral mesh with $3 \cdot 10^4$ spatial volumes. The steady state solution is found by time integration with a CFL number of 0.5. In Fig. 10 the flow field in terms of the local Mach number is shown for the mean value of the free stream Mach number $\mu_{Ma_\infty}$. Above the wing a large supersonic domain is present for which $Ma > 1$, which ends at a shock wave at $x = 0.6c$. Under the wing a small supersonic region and a weak shock wave at $x = 0.35c$ are present.

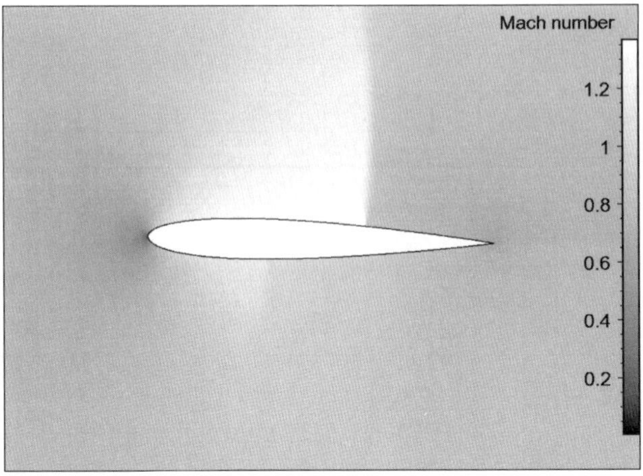


**Fig. 10** Transonic flow over a NACA0012 airfoil for the mean free stream Mach number $\mu_{\mathrm{Ma}_\infty}$ with the deterministic shock wave positions of $x = 0.6c$ and $x = 0.35c$ on the upper and lower surface, respectively

The effect of the uncertainty in $\mathrm{Ma}_\infty$ is given in Fig. 11 in terms of the Mach number along the airfoil surface. The mean Mach number and the 99 % uncertainty range are shown. Stochastic Finite Elements are applied with 4 elements, which results in nine deterministic solves. The output variables are the Mach numbers at all $6 \cdot 10^2$ volumes on the airfoil surface. Uniform grid refinement is applied, since refining adaptively based on a combination of $6 \cdot 10^2$ output variables effectively results in uniform refinement as well. The results are compared with Stochastic Collocation based on 5 deterministic solves.

Stochastic Finite Elements predict that the uncertainty smears the shock wave in the mean Mach number along the upper surface around its deterministic location, see Fig. 11a. The 99 % uncertainty range shows that the position of the shock wave is sensitive to the 1 % uncertainty in the free stream Mach number. The shock wave strength is nearly unaffected. Stochastic Collocation predicts less sensitivity of the shock position and a much larger effect on the possible Mach numbers near the shock wave, see Fig. 11b. The large uncertainty range given by Stochastic Collocation includes unrealistically high Mach numbers of up to 3 and unphysical negative Mach numbers. Increasing the number of collocation points in Stochastic Collocation increases the oscillatory behavior. In Fig. 11c, d qualitatively the same characteristics can be seen for the Mach number along the lower surface.

## 3 Method 2: Gradient-Enhanced Kriging with Adaptivity for Uncertainty Quantification

Any interpolation method can be used to perform UQ, and the form of the problem – in particular, dimensionality, smoothness of the response, and required accuracy – dictates the most appropriate method. We note that in our experience the UQ problem is typically characterized by relatively high-dimensional spaces (a large

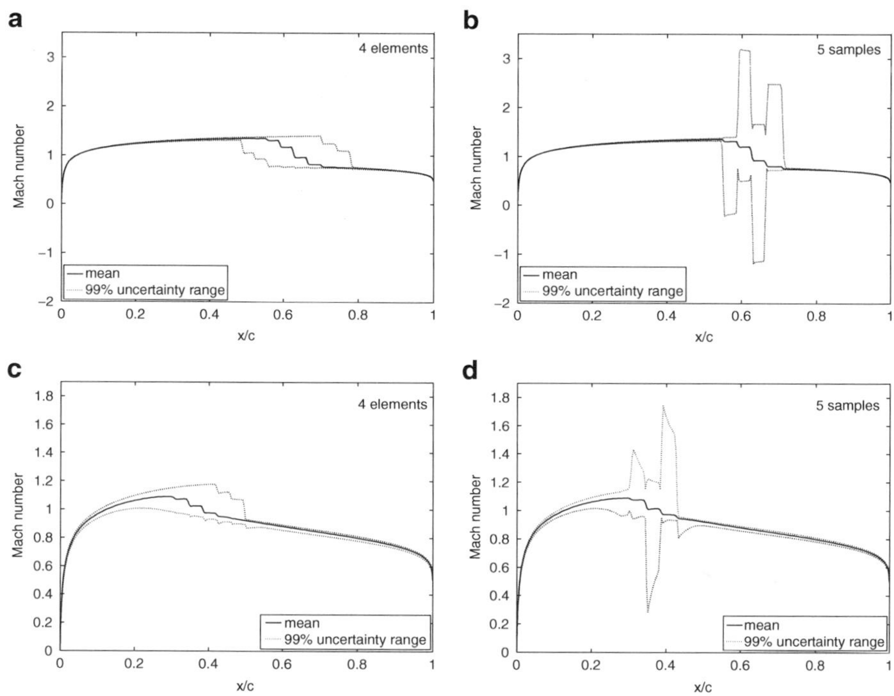


**Fig. 11** Mean Mach number and 99 % uncertainty range along the surface by Stochastic Finite Elements (SFE) and Stochastic Collocation (SC) for the uncertain free stream Mach number $Ma_\infty$ in the transonic flow over a NACA0012 airfoil. (**a**) SFE, upper surface. (**b**) SC, upper surface. (**c**) SFE, lower surface. (**d**) SC, lower surface

number of uncertain parameters), almost-linear response in many parameters with strong non-linearities in only a handful (few influential parameters), and a relatively low accuracy requirements. This latter is a consequence of *input* pdfs being of themselves of low accuracy – often they are not based on measurements, but are simply "best guesses". On the other hand we require our uncertainty estimates to be conservative.[2]

Given these criteria we propose an adaptive Kriging [21, 22] approach to UQ, using derivatives of the response where available. Kriging is a very flexible interpolation framework, allowing arbitrary sample locations, multiple correlated response variables, a tuneable level of regression, and possesses an inherent error estimate. In Sect. 3.3 it will be shown how these properties may be combined to create an adaptive response surface using gradient information suitable for UQ. By incorporating gradients (derivatives of the response w.r.t. input parameters) it

[2]Based on: R.P. Dwight and Z.-H. Han, *"Parametric Uncertainty Quantification with Adjoint Gradient-Enhanced Kriging"*, AIAA Paper, AIAA-2009–2276, 2011.

may be possible to perform UQ relatively cheaply in high-dimensions, especially if the response in many variables is almost linear. Gradients we obtained at a cost independent of dimension using an adjoint approach, see Sect. 3.2 – however this approach has some practical limitation discussed in Sect. 3.4. Numerical results for a model problem, and a transonic aerofoil are then presented in Sect. 3.5.

## *3.1 Uncertainty Quantification Problem*

We consider a slightly different UQ problem to previous chapters in order to exploit the availability of adjoint gradients: we assume a relatively low-dimensional functional $j(u)$ of the PDE solution $u$ that is of main interest.

Consider an arbitrary deterministic non-linear problem with unknowns $u$ and an $M$-dimensional deterministic parameter vector $\alpha$,

$$\mathcal{N}(u, \alpha) = 0, \tag{18}$$

where the computed solution $u$ is not of principle interest, but rather only an $N$-dimensional vector of functionals of the solution $j := j(u, \alpha)$. Such a situation arises often, e.g. in aerodynamics, where the forces and moments on a body, and not the entire flow field, are of primary engineering relevance.

Now assume that $A$ is a vector random variable with realization $\alpha$, and known probability density function (pdf) $f_A(\alpha)$. The objective is to determine $f_A(j)$ the pdf of the random variable $J$, and in particular the expectation, variance and higher moments of $J$, which may all be cast as integrals over the parameter space. For example the expectation $\mathbb{E}_A J$ is defined as

$$\mathbb{E}_A\left[J(u(\alpha), \alpha)\right] := \int_{\Omega} j(u(\alpha), \alpha) f_A(\alpha)\, \mathrm{d}\alpha, \qquad \Omega := \mathbb{R}^M. \tag{19}$$

If the individual components of $A$ are independent, then this integral over the parameter space $\Omega$, can be transformed into an integral over the *probability space* $\bar{\Omega} := [0, 1]^M$, using a change of variables. In particular let $f_{A,i}$ be the pdf of the $i$-th scalar component of $A$, and $F_{A,i} : \mathbb{R} \to [0, 1]$ the corresponding cumulative density function (cdf), with the property

$$\frac{\mathrm{d}F_{A,i}(\alpha)}{\mathrm{d}\alpha_i} = f_{A,i}(\alpha), \quad \forall i \in \{1, \ldots M\}.$$

Then a composite cdf can be defined which is invertible in each component:

$$\begin{aligned} F_A : \mathbb{R}^M &\to [0, 1]^M \\ (\alpha_1, \ldots, \alpha_M) &\mapsto (F_{A,1}(\alpha_1), \ldots, F_{A,M}(\alpha_M)), \end{aligned}$$

and can be used to change variables from $\alpha \in \mathbb{R}^M$ to $\xi \in [0,1]^M$. So integral (19) becomes

$$\mathbb{E}_A\left[J(u(\alpha),\alpha)\right] = \int_{\bar{\Omega}} j\left[u(F_A^{-1}(\xi)), F_A^{-1}(\xi)\right] \mathrm{d}\xi, \qquad \bar{\Omega} := [0,1]^M. \tag{20}$$

This change of variables is sometimes referred to as the *chaos transformation*, and transforms the problem from parameter space into probability space, reducing it to integration over an $M$-dimensional hypercube – a standard problem in quadrature – which is exploited by most UQ methods, including PC and Newton-Cotes triangulation. On the other hand the assumption of independence of components of $A$ is very restrictive, excluding e.g. multivariate normal distributions, and the integrand in (20) is likely to be much less regular than the integrand in (19), making quadrature more difficult. Therefore we prefer to work in the parameter space.

### 3.2 Gradient Evaluation via the Adjoint Method

One feature of the problem that may be exploited immediately if $N \ll M$ is that all gradients $\mathrm{d}j/\mathrm{d}\alpha$ may be obtained at a cost proportional to $N$ using a dual problem approach [4]. Briefly, let the *Lagrangian* $\mathscr{L}$, be defined as

$$\mathscr{L}(u,\alpha,\psi) = j(u,\alpha) + \psi^T \mathscr{N}(u,\alpha), \tag{21}$$

where $\psi$ is the adjoint state, a vector of the same dimension as $\mathscr{N}$. Now note that since $\mathscr{N} \equiv 0$ for all $\alpha$ and $u$ satisfying (18), $\mathscr{L} \equiv j$, $\forall\alpha$, and hence also

$$\frac{\mathrm{d}j}{\mathrm{d}\alpha} \equiv \frac{\mathrm{d}\mathscr{L}}{\mathrm{d}\alpha} = \left(\frac{\partial j}{\partial \alpha} + \frac{\partial j}{\partial u}\frac{\mathrm{d}u}{\mathrm{d}\alpha}\right) + \psi^T\left(\frac{\partial \mathscr{N}}{\partial \alpha} + \frac{\partial \mathscr{N}}{\partial u}\frac{\mathrm{d}u}{\mathrm{d}\alpha}\right) \tag{22}$$

$$= \left(\frac{\partial j}{\partial \alpha} + \psi^T\frac{\partial \mathscr{N}}{\partial \alpha}\right) + \left(\frac{\partial j}{\partial u} + \psi^T\frac{\partial \mathscr{N}}{\partial u}\right)\frac{\mathrm{d}u}{\mathrm{d}\alpha}. \tag{23}$$

Now if $\psi$ satisfies the linear *adjoint* equation:

$$\frac{\partial j}{\partial u} + \psi^T\frac{\partial \mathscr{N}}{\partial u} = 0, \tag{24}$$

then the unknown $\mathrm{d}u/\mathrm{d}\alpha$ term drops out of the expression for the gradient in (23), and we have

$$\frac{\mathrm{d}j}{\mathrm{d}\alpha} = \frac{\partial j}{\partial \alpha} + \psi^T\frac{\partial \mathscr{N}}{\partial \alpha}. \tag{25}$$

Since (24) is independent of $\alpha$, only $N$ such linear equations must be solved to obtain all gradients $\mathrm{d}j/\mathrm{d}\alpha$.

The main prerequisite for application of this technique is the availability of the partial derivatives of the (typically discrete) operator $\mathscr{N}$. For complex simulation codes, obtaining these is in principle a straight-forward process, but requires access to the source code, and represents a considerable amount of effort [4, 7]. The compressible Navier-Stokes solver examined in this paper, the DLR TAU-Code, has an adjoint solver mode available – originally developed for the purposes of gradient-based optimization and error estimation [5, 6] – and this method for determining gradients of $j$ is used in what follows.

## 3.3 Gradient-Enhanced Kriging for Uncertainty Quantification

We propose a method for adaptive uncertainty quantification based on a gradient-enhanced Kriging (GEK) response surface. Of the many response surface methods available – such as radial basis function models, neural networks, smoothing spline models, and support vector machines – Kriging is selected for its arbitrary sample locations, ability to incorporate gradients, sample error information (partial regression), as well as trend information from low fidelity models. The resulting surface comes equipped with error estimates in the form of standard deviations, suggesting natural adaptation indicators. Once a Kriging model of the parameter space is available, weighted integrals over $\mathbb{R}^M$ can be calculated numerically to obtain statistical moments of the output. If the full pdf of the output is needed, Monte-Carlo can be rapidly performed on the response surface.

### 3.3.1 Kriging Response Surfaces

A particularly lucid exposition of Kriging may be found in [11], here we present a brief overview of the main idea. Consider standard curve-fitting by regression for a deterministic scalar function $j$ sampled at $n$ parameter locations $\alpha_p$ with the index $p \in \{1, \ldots, n\}$, in $M$ dimensions. In this technique the observations are treated as if they were generated from the following model:

$$j(\alpha_p) = \sum_l \beta_l f_l(\alpha_p) + \epsilon_p, \tag{26}$$

where $f_l$ and $\beta_l$ are regression functions and coefficients respectively and the errors $\epsilon_p$ are assumed to be independently normally distributed with mean zero. Usually $f_l$ are pre-defined and a least-squares measure is used to determine the best values for $\beta_l$. However unless $f_l$ capture the full non-linear behaviour of $j$, the assumption of independence of errors is blatantly false for a deterministic $j$. Rather we would expect that errors are a function of position $\epsilon = \epsilon(\alpha)$, and errors at two close points are closely correlated.

Motivated by this argument the correlation between the errors at two points is modeled using a correlation function $r$:

$$\operatorname{corr}\left[\epsilon(\alpha_p), \epsilon(\alpha_q)\right] \approx r(\alpha_p, \alpha_q) := \exp\left\{-d(\alpha_p, \alpha_q)\right\}, \tag{27}$$

where $d$ is some weighted distance function

$$d(\alpha_p, \alpha_q) := \sum_{k=1}^{M} \theta_k \left|\alpha_{p,k} - \alpha_{q,k}\right|^{\phi}, \tag{28}$$

and $\alpha_{p,k}$ is the $k$th component of the point $\alpha_p$, the weights $\theta_k > 0$ account for different levels of correlation in each dimension, and $\phi = 2$ (resulting in a Gaussian correlation) in what follows. This model turns out to be sufficiently general that it is possible to dispense with the original regression, and use the sample model

$$j(\alpha_p) = \mu + \epsilon(\alpha_p) \tag{29}$$

where $\mu$ is the mean of the stochastic process, $\epsilon(\alpha_p)$ are normally distributed with zero mean and variance $\sigma^2$ and are no longer independent, but correlated according to (27).

The fitting is then performed by determining values for $\mu$, $\sigma^2$ and $\theta_i$ such that the likelihood of achieving the observed sample with this model is maximized. The likelihood is:

$$L(\theta, \mu, \sigma^2) := \frac{1}{(2\pi\sigma^2)^{n/2}|R|^{1/2}} \exp\left[-\frac{(\mathbf{j} - \mathbf{1}\mu)^T R^{-1}(\mathbf{j} - \mathbf{1}\mu)}{2\sigma^2}\right], \tag{30}$$

where $\mathbf{1}$ is a vector of ones of dimension $n$, $\mathbf{j}$ is the vector $(j(\alpha_1), \ldots, j(\alpha_n))$, and $R$ is the $n \times n$ correlation matrix generated by $r$:

$$R_{pq} := r(\alpha_p, \alpha_q).$$

Given $\theta$ there exist closed-form expressions for $\mu$ and $\sigma^2$ maximizing (30), namely

$$\hat{\mu} = \frac{\mathbf{1}^T R^{-1}\mathbf{j}}{\mathbf{1}^T R^{-1}\mathbf{1}}, \qquad \hat{\sigma}^2 = \frac{(\mathbf{j} - \mathbf{1}\hat{\mu})^T R^{-1}(\mathbf{j} - \mathbf{1}\hat{\mu})}{n}, \tag{31}$$

so that the $M$ dimensional optimization problem

$$\hat{\theta} := \operatorname*{argmax}_{\theta_k > 0} \left\{\log L(\theta, \hat{\mu}, \hat{\sigma}^2)\right\} \tag{32}$$

must be solved. The inversion of the positive definite matrix $R$ in (30) is performed with a Cholesky factorization.

Given values for $\hat{\theta}_i$ the model may be now used to predict the functional value at unsampled points. The best linear unbiased predictor of $j(\alpha^*)$ may be shown to be [19]

$$\hat{j}(\alpha^*) := \hat{\mu} + \mathbf{r}^T R^{-1}(\mathbf{j} - \mathbf{1}\hat{\mu}), \tag{33}$$

where the first term is the regression prediction, and the second term is a correction for the error correlation. The term $\mathbf{r}$ which represents the correlation between the error at the untried point $\alpha^*$ and the error at the sample points has components:

$$r_p := \text{corr}\left[\epsilon(\alpha^*), \epsilon(\alpha_p)\right] = \exp\left\{-d(\alpha^*, \alpha_p)\right\}. \tag{34}$$

An estimate of the error in the response surface at $\alpha^*$ is given by the predictor standard deviation:

$$\hat{\tau}(\alpha^*) := \hat{\sigma}\sqrt{1 - \mathbf{r}^T R^{-1}\mathbf{r}}, \tag{35}$$

whose accuracy is contingent on the correlation function and $\hat{\theta}$ being reasonable.

### 3.3.2 Gradient-Enhanced Kriging (GEK)

In order to incorporate gradients of $j$ obtained by an adjoint method, we employ GEK, for which there are two formulations: direct and indirect [10]. Both approaches have similar costs and give similar results. In indirect GEK the sample set is augmented with samples obtained by linear reconstruction a small distance $\delta$, in each coordinate direction from each sample point

$$j(\alpha_p \pm \delta) = j(\alpha_p) \pm \delta \left.\frac{\mathrm{d}j}{\mathrm{d}\alpha}\right|_{\alpha_p} + \mathcal{O}(\delta^2),$$

resulting in three times the number of samples (in 1d), and standard Kriging is performed on this extended sample set. The choice of $\delta$ is critical: it should be substantially smaller than the local distance between sample points, but not so small that the correlation matrix $R$ becomes stiff – as well as satisfying the usual requirements on truncation error and rounding error.

In direct GEK, the gradients in each dimension are treated as co-variables in a cokriging framework, and the gradient relationship between the original variable $j$ and its derivatives is established over the correlation function. In particular $\mathbf{j}$ is augmented with gradients, for example in 1d:

$$\mathbf{j}_c := \left( j(\alpha_1), \ldots, j(\alpha_n), \left.\frac{\mathrm{d}j}{\mathrm{d}\alpha}\right|_{\alpha_1}, \ldots, \left.\frac{\mathrm{d}j}{\mathrm{d}\alpha}\right|_{\alpha_n} \right)$$

and the correlation matrix is chosen as

$$R_c := \begin{pmatrix} R_{00} & R_{01} \\ R_{10} & R_{11} \end{pmatrix}$$

with

$$R_{00} := r(\alpha_p, \alpha_q) \qquad R_{01} := \frac{\partial r(\alpha_p, \alpha_q)}{\partial \alpha_p}$$

$$R_{10} := \frac{\partial r(\alpha_p, \alpha_q)}{\partial \alpha_q} \qquad R_{11} := \frac{\partial^2 r(\alpha_p, \alpha_q)}{\partial \alpha_p \partial \alpha_q}.$$

Kriging then proceeds as before with this new definition of $R$, except that the mean $\mu$, is computed for $j$ only, while the mean of $\mathrm{d}j/\mathrm{d}\alpha$ is assumed to be zero (for constant regression of $j$).

The left plot of Fig. 12 shows a comparison of direct and indirect GEK models for the function

$$j(x) = \frac{3}{10} + \sin\left(\frac{16}{15}\alpha - 1\right) + \sin^2\left[2\left(\frac{16}{15}\alpha - 1\right)\right],$$

with samples and gradient samples at $\{-3/4, 0, 3/4\}$. The indirect GEK uses a constant step size of 0.01. Correlation lengths are $\theta = 3.035$ for direct and $\theta = 3.384$ for indirect GEK. Only small differences are visible between direct and indirect GEK, and both are much better than Kriging without gradient information. The right-hand plot shows the design space of various GEK models, where single precision has been used to emphasise the influence of rounding error in the likelihood estimator for this low number of samples. The cost function for indirect GEK becomes noisy with decreasing $\delta$, as $R$ becomes increasingly ill-conditioned when sample points are almost collocated. Direct GEK suffers no such problems.

The computational cost of GEK can be significantly greater than Kriging however. Consider the cost of building Kriging and GEK models for $n$ samples in $M$-dimensions. For basic Kriging each evaluation of (30) requires inversion of the $n \times n$ dense matrix $R(\theta)$. The Cholesky decomposition applied to $R$ costs $\mathcal{O}(n^3)$. Assume that the optimization problem for $\theta$ may be solved at a cost linear in the dimension of the problem $\mathcal{O}(M)$, which is usually optimistic. Then the total cost of building the response surface is $\mathcal{O}(n^3 M)$. The predictor may then be evaluated at a cost of $\mathcal{O}(nM)$ per point.

In both direct- and indirect GEK the augmented value vector has size $n(M+1)$, so that the total cost of building the response surface is $\mathcal{O}(n^3 M^4)$ and the predictor

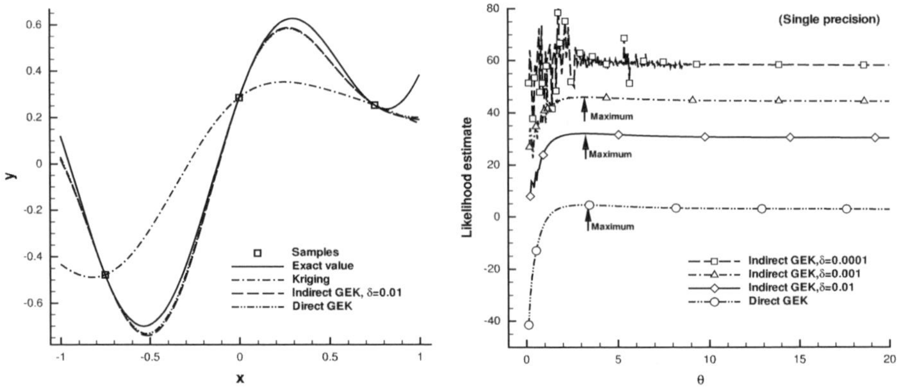


**Fig. 12** Comparison of Kriging, and direct and indirect gradient-enhanced Kriging for an analytic test function (*left*). Likelihood estimator versus $\theta$ for GEK models with single-precision arithmetic for the same data-set (*right*)

cost is $\mathcal{O}(nM^2)$. This and the difficulty of solving (32) are obstacles to the application of GEK in high-dimensional parameter spaces.

### 3.3.3 Adaptive Estimation of $\mathbb{E}_A J$

Given that the goal is to evaluate statistical moments of the output (such as the expectation (19)) using the Kriging response surface, we would like to ensure the response approximation of $j$ is accurate near regions of high probability in $\alpha$, and that effort is not wasted in zero-probability regions.

An adaptive scheme is based on the predictor standard-deviation $\hat{\tau}$ (35), which gives a measure of the error in the response surface. This suggests a simple error estimate:

$$\varepsilon := \int_{\Omega} \hat{\tau}(\alpha) f_A(\alpha)\, \mathrm{d}\alpha, \tag{36}$$

and corresponding adaptation indicator:

$$\eta(\alpha) := \hat{\tau}(\alpha) f_A(\alpha). \tag{37}$$

Adaptation is performed by adding a sample of $j$ at

$$\alpha_{\text{new}} := \underset{\alpha}{\operatorname{argmax}}\, \eta(\alpha),$$

and rebuilding the surface. I.e. we try to improve the response surface where the error in the integrand in (19) is estimated to be largest. Note that $\eta$ will be zero in regions of zero probability, and at existing sample points – precluding the possibility of $\alpha_{\text{new}}$ being an existing point.

The optimization problem for $\alpha_{\text{new}}$ is made challenging by the existence of very large numbers of local minima at $\alpha$ between sample locations. In fact it is likely we wish to add multiple new points in a single adaptation step – for the purposes of parallelizing evaluations of $j$, or avoiding repeated construction of the response. Therefore we use optimization methods which can provide approximations of multiple local minima, e.g. differential evolution.

### 3.3.4 Implementation of Correlation Parameter Optimization

A sensitive aspect in our Kriging implementation is the resolution of the optimization problem (32). If $\hat{\theta}$ is too large, the response surface will be constant at $\hat{\mu}$ almost everywhere, with narrow Gaussian humps at the data points. If $\hat{\theta}$ is too small the data is over-fit and oscillations and over-shoots occur. In high-dimensions these problems are difficult to diagnose. As seen in Fig. 12 the $\theta$-design space is often noisy due to ill-conditioning of the correlation matrix $R$, often has multiple local minima, and the computational cost of evaluating $L(\theta, \mu, \sigma^2)$ is high. In summary the optimization problem may be considered to be of substantial difficulty.

Our approach is as follows: firstly, in order to standardize the problem, the sample data and coordinates in each dimension are normalized using the sample mean and standard deviation. Secondly, by taking logs of $\theta$ the problem is transformed from a constrained optimization problem for $\theta_i > 0$, to an unconstrained problem for $\log(\theta_i)$. Since $\theta$ represents a scaling it has a exponential character, and taking logs serves to better condition the design space. Finally logs are taken of $L$.

To perform the optimization we employ the *Subplex* algorithm of Rowan [18]. It is a generalization of the Nelder-Mead Simplex (NMS) method [12, 15] intended for high-dimensional problems, and works by applying NMS on a sequence of low-dimensioned sub-spaces. Subplex retains the robustness of NMS with respect to noisy functions, but the number of function evaluations required for convergence typically increases only linearly with the problem dimension. Subplex is a local optimizer and therefore requires some globalization procedure. In this instance we apply the Subplex method repeatedly with $M + 2$ randomly selected initial conditions. These are chosen using Latin-Hypercube sampling [14] in order to encourage a wide spread of values.

## *3.4 Limitations of Gradients for Response Surfaces in CFD*

Before presenting numerical results, the accuracy of adjoint gradients for problems in aerodynamics, and the consequences for response surface construction are examined.

The adjoint methods described in Sect. 3.2 were developed in CFD in the context of optimization problems, where the aim is typically to minimize the drag on an aerofoil by modifying its shape (parameterized by a large number of design variables), subject to constraints lift and aerofoil thickness. This type of problem is characterized by regular attached flows: as the region of the optimum is approached, shocks and regions of separated flow have been eliminated as far as possible because these flow features represent the greatest contribution to the removable drag. Furthermore many standard optimization algorithms are quite robust to gradient error, for example conjugate-gradient method combined with a line-search [8].

When building a general-purpose response surface with gradients the situation is different. Regions of interest in the parameter space are likely to include "difficult" flows with separation, stall, and shocks, where gradients may not even be defined. Poor gradients might lead to spurious oscillations in the response surface. To illustrate these problems, we present two issues associated with obtaining accurate gradients in CFD, one associated with non-smooth resolution of shocks by the flow solver, and one associated with a standard approximation of freezing the turbulence equations in the linearization. The later may be resolvable with further development of numerical techniques, but appears at present intractable. The former is a more fundamental problem. In any case these results will show that strict enforcement of the gradient in a response-surface model is not always appropriate.

### 3.4.1 Oscillatory Gradients Due to Shocks

The presence of discontinuities in solutions of the Euler equations causes a variety of discretization difficulties which modern numerical techniques are capable of handling well. However shock-capturing methods often suffer from an oscillatory dependence of the solution on flow parameters. Such a phenomena was first observed by Giles et al. [9].

To demonstrate this effect we give the example of a NACA 0012 aerofoil at an angle-of-attack of $0.1^{\circ}$ on a sequence of three structured grids, and Mach number varying from a subsonic to a transonic regime. The lift coefficient is plotted in the top left, and (zoomed) top right of Fig. 13. For small Mach numbers the curve is smooth, but at a Mach number of about 0.78 a shock starts to form on the upper and lower surfaces and as the speed increases the shocks travel slowly along the aerofoil. The correct physical behaviour is a smooth Mach–lift curve, the oscillations seen in the figure are numerical artefacts caused by the varying position of the shocks in relation to the mesh points. If a shock occurs on the boundary of two mesh cells, it is likely to be better resolved than if it lies in interior of a cell, as the solution representation is continuous within cells and discontinuous on cell boundaries. This can be confirmed by noting that the frequency of the oscillations double when the spacing of the surface mesh points halves.

These oscillations are only visible because of the fine resolution of the curve, and are generally not considered a serious problem because they are aerodynamically not of significant amplitude, and as the mesh is refined their size decreases – if

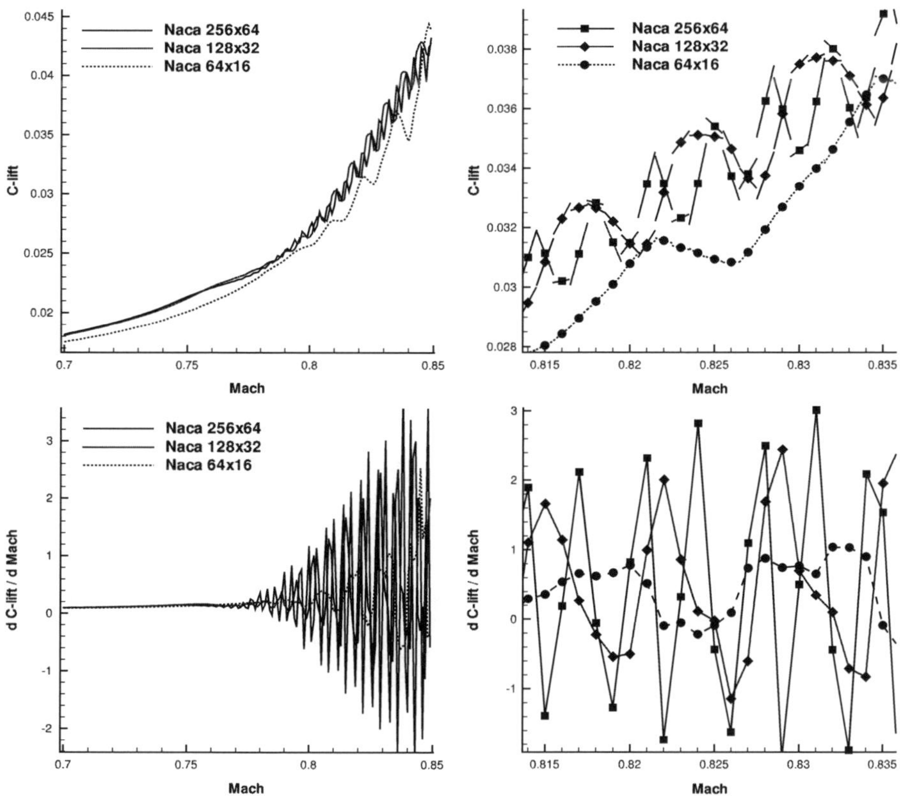


**Fig. 13** Lift coefficient and derivatives against Mach number for a transonic NACA0012 aerofoil on three grids. The bars in the *top* plots display the derivatives computed using the discrete adjoint

the discretization is consistent. However the linearization of the discretization is also aware of these oscillations, and the derivatives calculated from the linearization faithfully follow them. This can be seen again in the top two plots of Fig. 13, the bar attached to each point is based on the adjoint derivative of lift with respect to Mach number at that point.

Now the problem arises that although the *amplitude* of these oscillations reduces as the mesh is refined, so their *frequency* increases, and their overall *shape* remains roughly constant. Hence the magnitude of the first derivatives of these oscillations *does not* decrease with mesh resolution. This is shown in the bottom plots of Fig. 13, where the gradient obtained by adjoint is plotted directly. The oscillations are of large magnitude compared to the absolute value of the gradient, and their magnitude actually increases with mesh resolution. Clearly such gradients are of limited practical value despite being perfectly correct descriptions of the local behaviour of the discrete flow solver.

### 3.4.2 Gradient Error Due to a Frozen-Turbulence Approximation

Although it is possible to completely linearize the $\mathscr{N}$ of (18) for discretizations of the Reynolds Averaged Navier-Stokes (RANS) equation, including the turbulence model – if this is actually performed the resulting linear system is often so badly conditioned that it can not be solved using anything less than direct inversion [8]. For this reason very often the turbulence model is not linearized, rather *frozen* by treating eddy-viscosity, turbulence energy and any other turbulence quantities as constant with respect to the linearization. This is termed the frozen-turbulence approximation, and modifies the solution of the linearized problem, and hence the calculated gradients. In [8] the present author examined the influence of this approximation of the Jacobian on gradient accuracy with the conclusion that the magnitude of the error is strongly dependent on the flow under consideration.

To demonstrate the type of gradient errors that can result we consider an RAE 2822 single-element 2d aerofoil, modeled with the Spallart-Almaras-Edwards one-equation turbulence model at a Reynolds number of $\mathrm{Re} = 6.5 \times 10^6$, Mach number of $M_\infty = 0.73$ and varying angle-of-attack $\phi$. Lift and drag are plotted in Fig. 14. A variety of flow phenomena occur in this $\phi$ range. A shock develops at about $\phi = 2°$, flow begins to separate at about $\phi = 3°$, the flow is in the process of massive separation in the range $\phi = 4.5$–$8°$ and the flow solver can not obtain a stationary solution here. From $\phi = 8°$ the flow is fully separated from the upper surface, and the solver can again obtain a stationary solution.

The line segments plotted in Fig. 14 represent gradients obtained with the adjoint method at each angle-of-attack, with and without linearization of the turbulence model. Apart from the region in which no stationary solutions are obtained, the fully-linearized adjoint gives accurate gradients (judging by agreement with the polar). The frozen-turbulence approximation performs well provided separation is either not present, or has no significant effect on the integrated forces. However, as maximum lift is approached the approximation breaks down completely, presumably as the importance of variation of turbulence to the flow grows.

To see the effect such gradient errors have on a surrogate model consider Fig. 15, which shows Kriging and gradient-enhanced Kriging (GEK) response surfaces (to be described in the following sections) with exact and frozen-turbulence gradients. The surfaces are smooth, are required to pass through all the sample points, and in the case of GEK the surface is required to have the specified gradient at the sample point too. The surfaces are based on four sample locations deliberately chosen to lie outside the oscillatory region where the flow solver did not converge to a stationary state. Given accurate gradients GEK performs substantially better than basic Kriging. With poor quality gradients it is clear that the response surface is of no value, returning unphysical negative drag in some regions.

One chance of resolving these difficulties with linearization of turbulence models lies in the enforcement of *realizability conditions* to explicitly avoid the creation of un-physical states, and choosing variables in which the solution is smoother e.g. using the variable $\log \omega$ rather than $\omega$ directly in the implementation of the $k - \omega$ turbulence model [1]. The former however involves some modifications that change

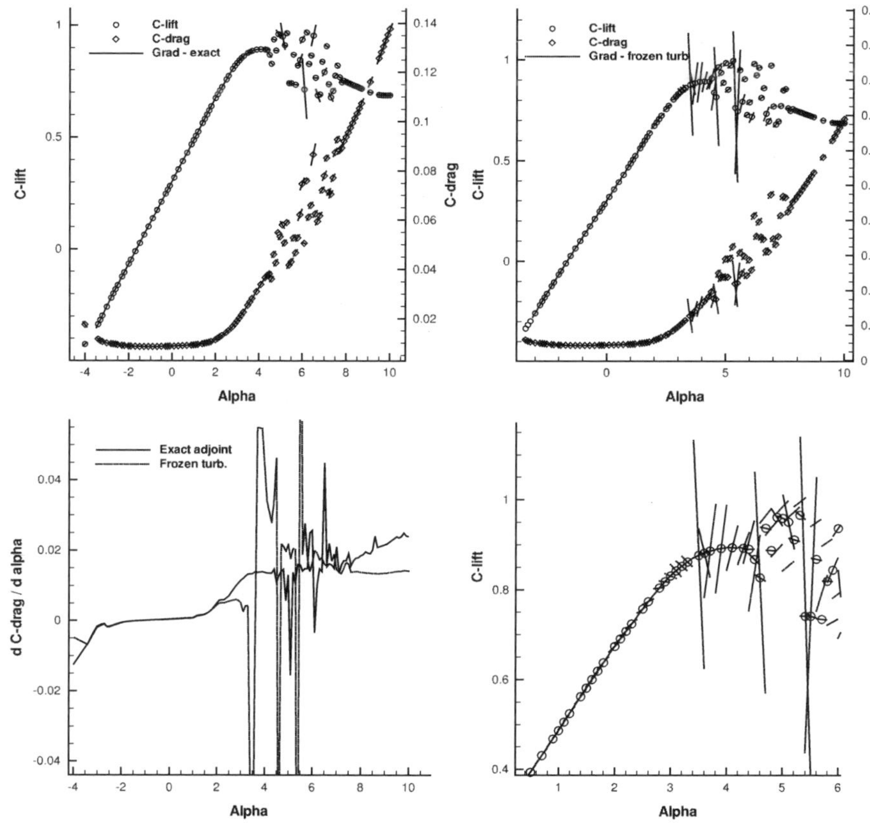


**Fig. 14** Lift and drag against angle-of-attack for an RAE 2822 aerofoil. *Line* segments represent gradients computed with a discrete adjoint code, with a full linearization of the turbulence model (*black*), and a frozen-turbulence approximation (*dotted*)

the physical behaviour of the model, and it remains to be seen if such changes will be accepted by the aerodynamics community.

In the following where CFD gradients are used they are obtained with full linearization of the turbulence model in order to observe the uncertainty quantification algorithm in the absence of gradient error. At present this is only feasible in 2d, where direct inversion of the linear system is feasible, using e.g. SuperLU [3].

## 3.5 Numerical Results

In order to compare the algorithms described so far, we resort initially to a model problem. Once general bounds on efficiency (in terms of number of functional

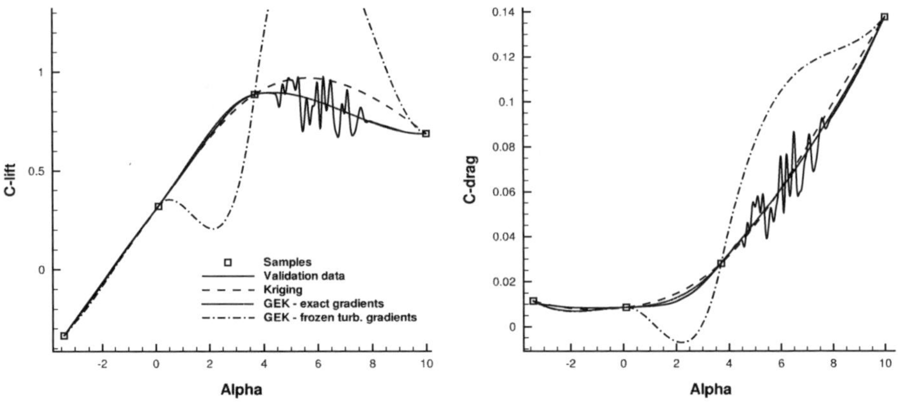


**Fig. 15** One-dimensional Kriging and GEK response surfaces for lift and drag based on four support points for the RAE 2822 test-case with varying angle-of-attack

evaluations required) have been established, the most successful techniques are applied to problems in CFD.

### 3.5.1 "Sandtimer" Model Problem

For the purposes of testing we define an $M$-dimensional model problem as follows:

$$j := \sum_{i=1}^{M-1} u_i^2 + S_q(\alpha_M)$$

where each $u_i$ for $1 \le i \le M-1$ is defined implicitly by

$$N(u_i, \alpha) = (\alpha_M^2 + 1)(u_i + u_i^3) - \alpha_i = 0,$$

and $S_q$ is a quartic with coefficients chosen such that it is slightly skewed, but still has a unique minimum at $\alpha = 0$ with value 0:

$$S_q(\alpha) = r\left(\tilde{\alpha}^4 + (\tilde{\alpha} - 1)^2 - q\right), \qquad \tilde{\alpha} = s\alpha + p,$$

$$p = 0.589754512301, \quad q = 0.289273423937, \quad r = 0.05, \quad s = 0.25.$$

The resulting surface in two dimensions can be seen in the upper left plot of Fig. 16, and will be denoted the *sandtimer* model problem. The goal is to evaluate the expected value of $J$ for $\alpha$ with specified pdf.

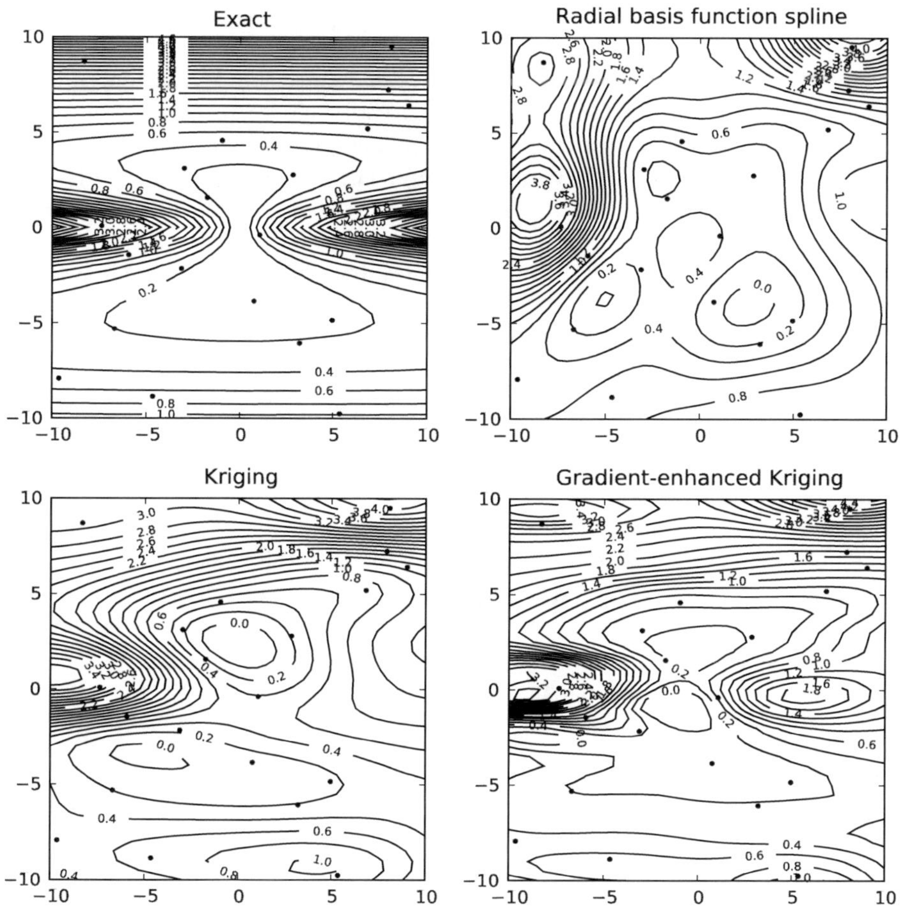


**Fig. 16** Reconstruction of a test surface using 20 random sample points

This choice of problem serves several purposes: firstly to model the situation in CFD where we have a cost-function $j$ which is an explicit function of a solution $u$, which is itself defined implicitly. Secondly $j$ can not be represented by a finite polynomial, otherwise the probabilistic collocation method would give an exact uncertainty result for sufficient order. Thirdly the resulting surface is sufficiently irregular that the uncertainty quantification problem is difficult if the variation of $\alpha$ is large enough. Finally we arc looking towards robust optimization problems: the optimization problem for $j$ has a unique minimum at $\alpha = 0$, while the optimization problem for $\mathbb{E}_A J$ has at least two local minima away from zero.

It should be emphasized that this problem is particularly demanding, containing local features and wide disparities in gradients. Typical uncertainty problems with relatively small parameter variations are unlikely to give such highly featured probability spaces, and will therefore generally require less samples for accurate reconstruction.

To begin we illustrate the effect of incorporating gradient information into the Kriging response surface for $j$, see Fig. 16. The response surface is built using Kriging, GEK and radial basis function (RBF) interpolation with 20 random Latin-hypercube samples. The RBF interpolation used Gaussian functions, the parameters of which were chosen by hand to produce reasonable results. With 20 samples, the lack of a sample within the peak on the left causes this feature to be absent in the RBF spline and Kriging model, though it is reproduced well when gradient information is added.

In order to estimate $\mathbb{E}_A J$ we employ the adaptive algorithm described in Sect. 3.3.3. As a starting point we sample $j$ and $\mathrm{d}j/\mathrm{d}\alpha$ at $M + 1$ points in the $M$-dimensional parameter space – 1-point at the mean of $A$ and an additional $M$ points taken from Latin-hypercube sampling weighted with the distribution of $A$. GEK is performed on this sample set in parameter space, and numerical integration is performed on the resulting response surface. This provides a first estimate of $\mathbb{E}_A J$, on which the error estimate (36) and the adaptation indicator (37) are calculated.

An example of the action of the method is shown for the 2d sandtimer problem with $A_1 \sim \text{Norm}(3,3)$ and $A_2 \sim \text{Norm}(2,3)$ in Fig. 17 after nine adaptation iterations. The upper left plot is again $j$, and upper right the GEK approximation to $j$ based on the samples shown by dots. Lower left is the Kriging error estimate, which is small near sample locations, and large far away – as expected. When this estimator is weighted with the pdf of $A$ the highly multi-modal adaptation indicator $\eta(\alpha)$ is obtained (lower right), which compromises extending sampling further away from the mean, and refining locally.

The above algorithm is applied to estimation of $\mathbb{E}J$ for the 2d and 4d sandtimer problem, with $A$ normally distributed with standard deviation $\sigma = \frac{1}{2}, 1, 3, 5,$ and mean $\bar{\alpha} = (3,2)$ in the 2d case, and $\bar{\alpha} = (3,2,1,2)$ in the 4d case. It is compared with Gauss rule integration on the probability space, probabilistic collocation and sparse grid integration using SMOLPACK [17,20]. Also considered adaptive Kriging without use of gradient information.

Results for the 2d case are plotted in Fig. 18, and for the 4d case in Fig. 19. The oscillatory behaviour of probabilistic collocation for larger $\sigma_A$ may be explained by the fact that a Taylor series is a poor approximation for $j$ over the larger range. In these cases the Gauss rule converges much more regularly, likely because the sample locations remain much more closely clustered around the mean, whereas with PC the sample points rapidly spread to the tails of the distribution and thereby the corners of the parameter space. This property of PC can be countered by using a *truncated* normal distribution as an input, in which case more regular convergence would expected to be observed. For small $\sigma_A$ PC beats all other methods convincingly.

The oscillations present in the Kriging and GEK convergence histories are partly due to the sensitivity of Kriging to the addition of new samples. The introduction of a single sample can cause the Kriging MLE to change smoothly, but in such a way that the location of the global maximum jumps from one location to another,

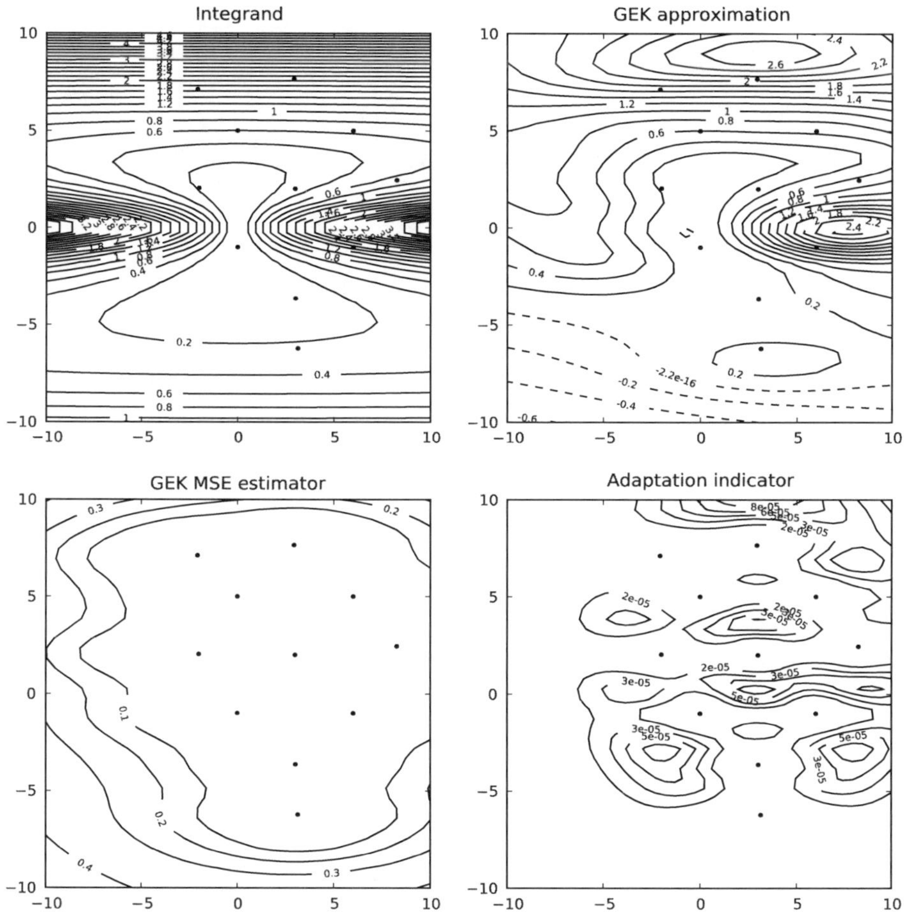


**Fig. 17** Adaptive sampling of $j$ when estimating $\mathbb{E}J$ for $A$ Norm$(\bar{\alpha}, 3)$ where $\bar{\alpha} = (3, 2)$. $j$ (*top left*), GEK reconstruction (*top right*), Kriging mean-squared error estimator (*lower left*), adaptation indicator (*lower right*)

changing the values of the correlation parameter $\theta$ and thereby the character of the entire surface. This leads to a jump or a spike in $\mathbb{E}J$. In an attempt to suppress this behaviour somewhat the choice of correlation parameter is under-relaxed from one iteration to the next, and is not taken as the optimum for the given data set. Even so oscillations are still highly in evidence, especially in the 4-d case.

Despite the corresponding difficulty in judging the convergence of these methods, they perform effectively, with GEK being of comparable or better efficiency than the other approaches in all cases. With comparable efficiency and additional flexibility, the Kriging-based approaches show promise as a replacement for the other methods in some circumstances.

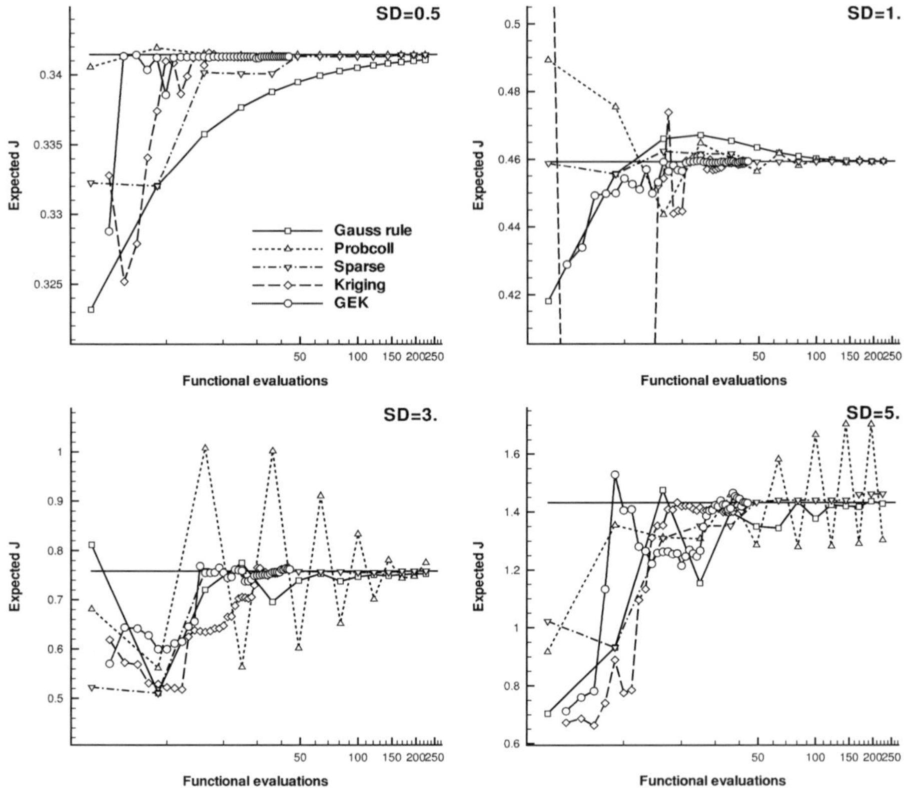


**Fig. 18** Convergence of estimates for $\mathbb{E}J$ for the 2-dimensional sandtimer model problem for $A$ normally distributed with various standard deviations

### 3.5.2 Piston Problem Redux

In order to draw a comparison between GEK and ASFE we now apply the GEK methodology to the piston problem of Sect. 2.3.1. In particular we consider the 2d case, with the same log-normal distributions for $u_{\text{piston}}$ and $p_{\text{pre}}$, and expect to observe the same mean and standard-deviation in the output. Unlike ASFE, GEK is not designed for discontinuous responses, it assumes a smooth surface, so we expect this to be a challenging test-case.

We choose to construct the response in the probability space (in which the shock becomes curved), and to use Cartesian sample points. In the case of discontinuities we observe that Kriging is substantially more stable when points are roughly equidistantly distributed. Samples from log-normal distributions are closely spaced near the mean, and sparse in the tails in the physical space, but of course uniform in the probability space – hence the choice of probability space to construct the GEK response. We also observe increased stability in the case of discontinuities on regular grids, and it is these best-case results that we present.

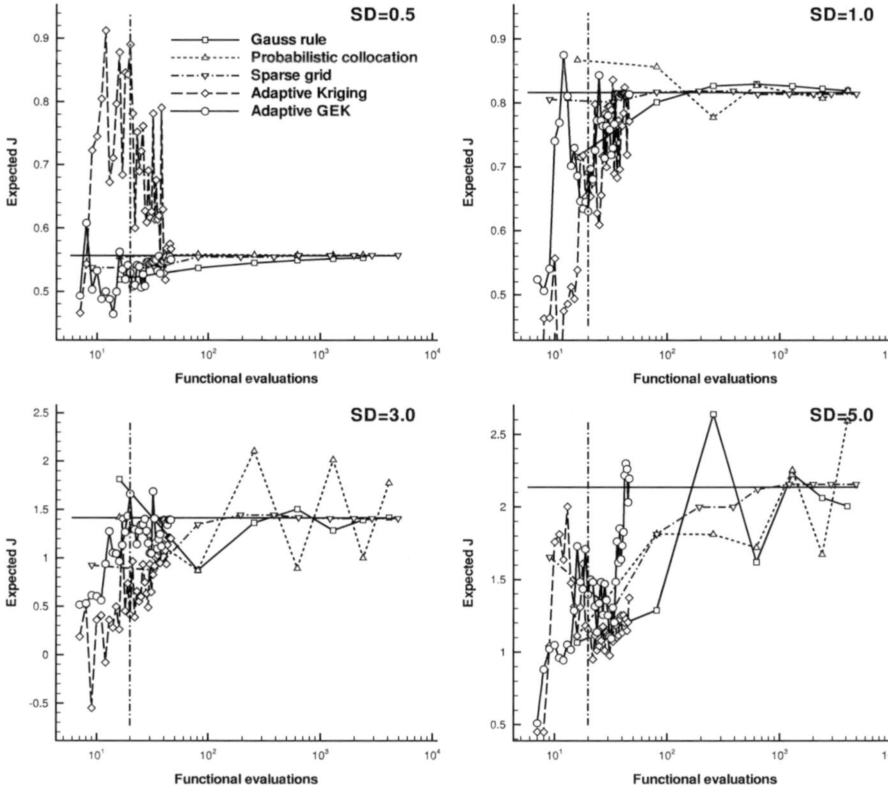


**Fig. 19** As for Fig. 18 for the 4-dimensional sandtimer problem

The GEK response surface for $16^2$ samples is plotted in Fig. 20 (right). The shock is resolved between two samples points with minimal oscillations, the region of zero response is flat, and the smooth region to the right of the shock contains details on the scale of the mesh-spacing. The key to this good shock resolution is a correlation length, $\hat{\theta}$, proportional to the mesh spacing. Near the shock no correlation length is short enough, and (32) gives roughly:

$$\hat{\theta} \approx 1.4N, \tag{38}$$

where $N$ is the total number of points, for a wide range of $N$. I.e. the "width" of the basis functions in (28) is roughly $1/\sqrt{N}$, and the method becomes very local, and thereby avoids oscillations. The correlation length itself remains a global quantity, and the large value due to the shock reduces the accuracy of GEK in smooth regions (see e.g. Fig. 20, and the contours in top-right corner of the probability space).

The convergence of mean and standard-deviation for GEK is given in Table 2. The accuracy is comparable to ASFE: slightly better for the mean and slightly worse

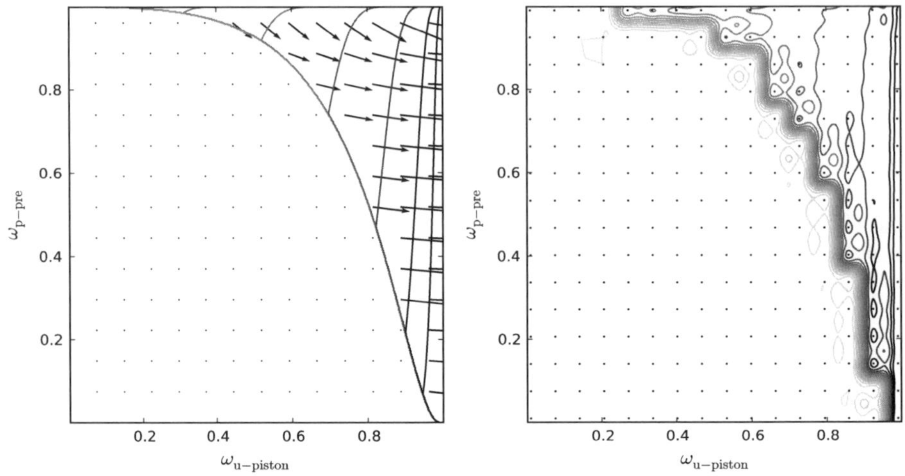


**Fig. 20** The piston problem response in the probability space. *Left*: the exact response (*arrows* indicate gradients, *dots* zero gradients). *Right*: the GEK response with $16^2$ samples, *dots* indicate sample locations

for the standard-deviation. The variance is consistently underestimated, testifying to the non-oscillatory reconstruction. Based on this study the method appears to converge, and it is easy to show that it will do so provided $\hat{\theta}$ continues to satisfy (38) as $N \to \infty$.

In conclusion GEK detects the presence of the shock when estimating the correlation length, and responds appropriately – testifying to the flexibility of the approach. As a caveat, note that for highly irregular meshes it is impossible to find a single global $\hat{\theta}$ that will give both shock resolution, and acceptable reconstruction in smooth regimes. A locally varying $\hat{\theta}$ is called for here.

### 3.5.3 Shape Uncertainty for the RAE 2822 Aerofoil

Now we consider the above algorithms applied to estimation of expected drag, $\mathbb{E}c_D$ for CFD simulation of the RAE 2822 aerofoil previously mentioned. The flow conditions are deterministic, and identical to those of Sect. 3.4.2. However the camber-line of the aerofoil is parameterized with $(a_1, \ldots, a_4)$ multiplying four Hicks-Henne bump functions. The aerofoil thickness is held constant. This parameterization is of relevance in optimization, where $c_D$ is to be minimized, and the lift $c_L$ is held constant by varying the angle-of-attack, $\phi$. Without this constraint, minimum drag would be achieved by eliminating lift and therefore also the induced drag. The constraint on the lift modifies the expression for the derivative of $c_D$ as follows:

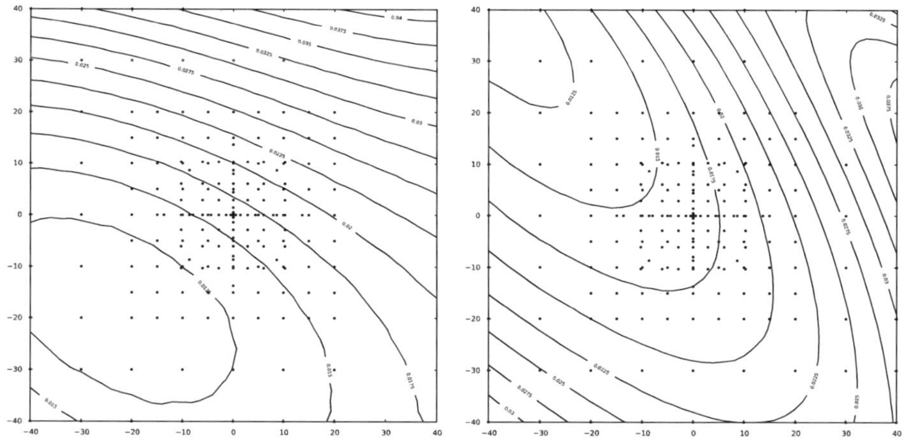

**Fig. 21** Slices through the design space of RAE 2822 test case, for the $a_2 = a_3 = 0$ plane (*left*), and the $a_0 = a_1 = 0$ plane (*right*). Contours plotted from a Kriging reconstructed surface, samples plotted as *black dots*

$$\nabla_L c_D = \nabla_\phi c_D - \left( \frac{\mathrm{d}c_D}{\mathrm{d}\phi} \Big/ \frac{\mathrm{d}c_L}{\mathrm{d}\phi} \right) \cdot \nabla_\phi c_L,$$

where $\nabla_L$ specifies the derivative at constant lift, and $\nabla_\phi$ at constant angle-of-attack. Therefore two adjoint solutions, one of lift and one for drag, are required for the gradient at each sample location.

Here this same parameterization is used as a test case for uncertainty in aerofoil geometry. It it is not intended to represent variation that might occur in practice, however the constraint on lift is still reasonable, as a pilot will always attempt to maintain constant lift, not constant angle-of-attack.

Two mean states are considered, the $a = 0$ case corresponding to the original RAE 2822, and $a = a_{\mathrm{opt}}$, which minimizes drag at the given flow conditions. The design space near the former is likely to be locally linear, while near the latter it should be locally quadratic – providing two cases of varying character. For the former we consider $a$ normally distributed with standard deviation 5, and for the latter standard deviations of 5 and 10.

Slices through the parameter space are plotted in Fig. 21, where a Kriging response surface has been used to compute contours, the samples of $c_D$ come from calculations used in the various uncertainty approximation techniques applied to this case. The parameter space appears to have a simple structure, but more slices would be needed to verify this.

Estimates of $\mathbb{E}c_D$ for various methods are plotted in Fig. 22, as well as $c_D$ for the zero and optimal states. In all three cases probabilistic collocation achieves an accurate solution with the least possible number of samples (16). Sparse grid integration obtains similar accuracy with 9 samples for the $\sigma = 5$ cases, but fails for the larger standard deviation. The adaptive GEK approach described here

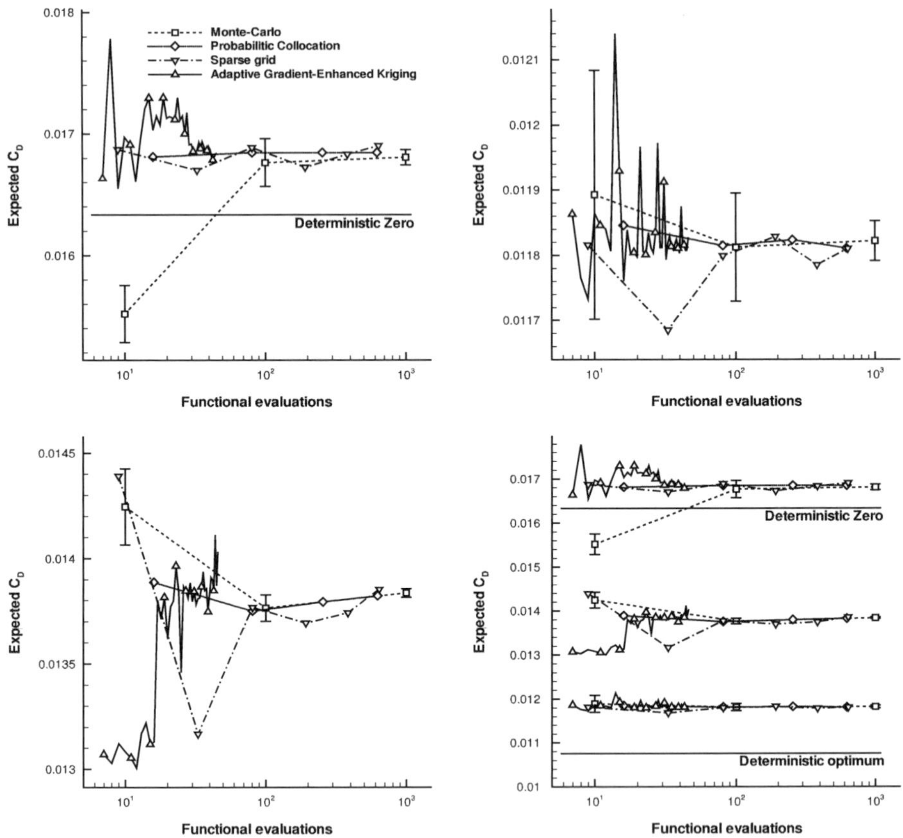


**Fig. 22** Convergence of various uncertainty quantification methods for the 4-parameter RAE 2822 problem. Three cases are considered: $\mu = 0, \sigma = 5$ (*top left*); $\mu = a_{\text{opt}}, \sigma = 5$ (*top right*); $\mu = a_{\text{opt}}, \sigma = 10$ (*bottom left*), where $a_{\text{opt}}$ is the deterministic minimum drag shape. *Bottom right* is a composite of the three cases – making the relative sizes of errors visible

performs comparably well with the more traditional methods in all three cases. Again the value of the Kriging correlation parameter was under-relaxed in order to smooth the convergence behaviour, and in this case this proved very effective. In particular the under-relaxation did not prevent the response surface in the case $a = a_{\text{opt}}, \sigma_a = 10$, jumping from an apparently incorrect state (before about 12 functional evaluations) to a correct state (thereafter).

In all cases the magnitudes of the errors made in $\mathbb{E}c_D$ are small in comparison to the variation of $c_D$ between the different test cases, suggesting that the 4-parameter uncertainty quantification may be accurate enough for many purposes, e.g. robust design in this case.

## References


1. Bassi, F., Crivellini, A., Rebay, S., Savini, M.: Discontinuous galerkin solution of the Reynolds-averaged Navier-Stokes and k-w turbulence model equations. Computers and Fluids **34**(4–5), 507–540 (2005). DOI DOI:10.1016/j.compfluid.2003.08.004
2. Chorin, A., Marsden, J.: A mathematical introduction to fluid mechanics. Springer-Verlag, New York (1979)
3. Demmel, J.W., Eisenstat, S.C., Gilbert, J.R., Li, X.S., Liu, J.W.H.: A supernodal approach to sparse partial pivoting. SIAM J. Matrix Analysis and Applications **20**(3), 720–755 (1999)
4. Dwight, R.: Efficiency improvements of RANS-based analysis and optimization using implicit and adjoint methods on unstructured grids. Ph.D. thesis, School of Mathematics, University of Manchester (2006)
5. Dwight, R.: Goal-oriented mesh adaptation using a dissipation-based error indicator. International Journal of Numerical Methods in Fluids **56**(8), 1193–1200 (2008). DOI: 10.1002/fld.1582
6. Dwight, R.: Heuristic a posteriori estimation of error due to dissipation in finite volume schemes and application to mesh adaptation. Journal of Computational Physics **227**(5), 2845–2863 (2008). DOI: 10.1016/j.jcp.2007.11.020
7. Dwight, R., Brezillon, J.: Effect of approximations of the discrete adjoint on gradient-based optimization. AIAA Journal **44**(12), 3022–3071 (2006)
8. Dwight, R., Brezillon, J.: Effect of various approximations of the discrete adjoint on gradient-based optimization. In: Proceedings of the 44th AIAA Aerospace Sciences Meeting and Exhibit, Reno NV, AIAA-2006-0690 (2006)
9. Giles, M., Duta, M., Muller, J.D., Pierce, N.: Algorithm developments for discrete adjoint methods. AIAA Journal **41**(2), 198–205 (2003)
10. H.-S., C., Alonso, J.: Using gradients to construct cokriging approximation models for high-dimensional design optimization problems. In: AIAA Paper Series, Paper 2002-0317 (2002)
11. Jones, D., Schonlau, M., Welch, W.: Efficient global optimization of expensive black-box functions. Journal of Global Optimization **13**, 455–492 (1998)
12. Lagarias, J., Reeds, J., Wright, M., Wright, P.: Convergence properties of the Nelder-Mead Simplex method in low dimensions. SIAM Journal on Optimization **9**(1), 112–147 (1998)
13. Maitre, O.L., Najm, H., Ghanem, R., Knio, O.: Multi-resolution analysis of Wiener-type uncertainty propagation schemes. J. Comput. Phys. **197**, 502–531 (2004)
14. McKay, M., Conover, W., Beckman, R.: A comparison of three methods for selecting values of input variables in the analysis of output from a computer code. Technometrics **21**, 239–245 (1979)
15. Nelder, J., Mead, R.: A simplex method for function minimization. Computer Journal **7**(4), 308–313 (1965)
16. O'Hagan, A., Oakley, J.E.: Probability is perfect, but we can't elicit it perfectly. Reliability Engineering and System Safety **85**(1–3), 239–248 (2004). DOI 10.1016/j.ress.2004.03.014
17. Petras, K.: Fast calculation of coefficients in the smolyak algorithm. Numerical Algorithms **26**(2), 93–109 (2001)
18. Rowan, T.: The subplex method for unconstrained optimization. Ph.D. thesis, Department of Computer Sciences, Univ. of Texas (1990)
19. Sacks, J., Welch, W., Mitchell, T., Wynn, H.: Design and analysis of computer experiments (with discussion). Statistical Science **4**, 409–435 (1989)
20. Smolyak, S.: Quadrature and interpolation formulas for tensor products of certain classes of functions. Doklady Akademii Nauk SSSR **4**, 240–243 (1963)
21. Webster, R., Oliver, M.: Geostatistics for Environmental Scientists, second edn. Wiley (2007). ISBN 0470028580
22. Wikle, C., Berliner, L.: A Bayesian tutorial for data assimilation. Physica D **230**, 1–16 (2007)
23. Xiu, D., Hesthaven, J.: High-order collocation methods for differential equations with random inputs. SIAM J. Sci. Comput. **27**, 1118–1139 (2005)

# Implementation of Intrusive Polynomial Chaos in CFD Codes and Application to 3D Navier-Stokes

**Chris Lacor, Cristian Dinescu, Charles Hirsch, and Sergey Smirnov**

## 1 Introduction

In present day, the technology reduction of product development costs and design cycle time are essential ingredients in a competitive industrial environment. E.g. in aeronautics the objectives set by the EU are to reduce aircraft development costs by resp. 20 and 50 % in the short and long term. Virtual prototyping and advanced design optimization, which largely depend on the predictive performance and the reliability of simulation software, are essential tools to reach this goal.

The quality assurance of the software tools is therefore essential. This includes the verification and the validation of the software as well as a control of the related simulation uncertainties.

In the field of aeronautics, this has led to a large number of EU funded projects over the last few decades, in the area of CFD with the goal of improving the reliability of models and algorithms, such as turbulence or combustion models and their validation, hereby aiming at the reduction of various sources of numerical errors and numerical uncertainties.

Apart from the numerical uncertainties, also uncertainties in operating conditions have to be considered as well as geometrical uncertainties due to imprecise geometrical definitions and/or fabrication variability. The main consequence is that, although actions towards the reduction of the uncertainties and towards extension

---

C. Lacor (✉)
Vrije Universiteit Brussel, Pleinlaan 2, 1050 Brussel, Belgium
e-mail: chris.lacor@vub.ac.be

C. Dinescu · C. Hirsch
NUMECA Int., Terhulpsesteenweg 189, 1170 Brussel, Belgium
e-mail: cristian.dinescu@numeca.be; charles.hirsch@numeca.be

S. Smirnov
Formerly at Vrije Universiteit Brussel, Pleinlaan 2, 1050 Brussel, Belgium

H. Bijl et al. (eds.), *Uncertainty Quantification in Computational Fluid Dynamics*, Lecture Notes in Computational Science and Engineering 92,
DOI 10.1007/978-3-319-00885-1_5, 

of the validation base are still required, most sources of uncertainties cannot be eliminated and have to be taken into account in the simulation process.

New methodologies are therefore required to incorporate the presence of uncertainties at the level of the simulation tools in order to

- Improve the predictive reliability of the simulation process
- Introduce the existence of these uncertainties in the decision process related to industrial design to come to so-called **robust designs**, reducing cost and failure risks.

This requires the following steps

- Identification, quantification and description of the uncertain parameters. Once the uncertainties are identified their description can be based on interval bounds, membership functions or probability density functions (PDF)
- Development of a non-deterministic methodology
- Integration of the non-deterministic approach into the design process

In the present contribution we will focus on the second item in the framework of CFD.

## 2 Polynomial Chaos Methodology

The Polynomial Chaos Methodology (PCM) is a very recent approach, which offers a large potential for CFD related non-deterministic simulations, as it allows the treatment of a large variety of stochastic variables and properties that can be described by probability density functions (PDF). The method is based on a spectral representation of the uncertainty where the basis polynomials contain the randomness, described by random variables $\xi$, and the unknown expansion coefficients are deterministic, resulting in deterministic equations. The methodology was originally formulated by Wiener [1], and was recently rediscovered and used for CFD applications by several groups, e.g. Xiu and Karniadakis [2], Lucor et al. [3], Le Maître et al. [4], Mathelin et al. [5], and Walters and Huyse [6].

In the original method of Wiener, Hermite polynomials are used in the expansion. However, other polynomials can also be used, see [2]. Each polynomial however has an optimal random distribution associated, which results in a fast convergence rate, according to the so-called **Askey scheme**. For example, Hermite polynomials for Gaussian distributions, Charlier polynomials for Poisson distributions, Laguerre polynomials for Gamma distributions, Jacobi polynomials for Beta distributions, etc. The appropriate polynomials are such that they are orthogonal with respect to the PDF of the distribution. Therefore, in case of less common distributions, an optimal PCM can always be found by constructing the polynomials via a Gram-Schmidt procedure, see Witteveen and Bijl [7]. If a random distribution is combined with a non-optimal polynomial, the projection of the uncertain input variables, say $k$, requires a transformation to transform the fully correlated random variables $k$ and $\xi$ to the same probability space, [8].

The dimension of the problem is determined by the number of independent random variables. In case of a random process (as opposed to random variable) a Karhunen-Loève expansion, [9, 10] can be applied to the correlation function to decompose the random input process in a set of independent random variables. The stochastic dimension depends on the correlation length, resulting in a high dimensional chaos expansion for processes with a very short correlation length (such as e.g. white noise), making the PCM more expensive. Non-Gaussian random processes are much more difficult to treat than Gaussian, [11]. In the former case mean and covariance are far from sufficient to completely specify the process. This remains an active area of research.

In the present contribution we will not go into the theoretical details of the PCM. This is already covered in other contributions. The main emphasis will be on the implementation issues of the PCM in the 3D turbulent Navier-Stokes equations, Lacor and Smirnov [12, 13], Dinescu et al. [14], and Onorato et al. [15]. This work was done in a recent EU STREP project, NODESIM-CFD 1/11/2006-28/2/2010, see www.nodesim.eu.

The PCM we consider is **intrusive**. This means that the deterministic CFD code has to be modified (significantly).

**Non-intrusive** PCM was also developed during the years. Basically two different approaches have been formulated: (i) the so-called projection method, which is based on a numerical evaluation of the Galerkin integrals, Le Maître et al. [4, 16] and Nobile et al. [17]; (ii) using a linear system or regression based on a selected set of points, Berveiller et al. [18] and Hosder et al. [19, 20]. The latter method is also known as point collocation or stochastic response surfaces.

In general the following methods can be used for the numerical quadrature, see [21]

- Full tensorization of 1D Gaussian quadrature, [16]
- Sampling with Monte Carlo simulation, [22, 23], or Latin Hypercube sampling, [24]
- Smolyak sparse grid, [25]. Adaptive algorithms have been developed recently that further reduce cost, [26, 27]

In the dimension reduction method an additive decomposition of a response is used that simplifies a multi-dimensional integration to multiple 1D integrations (Univariate Dimension Reduction, [28]) or to multiple 1D and 2D integrations (Bivariate Dimension Reduction, [29]).

Mathelin et al. [30], compare the number of operations required within the numerical quadrature approach with those of the intrusive PC. Triple products are always more costly to calculate with intrusive PC, whereas for double products it is the inverse. Xiu [11] mentions all of the existing collocation methods require solutions of (much) larger number of equations than that of intrusive PC, especially for higher dimensional random spaces. Furthermore, the aliasing errors in collocation methods can be significant, especially for higher dimensional random spaces. This indicates that the intrusive method offers the most accurate solutions involving the least number of equations in multi-dimensional random spaces, even though the equations are coupled, [11]. This is in accordance with Alekseev et al. [31], who mention intrusive methods are more flexible and in general more precise than non-intrusive methods.

Uncertainties can be classified in **epistemic** and **aleatory**, see Oberkampf et al. [32]. An epistemic uncertainty is defined as any lack of knowledge or information in any phase or activity of the modeling process. Aleatory uncertainty is the inherent variation associated with the physical system or the environment under consideration. The sources of aleatory uncertainty can commonly be singled out from other contributors to uncertainty by their representation as randomly distributed quantities. The mathematical representation most commonly used is then a probability distribution.

We assume that the uncertainties mainly come from the operating conditions. As an example, and to fix thoughts, think of a turbomachinery application e.g. a 3D compressor. Uncertainties can come from the Mach number at inlet, the flow angle at inlet, the total pressure at inlet, the static pressure at outlet, etc.

Additional uncertainties can come from the fluid properties (e.g. the viscosity or the conductivity) or from model constants (e.g. the constants used in various turbulence models).

A separate class are geometrical uncertainties, e.g. uncertainty on the shape of the compressor blade due to manufacturing tolerances. The treatment of such uncertainties needs a different approach than for uncertainties coming from operation conditions or fluid properties. Within an intrusive PCM approach the idea is to use a transformation such that the deterministic problem in a stochastic domain becomes a stochastic problem in a deterministic domain, e.g. Xiu and Tartakovsky [33]. An alternative is the use of a so-called fictitious domain method, [34, 35], or by introducing the uncertainty directly in the surface normals within a control volume approach, [36, 37].

A geometrical uncertainty is usually a random process. As mentioned above, a Karhunen-Loève expansion will therefore be needed.

We will assume that the PDF of the uncertainties, which are treated as random variables, is known. In view of the Askey scheme several distributions can be treated in an efficient way by proper choice of the corresponding polynomials. In the following, if not mentioned otherwise, Gaussian distributions will be assumed.

We will assume that we have in total $N$ **independent uncertain variables** related to the operating conditions, the fluid properties or to model constants. We will refer to these as **uncertain input variables**.

The case of random fields (or processes) will not be further discussed here, but as mentioned above, it can be reduced to the problem of independent random variables via the Karhunen-Loève expansion. An example will be given in Sect. 6.1.2.

The problem considered is then the propagation of the input uncertainties through the CFD model to evaluate the effect on the output. This is sometimes referred to as **the forward uncertainty problem** as opposed to the **inverse uncertainty problem**, where the unknown model parameters have to be estimated based on observational data of the output, which can be of experimental and computational nature.

The required effort for extending a deterministic CFD code with intrusive PC depends on the characteristics of the code: computer language, structured/unstructured, handling of data storage etc.

In the present case we started from the commercial code Fine/Turbo of NUMECA. It is written in Fortran and based on structured grids within a multiblock-multigrid approach. The data storage is very flexible so that by adapting the number of unknowns, space for the additional unknowns is automatically generated. Since the loop over the unknowns is inside the grid (for multigrid) and block loop, only minor changes are needed to have multiblock-multigrid available, once the single grid, single block works.

The main effort goes therefore into the adaptation of the calculation of convective and viscous terms and of the routines for the boundary conditions. Introducing additional uncertainties only requires adaptation of the related boundary condition routines; the calculation of convective and viscous terms as well as of the matrices $M_{ijk}$ and $L_{ijkl}$ will be automatically adapted. Note that, for computational efficiency, the (non-zero) elements of these matrices are stored; by using the pseudo-spectral approach the use of the $L$ matrix can be avoided saving a lot of memory.

All implementations were done by the last author. The number of additional lines of code is very limited, compared to the length of the original, deterministic code. However, changes are not restricted to a local part of the code. This increases the risk of introducing bugs and requires someone who is very familiar with all aspects of the code. This is a big disadvantage compared to non-intrusive PC and the main reason that the application of intrusive PC in commercial codes is very limited.

## 3 Mathematical Formulation

The $N$ uncertain input variables will cause all flow variables to be uncertain too. In PCM all uncertain variables are decomposed in a basis of complete orthogonal polynomials, the so-called **polynomial chaos expansion** (PCE). E.g. for the x-component of the velocity $u$ the PCE gives

$$u(\mathbf{x}, t, \theta) = \sum_{i=0}^{P} u^i(\mathbf{x}, t)\psi_i(\boldsymbol{\xi}(\theta)) \tag{1}$$

$\theta$ is the outcome of an experiment related to a random event. The unknown $u^i$ are deterministic coefficients and represent the random mode $i$ of the velocity component $u$. $\psi_i$ are the orthogonal polynomials which are function of $\boldsymbol{\xi} \equiv (\xi_1, \xi_2, \cdots, \xi_N)$ where $\xi_j(\theta)$ is a random variable. $N$ is the number of input uncertainties which is also the **number of random dimensions**. The total number of terms $P+1$, used in (1) depends on the highest order of the polynomial that is used (denoted $p$) and on the number of random dimensions. One has, cf. [8]

$$P+1=\frac{(N+p)!}{N!p!} \tag{2}$$

Orthogonality means that

$$<\psi_i\psi_j>=<\psi_i^2>\delta_{ij} \tag{3}$$

with <> denoting the inner product

$$<\psi_i\psi_j>\equiv\int W(\boldsymbol{\xi})\psi_i(\boldsymbol{\xi})\psi_j(\boldsymbol{\xi})d\boldsymbol{\xi} \tag{4}$$

with $W$ the weighting function.

In the original method of Wiener, the trial basis are **Hermite** polynomials. These are optimal for random variables with **Gaussian** distribution. The reason is that the weighting function $W$ in the orthogonality condition for Hermite polynomials is Gaussian i.e.

$$W_{Herm}(\boldsymbol{\xi})\equiv\frac{1}{\sqrt{(2\pi)^N}}exp(-\frac{1}{2}\boldsymbol{\xi}^{\tau}\boldsymbol{\xi}) \tag{5}$$

# 4 Application to 3D Compressible Navier-Stokes

The deterministic 3D compressible Navier-Stokes system can be written as

$$\frac{\partial\rho}{\partial t}+\frac{\partial}{\partial x_i}(\rho u_i)=0 \tag{6}$$

$$\frac{\partial(\rho u_i)}{\partial t}+\frac{\partial}{\partial x_j}\left(\rho u_i u_j\right)=-\frac{\partial p}{\partial x_i}+\frac{\partial\tau_{ij}}{\partial x_j} \tag{7}$$

$$\frac{\partial(\rho E)}{\partial t}+\frac{\partial}{\partial x_j}\left(\rho u_j E\right)=-\frac{\partial}{\partial x_j}\left(u_j p\right)+\frac{\partial}{\partial x_j}\left(\tau_{ji}v_i\right) \tag{8}$$

where $\rho$ is the density, $u_i$ the i-th velocity component, $p$ the pressure, $E$ the total energy and $\tau_{ij}$ component $ij$ of the stress tensor. The spectral expansion of type (1) is now introduced. Since the unknowns can be considered as $\rho$, $u_i, i=1,2,3$ and $E$ one would, at first sight, apply the decomposition to these variables. However looking at the energy equation this would introduce a coupling between the PCE coefficients $\rho^i$ of the density and those of the total energy $E^i$. Indeed Eq. (8) becomes:

$$\sum_k \sum_l \psi_k \psi_l \frac{\partial \rho^k E^l}{\partial t} + \sum_k \sum_l \sum_m \psi_k \psi_l \psi_m \frac{\partial}{\partial x_j} \left( \rho^k u_j^l E^m \right) =$$
$$- \sum_k \sum_l \psi_k \psi_l \frac{\partial}{\partial x_j} \left( u_j^k p^l \right) + \sum_k \sum_l \psi_k \psi_l \frac{\partial}{\partial x_j} \left( \tau_{ji}^k v_i^l \right) \tag{9}$$

Note that all summations, also in the following equations, unless mentioned otherwise, always go from 0 to $P$.

To get equations for the spectral components, a Galerkin projection is used. Multiplication of Eq. (9) with $\psi_n$ and performing the inner product of Eq. (4) gives

$$\sum_k \sum_l M_{kln} \frac{\partial \rho^k E^l}{\partial t} + \sum_k \sum_l \sum_m L_{klmn} \frac{\partial}{\partial x_j} \left( \rho^k u_j^l E^m \right) =$$
$$- \sum_k \sum_l M_{kln} \frac{\partial}{\partial x_j} \left( u_j^k p^l \right) + \sum_k \sum_l M_{kln} \frac{\partial}{\partial x_j} \left( \tau_{ji}^k v_i^l \right) \tag{10}$$

with

$$M_{kln} \equiv \frac{\langle \psi_k \psi_l \psi_n \rangle}{\langle \psi_n \psi_n \rangle} \tag{11}$$

and

$$L_{klmn} \equiv \frac{\langle \psi_k \psi_l \psi_m \psi_n \rangle}{\langle \psi_n \psi_n \rangle} \tag{12}$$

Equation (10) represents $P+1$ equations (if $n$ varies from 0 to $P$) for the unknowns $E^l, l = 0, \cdots, P$. Note that these $E^l$ are coupled to $\rho^k$ which complicates the solution of this equation.

It is therefore more convenient to consider $\rho$, $u_i, i = 1, 2, 3$ and $\rho E$ as unknowns. The energy equation, after introducing the PCE and the Galerkin projection, reads

$$\frac{\partial (\rho E)^m}{\partial t} + \sum_k \sum_l M_{klm} \frac{\partial}{\partial x_j} \left( (\rho E)^k u_j^l \right) =$$
$$- \sum_k \sum_l M_{klm} \frac{\partial}{\partial x_j} \left( u_j^k p^l \right) + \sum_k \sum_l M_{klm} \frac{\partial}{\partial x_j} \left( \tau_{ji}^k v_i^l \right) \tag{13}$$

For $l = 0, \cdots, P$ this gives $P$ equations which can directly be solved for the unknowns $(\rho E)^l$. Also note that the convection term contains far less terms than in Eq. (10); instead of a triple summation we only have a double summation.

At this stage one would be tempted to try the same approach for the momentum equation and thus use a PCE of $\rho u_i$ rather than of $u_i$. However the convection terms of momentum contain products $\rho u_i u_j$ which should then be rewritten as $\frac{\rho u_i \rho u_j}{\rho}$ to do the PCE. This brings in a division which is more difficult to deal with, see e.g. further down with the determination of the PCE coefficients of temperature. Therefore this avenue is not further explored. Hence, applying the PCE to $\rho$ and $u_i$ the momentum equations become, after Galerkin projection:

$$\sum_k \sum_l M_{kln} \frac{\partial \rho^k u_i^l}{\partial t} + \sum_k \sum_l \sum_m L_{klmn} \frac{\partial}{\partial x_j} \left( \rho^k u_i^l u_j^m \right) = -\frac{\partial p^n}{\partial x_i} + \frac{\partial \tau_{ij}^n}{\partial x_j} \tag{14}$$

Finally for the continuity equation one obtains

$$\frac{\partial \rho^m}{\partial t} + \sum_k \sum_l M_{klm} \frac{\partial}{\partial x_i} \left( \rho^k u_i^l \right) = 0 \tag{15}$$

Equations (15), (14) and (13) constitute the non-deterministic Navier-Stokes system. To solve it, it will be discretized using the Finite Volume method.

The coefficients $p^m$ of the PCE of the pressure can be found from the relation linking pressure to total and kinetic energy:

$$p = (\gamma - 1) \left( \rho E - \frac{1}{2} \rho u_i u_i \right) \tag{16}$$

Using PCE in both sides of this equation and applying Galerkin projection, yields

$$p^n = (\gamma - 1) \left( (\rho E)^n - \sum_k \sum_l \sum_m L_{klmn} \frac{1}{2} \rho^k u_i^l u_i^m \right) \tag{17}$$

Similarly the components $\tau_{ij}^m$ of the stress tensor follow from its definition:

$$\tau_{ij} = \mu ( \frac{\partial u_i}{\partial x_j} + \frac{\partial u_j}{\partial x_i} - \frac{2}{3} \frac{\partial u_k}{\partial x_k} \delta_{ij} ) \tag{18}$$

One obtains

$$\tau_{ij}^m = \mu ( \frac{\partial u_i^m}{\partial x_j} + \frac{\partial u_j^m}{\partial x_i} - \frac{2}{3} \frac{\partial u_k^m}{\partial x_k} \delta_{ij} ) \tag{19}$$

Note that it is assumed here that the viscosity $\mu$ is deterministic. If this is not the case, it has to be expanded also and the equation for $\tau_{ij}^n$ now becomes

$$\tau_{ij}^n = \sum_l \sum_m M_{lmn} \mu^l ( \frac{\partial u_i^m}{\partial x_j} + \frac{\partial u_j^m}{\partial x_i} - \frac{2}{3} \frac{\partial u_k^m}{\partial x_k} \delta_{ij} ) \tag{20}$$

The static temperature can be obtained from pressure and density via the perfect gas law:

$$p = \rho r T \tag{21}$$

with $r$ the gas constant. Introducing the spectral decomposition and applying the Galerkin projection one obtains

$$p^k = r \sum_j \left( \sum_i \rho^i M_{ijk} \right) T^j \tag{22}$$

Knowing the PCE coefficients $p^k$ and $\rho^i$, Eq. (22) is a linear system that can be solved for the unknown $T^j$.

# 5 Simplifications of the Non-deterministic Navier-Stokes Equations

Several simplifications can be introduced to make the solution more efficient.

## *5.1 Pseudo Spectral Approach*

The terms in the non-deterministic equations containing the $M$ and the $L$ matrices contain resp. double and triple summations where the indices vary from 0 to $P$. The number of terms therefore grows quickly with $P$ especially for the triple summations.

On the other hand the matrices contain also many zero entries. Since the matrices depend only on the polynomials they can be calculated upfront. In order to save CPU time zero entries are ignored and the summation is only over the non-zero entries.

Even then many terms have to calculated. Table 1 shows the number of non-zero entries in $M$ and $L$ for the case of 3 uncertainties ($N = 3$) and an increasing polynomial order $p$.

It is observed that the number of non-zero entries of the $L$ matrix gets quickly very high. The calculation of the terms containing this matrix, i.e. in the momentum equation (14) and in the calculation of the PCE coefficients of p, Eq. (17), becomes therefore quite CPU intensive (unless $P$ is small).

An alternative is then to use a **pseudo spectral approach** for the triple products. The latter are then calculated in two steps, avoiding the use of $L$. To fix thoughts consider the calculation of $\rho u^2$ where $u$ is the x-component of velocity, $u \equiv u_1$. In a first step the PCE expansion of $\rho u$ is calculated:

**Table 1** Non-zero entries in matrices $M$ and $L$ for $N = 3$ and varying polynomial order $p$

| Order | No. polynomials | No. non-zero elements $M$ | No. non-zero elements $L$ |
|---|---|---|---|
| 1 | 4 | 10 | 40 |
| 2 | 10 | 82 | 820 |
| 3 | 20 | 382 | 9,584 |
| 4 | 35 | 1,525 | 75,416 |
| 5 | 56 | 4,783 | 443,888 |
| 6 | 84 | 13,592 | 2,092,888 |
| 7 | 120 | 33,752 | 8,296,960 |

$$\rho u = \sum_k \sum_l \rho^k u^l \psi_k \psi_l \tag{23}$$

A simplified expression will now be used, i.e.

$$\rho u = \sum_k (\rho u)^k \psi_k \tag{24}$$

Equating the Galerkin projection of both equations leads to

$$(\rho u)^m = \sum_k \sum_l M_{klm} \rho^k u^l \tag{25}$$

Next the PCE expansion of the triple product can be calculated

$$\rho u^2 = (\rho u) u = \sum_k \sum_l (\rho u)^k u^l \psi_k \psi_l \tag{26}$$

Again a simplified expression will be used

$$\rho u^2 = \sum_k (\rho u^2)^k \psi_k \tag{27}$$

Equating the Galerkin projection of both equations, one obtains

$$(\rho u^2)^m = \sum_k \sum_l M_{klm} (\rho u)^k u^l \tag{28}$$

The momentum equation (14) is then simplified to

$$\sum_k \sum_l M_{kln} \frac{\partial \rho^k u_i^l}{\partial t} + \sum_k \sum_l M_{kln} \frac{\partial}{\partial x_j} \left( (\rho u_i)^k u_j^l \right) = -\frac{\partial p^n}{\partial x_i} + \frac{\partial \tau_{ij}^n}{\partial x_j} \tag{29}$$

A similar approach can be used to calculate $p^k$. Equation (17) becomes

$$p^n = (\gamma - 1) \left( (\rho E)^n - \sum_k \sum_l M_{kln} \frac{1}{2} (\rho u_i)^k u_i^l \right) \tag{30}$$

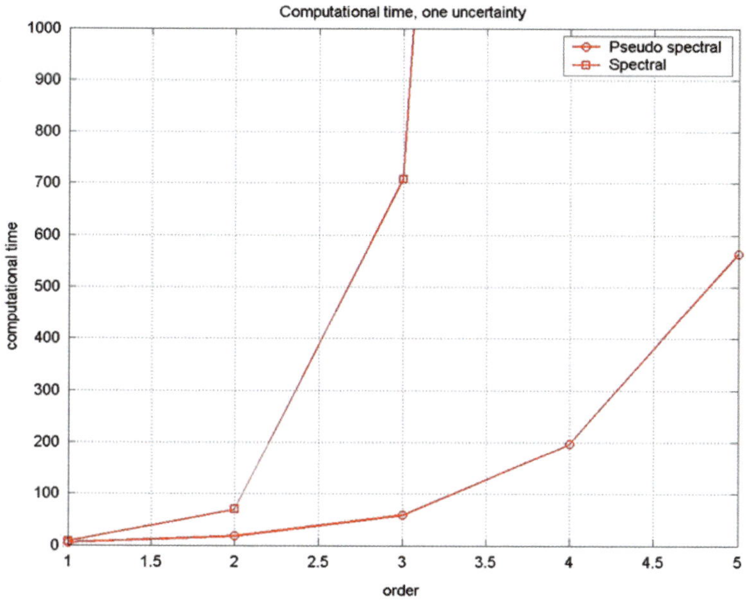


**Fig. 1** Comparison of CPU time for full and pseudo spectral approach for $N = 3$ and $p$ varying. Application is a quasi-1D nozzle

Figure 1 compares the CPU effort using the full spectral and the pseudo spectral approach for the application of a quasi-1D nozzle with 3 uncertainties and this for an increasing polynomial order $p$.

It is observed that the CPU of the full spectral approach grows exponentially fast. E.g. for $p = 3$ the CPU is about ten times higher than for the pseudo spectral approach.

## *5.2 Steady State Solutions of the Navier-Stokes Equations*

In many applications we are interested in the steady solution only. In this case a **preconditioning** of the system can be used. A very simple preconditioning is obtained by replacing the time derivatives in the momentum equations (14) with the time derivative of the PCE coefficients $u_i^n$ of the velocity components $u_i$. Equation (14) is then reduced to

$$\frac{\partial u_i^n}{\partial t} + \sum_k \sum_l \sum_m L_{klmn} \frac{\partial}{\partial x_j} \left( \rho^k u_i^l u_j^m \right) = -\frac{\partial p^n}{\partial x_i} + \frac{\partial \tau_{ij}^n}{\partial x_j} \tag{31}$$

The above equation can then be interpreted as an equation for $u_i^n$ i.e. the different PCE coefficients are not coupled anymore as in (14), largely simplifying the solution.

## 5.3 Truncation of Higher Order Terms

To further reduce CPU time one can truncate the PCE of the nonlinear terms. To fix thoughts, consider for instance the convection term in the non-deterministic energy equation, (13):

$$\sum_{k=0}^{P}\sum_{l=0}^{P} M_{klm}\frac{\partial}{\partial x_j}\left((\rho E)^k u_j^l\right) \tag{32}$$

Here we propose to simplify it to

$$\sum_{k+l\leq P} M_{kln}\frac{\partial}{\partial x_j}\left((\rho E)^k u_j^l\right) \tag{33}$$

and similar for the other summation terms in the Navier-Stokes equations.

This is motivated by the sparsity of the PC expansion: the size of the expansion coefficients quickly goes down with increasing order. This feature is also exploited in the non-intrusive compressive sampling approach of Doostan and Owhadi [38].

This approach was tested for efficiency and accuracy on a 1D Burgers equation in the domain $0 < x < 5$:

$$\frac{\partial u}{\partial t} + \frac{1}{2}\frac{\partial u^2}{\partial x} = \nu\frac{\partial^2 u}{\partial x^2} \tag{34}$$

The initial and boundary conditions are

$$\begin{aligned} u(x,0) &= 1 + exp\left[-5(x-1.5)^2\right] \\ u(0,t) &= 0 \\ u(5,t) &= 0 \end{aligned} \tag{35}$$

The viscosity is considered as uncertain (n $=$ 1) with a Gaussian distribution and with mean 0.05 and standard deviation $\sigma = 0.0025$.

Figure 2 shows the comparison in CPU time between the full expansion and the truncated expansion for increasing order $p$. It is noticed that the gain increases with the order. It is to be expected that for an increased number of uncertainties the gain will be much more.

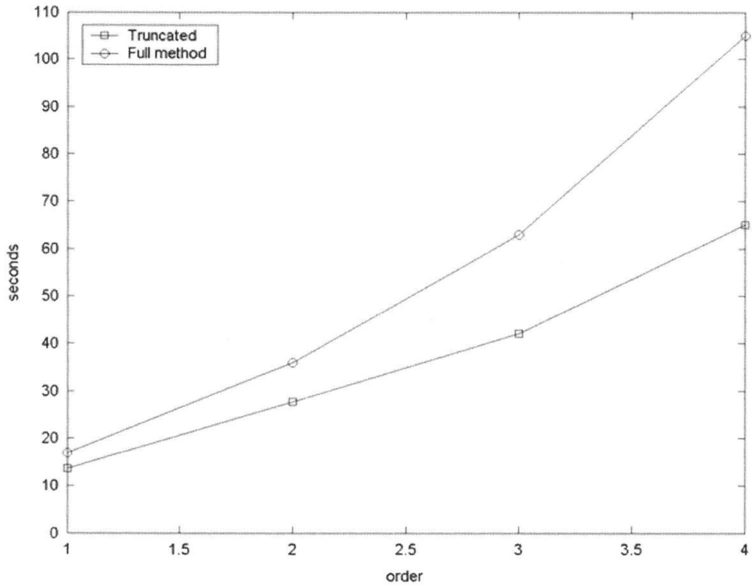


**Fig. 2** Comparison of CPU time for full and truncated spectral expansion for $N = 1$ and $p$ varying. Application is Burgers equation

Of course it has to be checked if the truncation does not have a negative impact on the quality of the results. To this end, the variance and the skewness of the output were checked. It is easy to show that for any quantity $a$ with PCE coefficients $a^k$ i.e.

$$a = \sum_k a^k \psi_k \tag{36}$$

the average $\overline{a}$, variance $\sigma_a^2$, and skewness can be obtained as

$$\overline{a} = E[a] = a^0 \tag{37}$$

$$\sigma_a^2 \equiv E\left[(a - \overline{a})^2\right] = \sum_k (a^k)^2 \langle \psi_k \psi_k \rangle - \overline{a}^2 \tag{38}$$

$$E\left[(a - \overline{a})^3\right] = \sum_i \sum_j \sum_k \langle \psi_i \psi_j \psi_k \rangle a^i a^j a^k - 3\overline{a}\sigma_a^2 - \overline{a}^3 \tag{39}$$

Figures 3 and 4 compare the results for the variance of the output resp. when the non-deterministic solution was obtained without and with truncation. Also the solution obtained with a Monte Carlo method is indicated. It is observed that all solutions are identical.

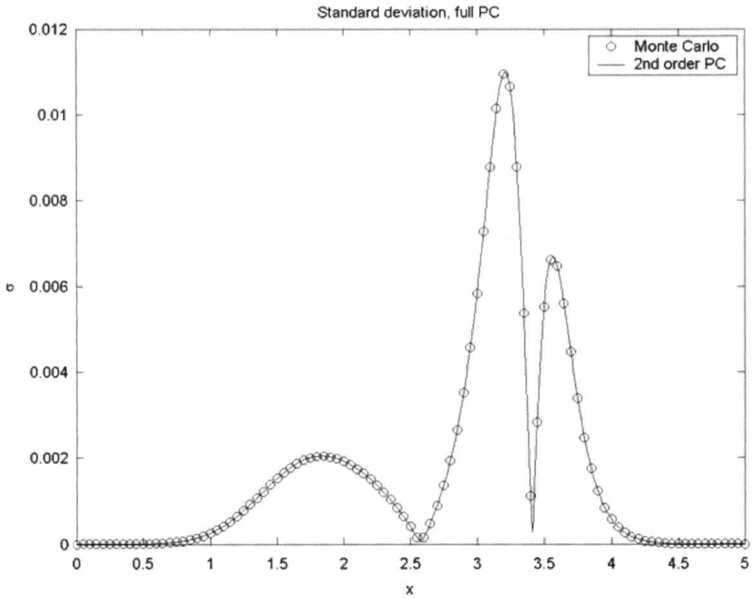


**Fig. 3** Comparison of variance of the output for full spectral expansion and comparison with Monte Carlo result. Application is Burgers equation

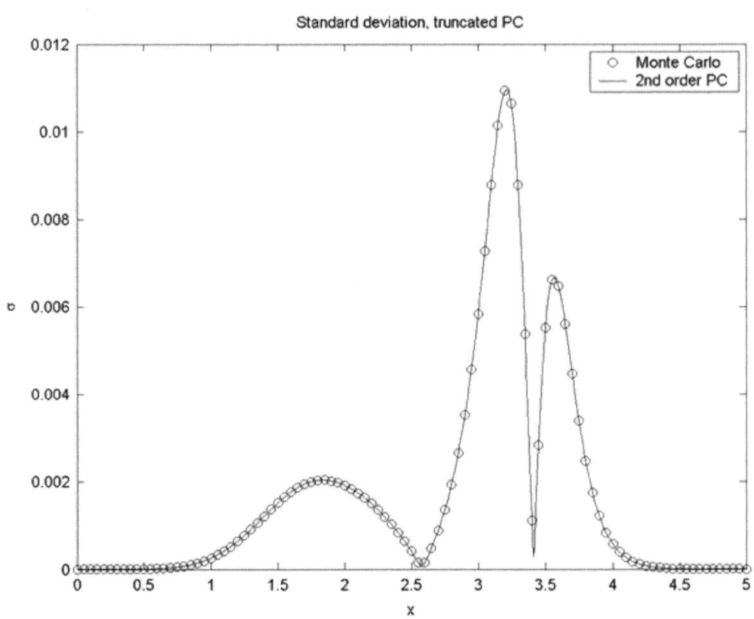


**Fig. 4** Comparison of variance of the output for truncated spectral expansion and comparison with Monte Carlo result. Application is Burgers equation

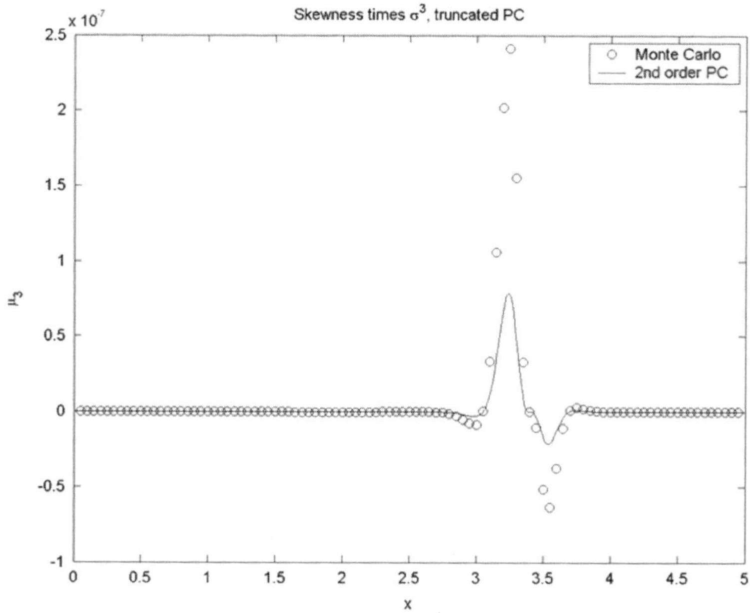


**Fig. 5** Skewness of the output calculated with truncated spectral expansion and comparison with Monte Carlo result. Application is Burgers equation

Figure 5 shows the skewness of the output, see Eq. (39), but with the truncation also applied in the skewness formula. The results are now completely off. It is therefore essential that truncation is not used in the postprocessing.

# 6 Applications

## *6.1 Supersonic 1D Nozzle*

A supersonic flow in a nozzle is considered. The cross-sectional area of the nozzle is given by the following relations

$$
\begin{aligned}
A &= 1 \quad 0 \leq x < 1 \\
A &= 1 + \Delta A \frac{1 - cos\left(\pi \frac{x-1}{3}\right)}{2} \quad 1 \leq x < 4 \\
A &= 1 + \Delta A \quad 4 \leq x \leq 5
\end{aligned}
\tag{40}
$$

with $\Delta A = 0.2$.

The flow is described by the quasi-1D Euler equations.

### 6.1.1 Uncertainty on Inlet Variables

All three inlet variables are considered uncertain:

$$
\begin{aligned}
u_{in} &= u^0 + u^1\xi_1 \\
\rho_{in} &= \rho^0 + \rho^1\xi_2 \\
p_{in} &= p^0 + p^1\xi_3
\end{aligned}
\tag{41}
$$

and

$$u^0 = 450\,\mathrm{m/s} \quad u^1 = 9\,\mathrm{m/s} \tag{42}$$

$$p^0 = 101{,}325\,\mathrm{N/m^2} \quad p^1 = 2{,}026,5\,\mathrm{N/m^2} \tag{43}$$

$$\rho^0 = 1\,\mathrm{kg/m^3} \quad \rho^1 = 0.02\,\mathrm{kg/m^3} \tag{44}$$

As the flow is supersonic, these values are imposed at inlet. Since a purely Gaussian distribution is assumed the PCE coefficients $u^i$, $p^i$, $\rho^i$, $i > 1$ are assumed to be zero at inlet. From relation (17) the PCE coefficients $(\rho E)^k$ can be calculated. At outlet, all variables are extrapolated.

The computational domain $0 \leq x \leq 5$ is meshed with a uniform grid of 500 equally sized cells of size $\Delta x = 0.01$.

Figure 6 shows the deterministic solution for the velocity. As the nozzle is divergent and the flow supersonic, the flow accelerates in the nozzle.

Figures 7 and 8 show the calculated standard deviation of the velocity with first and second order PC and compared to results of a Monte Carlo simulation. As can be observed, the standard deviation remains unchanged in the region $0 \leq x \leq 1$ . The reason is that the nozzle section is constant in this region and hence the flow is uniform. Note that in the simulations either the conservative variables $\rho$, $\rho u$, $\rho E$ or the so-called mixed variables $\rho.u$, $\rho E$ were stored. This can make a difference in CPU time but is irrelevant for the final results.

Though the results seem the same, there are some minor differences as can be observed in Fig. 9 where the error of the variance (as compared to the Monte Carlo result) is plotted for increasing PC order $p$. Again the difference between conservative and mixed variables is irrelevant here.

### 6.1.2 Geometrical Uncertainty

Since in the quasi-1D Euler equations the nozzle cross section $A(x)$ appears explicitly, it can be treated as an uncertain variable much in the same way as the uncertain flow variables at inlet.

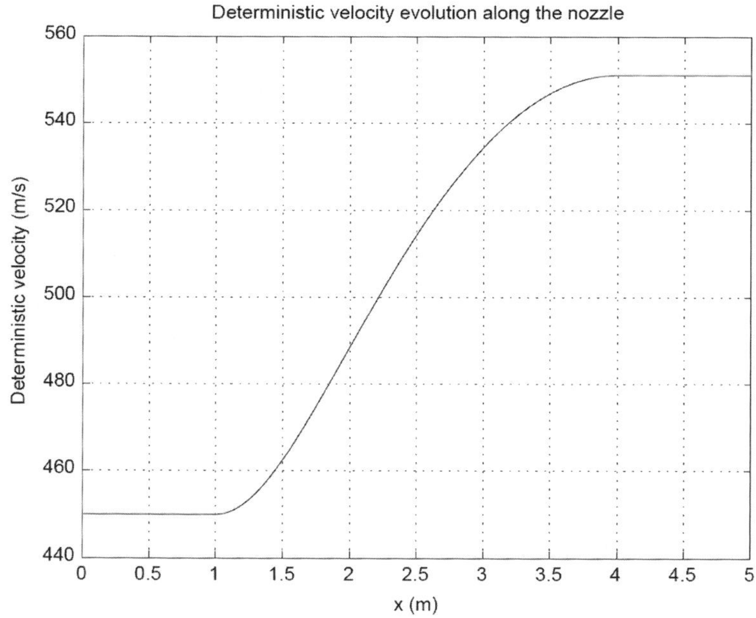


**Fig. 6** Deterministic solution for the velocity in the 1D nozzle)

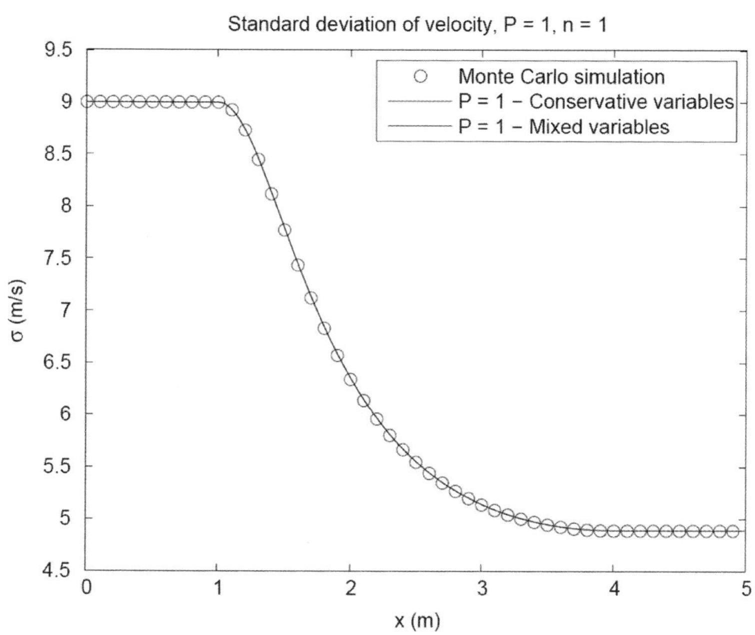


**Fig. 7** Standard deviation of velocity obtained with first order PC and a Monte Carlo method (1D nozzle with 3 uncertainties)

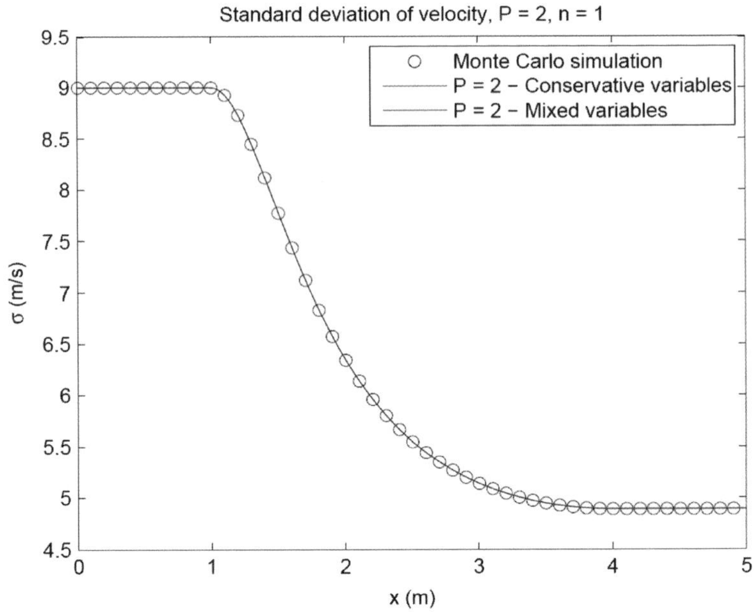


**Fig. 8** Standard deviation of velocity obtained with second order PC and a Monte Carlo method (1D nozzle with 3 uncertainties)

The flow variables are now considered deterministic. The nozzle cross section is considered as an uncertain Gaussian field with the following covariance function

$$C_{AA}(x_1, x_2) = \sigma_A^2 exp\left(\frac{-|x_1 - x_2|}{b}\right) \tag{45}$$

The correlation length $b$ is set to 20. Different values for the variance $\sigma_A^2$ will be considered ranging from 0.01 to 0.1. A Karhunen-Loève (K-L) expansion is used to describe the uncertain field $A(x)$ with a number of independent random variables. Since the eigenvalues in the K-L expansion rapidly decrease only few random variables are needed. In the simulations the number $N$ was varied from 2 to 5. In all cases a PC expansion of order $p = 1$ is used. For more details see Raets [39].

Figure 10 shows the influence of varying the variance of the correlation on the mean velocity. The number of random variables in the K-L expansion is limited to 2, i.e. $N = 2$. A clear difference with the deterministic solution is observed when $\sigma_A = 0.1$. For the smaller values 0.01 and 0.02 there is only a minor difference, which is only visible on a zoom (not shown).

The effect on the standard deviation of the velocity is shown in Fig. 11.

The effect of the number of independent random variables $N$ in the K-L expansion is investigated below.

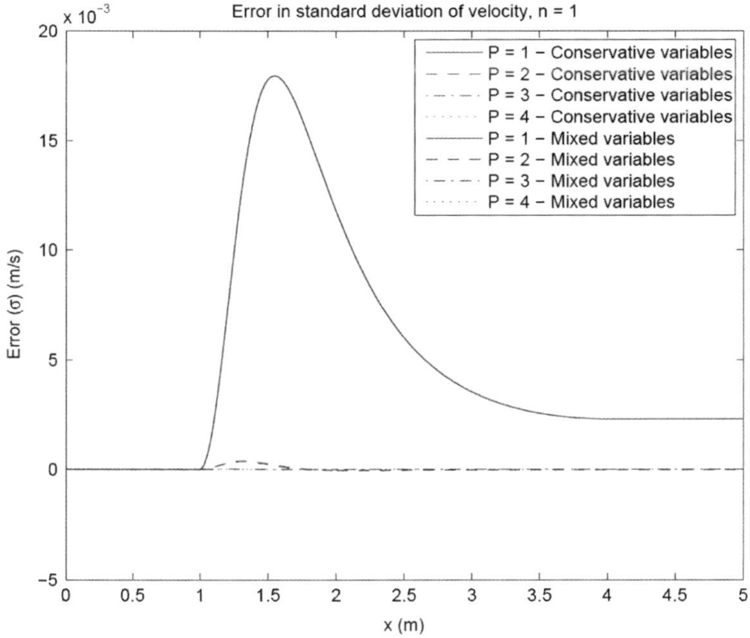


**Fig. 9** Error in standard deviation of velocity for PC of order $p = 1$ to $p = 4$ as compared to Monte Carlo (1D nozzle with 3 uncertainties)

Figures 12 and 13 show the mean velocity for $N$ varying from 2 to 5, $\sigma_A = 0.02$ and $p = 1$. The solutions are compared with a reference Monte Carlo (MC) solution. The differences are very small, even if $N = 1$, and can only be observed in the zoom on the right. With increasing $N$ the PC solution comes closer to the MC solution.

Finally the effect of $N$ on the standard deviation of the velocity is shown in Fig. 14. The solution for $N = 1$ is completely off. For increasing $N$ the PC solution converges towards the reference MC solution.

## 6.2 *Lid Driven Cavity Flow*

The second test case is a lid driven cavity flow. The geometry is a square box $0 < x, y < 1$ of which the upper wall ($y = 1$) is moving with a velocity $U = 1$ m/s. The Reynolds number is set to 100 so that the flow can be considered laminar. The 2D laminar Navier-Stokes equations are therefore solved with the PC method. The mesh consists of $40 \times 40$ uniform cells. The viscosity is considered as uncertain variable with a standard deviation of 10 % of its mean value. A second order pseudo-spectral PC approach is used. Figures 15 and 16 show the calculated mean velocity

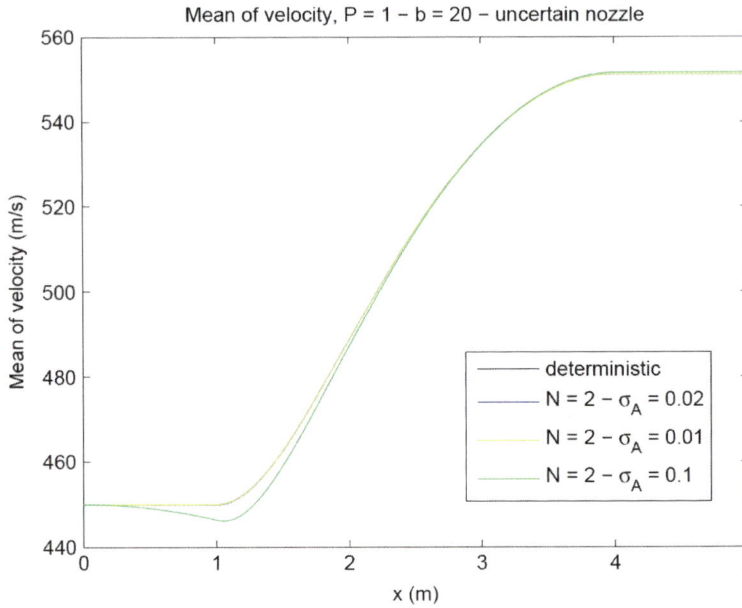


**Fig. 10** Nozzle with geometric uncertainty: mean velocity for covariance function with different variance $\sigma_A$. PC order $p = 1$, number of random variables $N = 2$

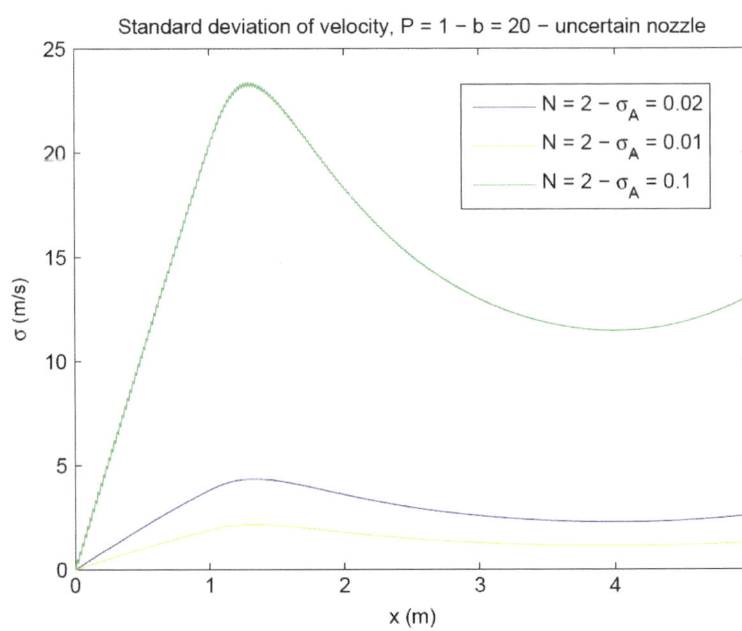


**Fig. 11** Nozzle with geometric uncertainty: standard deviation on the velocity for covariance function with different variance $\sigma_A$. PC order $p = 1$, number of random variables $N = 2$

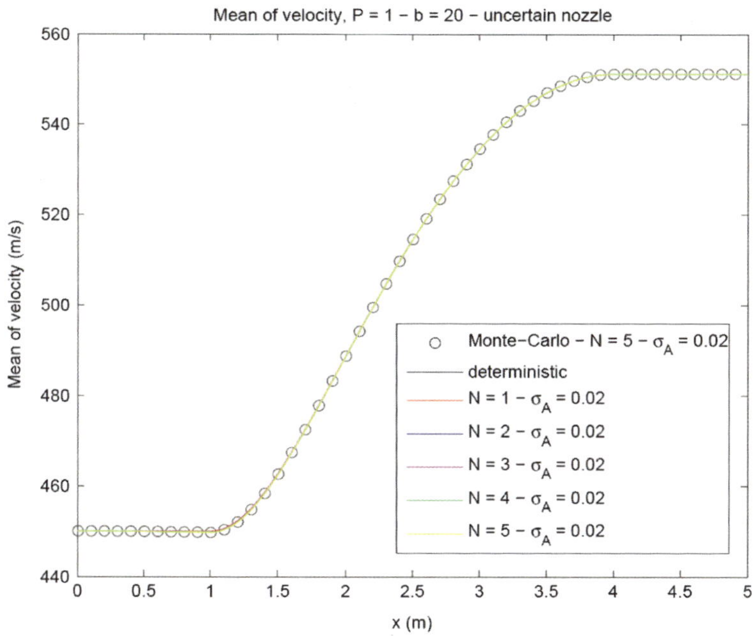


**Fig. 12** Nozzle with geometric uncertainty: mean velocity for covariance function with variance $\sigma_A = 0.02$ and varying number of random variables $N$. PC order $p = 1$ and comparison with MC solution. Global view

components along the centerlines x $= 0.5$ and y $= 0.5$. On the plot the associated uncertainty is indicated using an error bar. This error bar corresponds to the 90 % interval resulting from the local PDF of resp. u and v. Note that the error bar is not necessarily symmetric with respect to the mean value; this corresponds to a non-symmetric PDF.

## *6.3 NASA Rotor 37*

The transonic flow in the NASA Rotor 37 axial compressor was chosen as a representative test case within the NODESIM-CFD EU project, see Hirsch and Dinescu [40]. The experimental set-up and the validation data are described in detail in Dunham [41].

The deterministic version of the flow environment FINE/Turbo of NUMECA Int. has been used extensively to investigate this flow as emphasized in Tartinville and Hirsch [42]. It was shown that the numerical simulations are extremely sensitive to the quality of mesh, the employed turbulence model and the values of inlet mass flow rate. Based on the experience achieved with the deterministic flow solver meshes with more than 600,000 cells with proper clustering near solid boundaries ($y^+ \approx 12$) and the Spalart-Allmaras turbulence model are used for 3D computations.

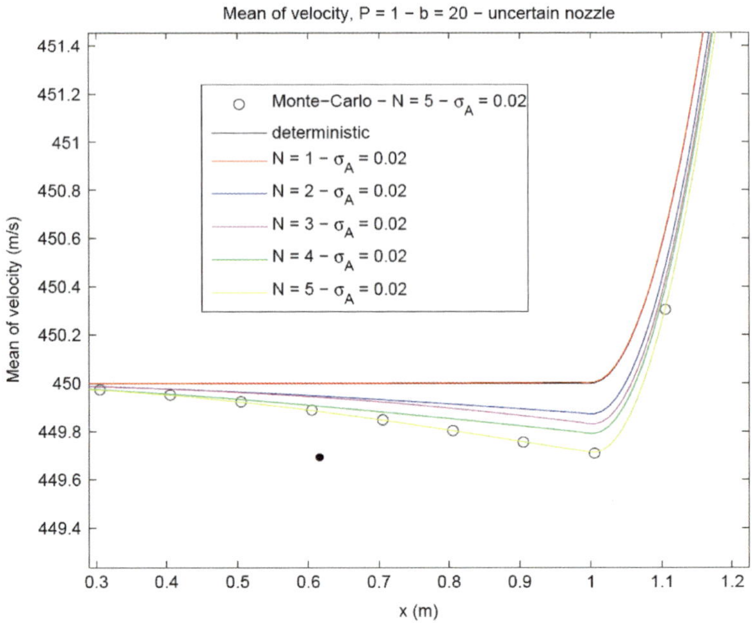


**Fig. 13** Nozzle with geometric uncertainty: mean velocity for covariance function with variance $\sigma_A = 0.02$ and varying number of random variables $N$. PC order $p = 1$ and comparison with MC solution. Zoom

The intrusive polynomial chaos methodology (IPCM) was implemented in the FINE/Turbo commercial solver and subjected to a verification and validation (V&V) program. In what follows, the outcomes of this V&V program are sketched hereafter, relying mostly on the computational results obtained for the operational uncertainty imposed on the outlet static pressure $p_{out}$. For more details and the context of the V&V program we refer the reader to Dinescu et al. [14] and Hirsch and Dinescu [40].

The operational uncertainty is modeled as a normal distribution with an imposed standard deviation $\sigma$ of 4,400 Pa.

For the verification phase, the computational accuracy of the IPCM has been verified by comparing it with that of the results obtained with the non-intrusive Probabilistic Collocation method (NIPCM) of Loeven et al. [43]. The investigated operating regime was that corresponding to $p_{out} = 110{,}000$ Pa and the monitored quantities of interest have been the mean and standard deviation of the pressure on the middle grid line of the mid-span section S2, shown in Fig. 17.

The discontinuous character of the solution, due to the bow shock formed near the leading edge, which impinges on the suction surface on the blade (see Fig. 18), is challenging for accurate computation of its statistical moments.

It is to be noted that results near discontinuities, obtained with global PC expansion, might be less robust than those obtained with local methods such as the

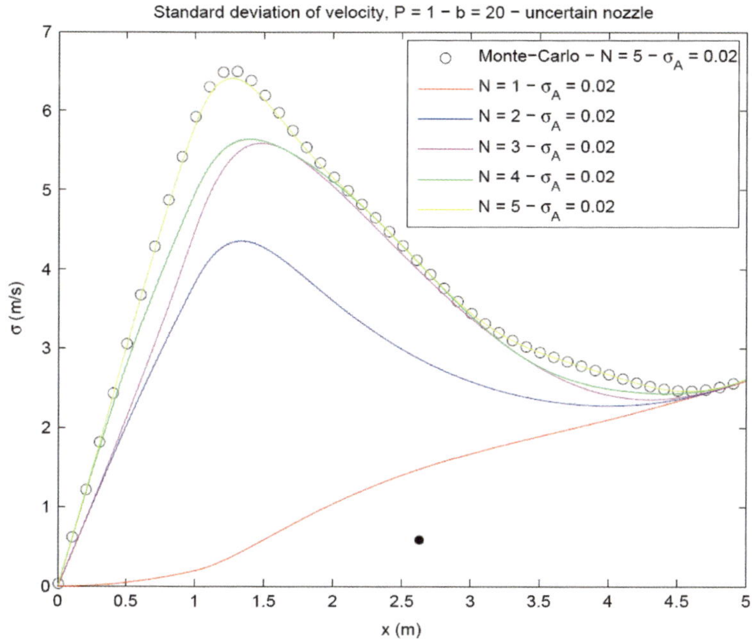


**Fig. 14** Nozzle with geometric uncertainty: standard deviation on the velocity for covariance function with variance $\sigma_A = 0.02$ and varying number of random variables $N$. PC order $p = 1$ and comparison with MC solution

multi-element PC, [44] or the Wiener-Haar wavelet PC, [45]. The convergence of the results is not guaranteed, though in the present case the second- and third-order results seem quite similar, suggesting such convergence.

In the validation phase, the experimental results reported in Dunham [41] were employed, the total pressure ratio versus mass flow rate and the efficiency versus mass flow rate maps. The five investigated operating regimes are listed in Table 2.

Concerning the verification of the mean pressure values recorded on the middle gridline of section S2, one observes in Fig. 19 a good agreement between the second order (p = 2) non-deterministic solutions computed with IPCM and NIPCM. For this operating regime the coefficient of variation (CoV), i.e. the ratio of standard deviation and mean, is 4 %. When switching to third order (p = 3) the IPCM mean solution only changes a little, see Fig. 20.

The standard deviation on the pressure is given in Figs. 21 and 22 for second and third order polynomial chaos, respectively. The intrusive result is again compared with the non-intrusive one. It can be observed that the intrusive results change little when going from p = 2 to p = 3. Only in the shock wave regions there is a small difference, with third order intrusive solution (p = 3) predicting a slightly smaller standard deviation.

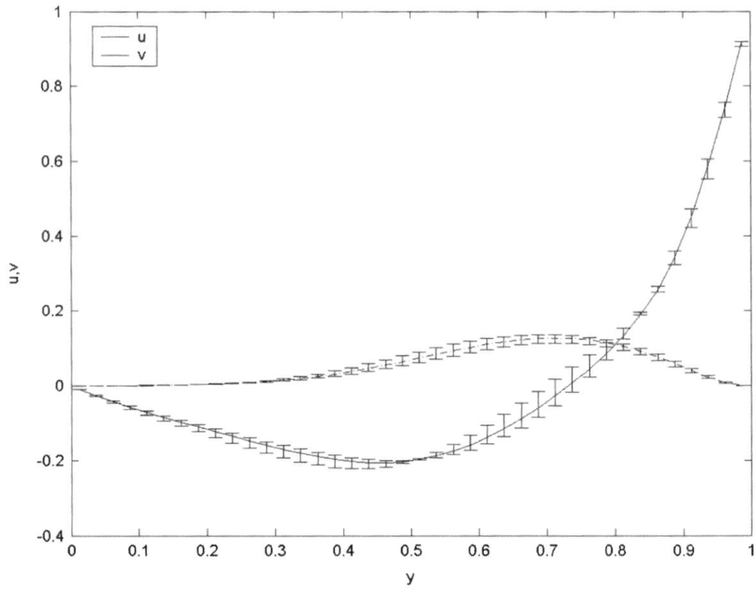


**Fig. 15** Lid driven cavity: mean values of u, v velocity components along the centerlines and the associated uncertainty indicated with error bars: $x = 0.5$

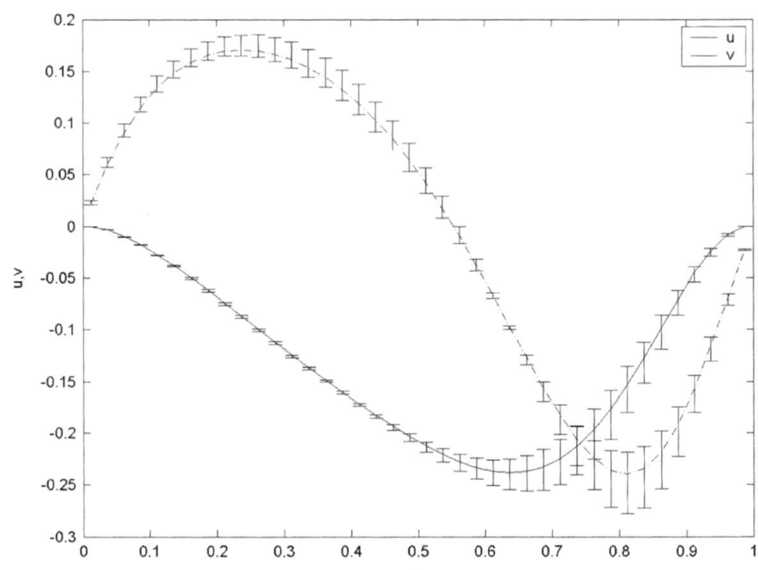


**Fig. 16** Lid driven cavity: mean values of u, v velocity components along the centerlines and the associated uncertainty indicated with error bars: $y = 0.5$

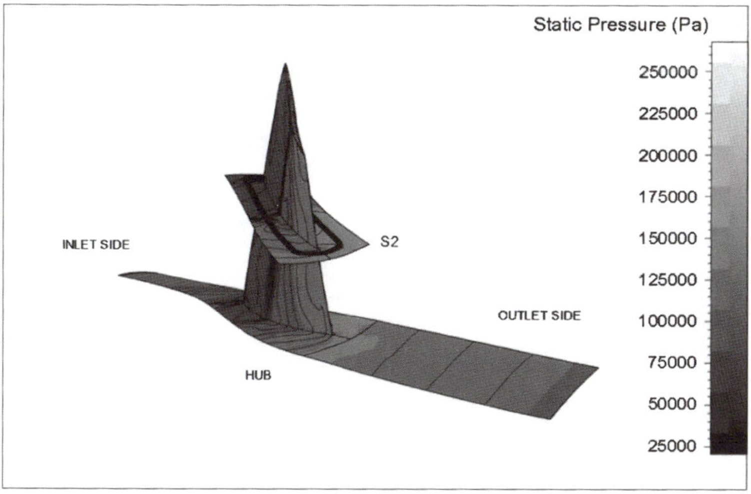


**Fig. 17** NASA Rotor 37: blade-to-blade section S2 with the circumferential line along which results are monitored, indicated as *thick black line*

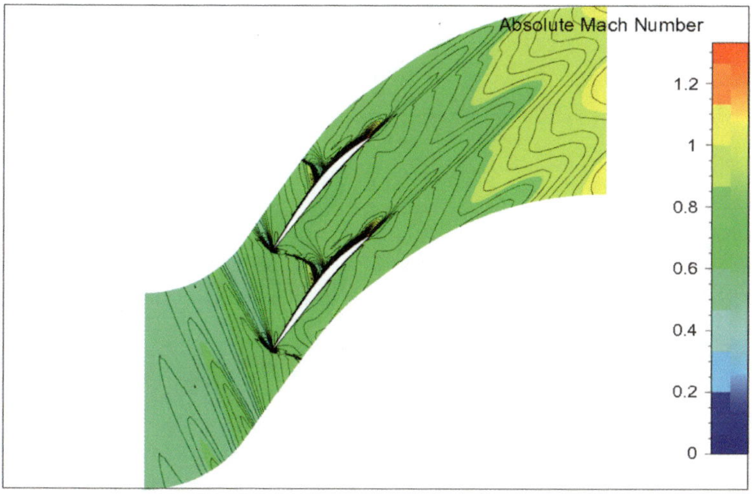


**Fig. 18** NASA Rotor 37: Mach isolines in section S2 for the operating regime 2 ($p_{out}$ = 110,000 Pa)

The comparison with the non-intrusive method is good, except in the shock wave region where the intrusive method predicts a larger standard deviation. The discrepancies recorded in the shock wave regions, among the predictions performed by each method, emphasize the challenge in evaluating accurately statistical moments of the non-deterministic solutions. Supplementary details on this topic can be found in Hirsch and Dinescu [40].

**Table 2** NASA Rotor 37: static outlet pressure for the five operating regimes

| Operating regime | Outlet static pressure (Pa) |
|---|---|
| 1 | 99,215 |
| 2 | 110,000 |
| 3 | 114,074 |
| 4 | 119,035 |
| 5 | 121,033 |

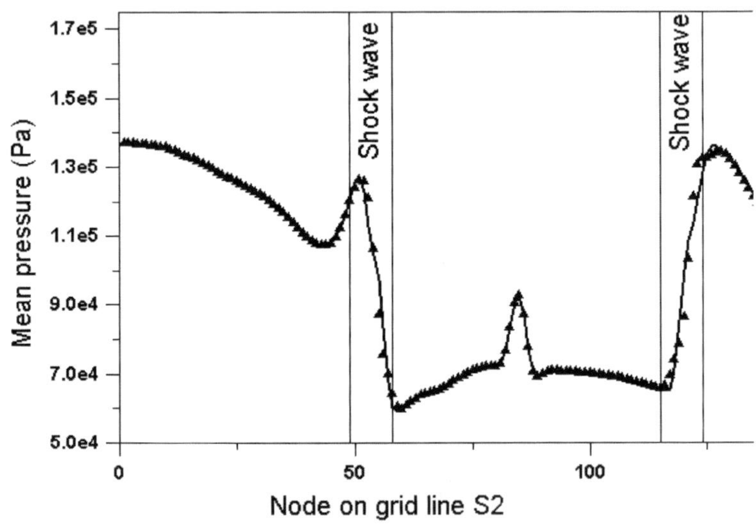


**Fig. 19** NASA Rotor 37: distribution of the mean static pressure along the circumferential line of Fig. 17 in plane $S2$. Comparison intrusive (ipcm) and non-intrusive (nipcm) polynomial chaos of order $p = 2$. Outlet pressure is uncertain (Adapted from Dinescu et al. [14])

In the validation phase, on each outlet static pressure of the operating regimes from Table 2 a normally distributed uncertainty with a standard deviation of 4,400 Pa has been imposed.

Figure 23 shows the resulting map of total pressure ratio versus mass flow rate, while Fig. 24 shows the efficiency versus the mass flow rate map. The uncertainties are indicated with uncertainty bars corresponding to the interval $[-\frac{\sigma}{2}, +\frac{\sigma}{2}]$. A second order polynomial chaos was used.

The main conclusion is that the nondeterministic simulation predicts a range, marked in Figs. 23 and 24 by the uncertainty bars centered on each predicted point, while the deterministic simulation can predict only a single point, e.g. in Tartinville and Hirsch [42]. The coordinates of the range centers are the predicted mean values of mass flow rate and efficiency or total pressure ratio, while the extent of the range is controlled by the predicted standard deviations of the same output functionals.

Moreover, the deterministically computed points in the compressor maps for the operating regimes from Table 2, not shown here, are different from the

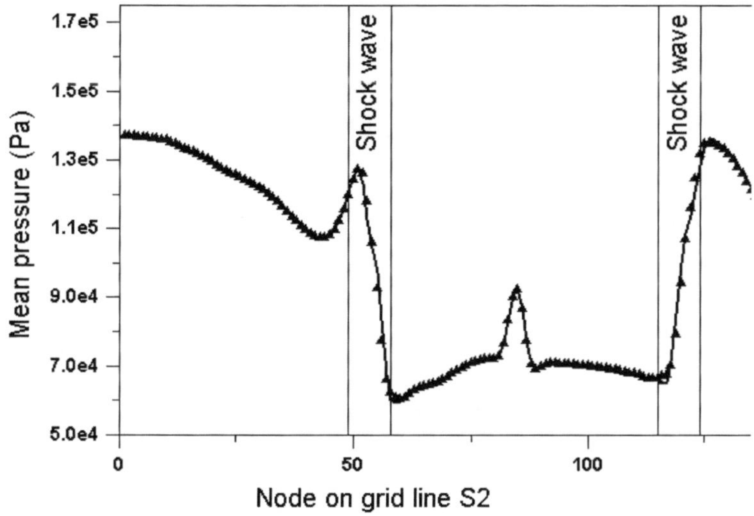


**Fig. 20** NASA Rotor 37: distribution of the mean static pressure along the circumferential line of Fig. 17 in plane $S2$. Comparison intrusive (ipcm) and non-intrusive (nipcm) polynomial chaos of order $p = 3$. Outlet pressure is uncertain (Adapted from Dinescu et al. [14])

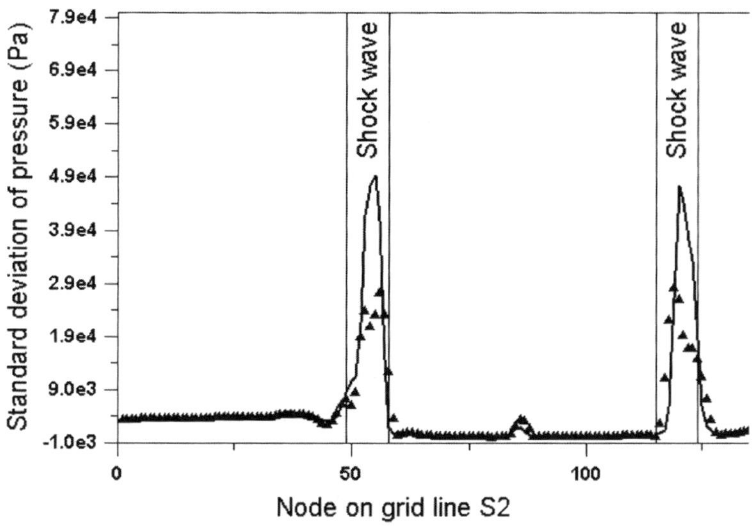


**Fig. 21** NASA Rotor 37: distribution of the standard deviation of static pressure along the circumferential line of Fig. 17 in plane $S2$. Comparison intrusive (ipcm) and non-intrusive (nipcm) polynomial chaos of order $p = 2$. Outlet pressure is uncertain (Adapted from Dinescu et al. [14])

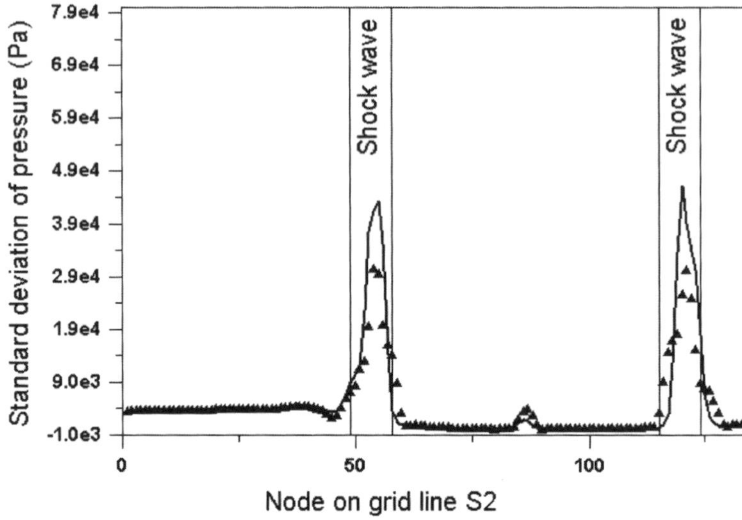


**Fig. 22** NASA Rotor 37: distribution of the standard deviation of static pressure along the circumferential line of Fig. 17 in plane $S2$. Comparison intrusive (ipcm) and non-intrusive (nipcm) polynomial chaos of order $p = 3$. Outlet pressure is uncertain (Adapted from Dinescu et al. [14])

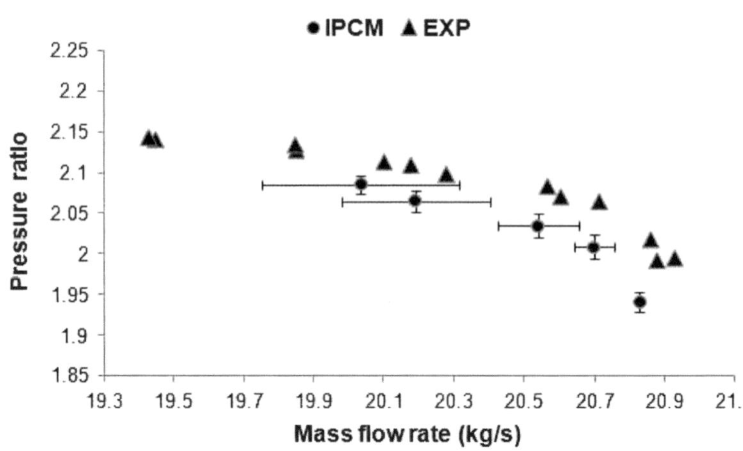


**Fig. 23** NASA Rotor 37: map of total pressure ratio versus mass flow indicating the five simulated running points with error bars as well as experimental data. $p = 2$

non-deterministic points, i.e. centers of the predicted ranges, shown in Figs. 23 and 24. Indeed, focusing for example only on the operating regime 2, one finds in Tables 3–5 that the predicted deterministic values of the mass flow rate, efficiency and total pressure ratio differ from the mean values of the second and third order intrusive results.

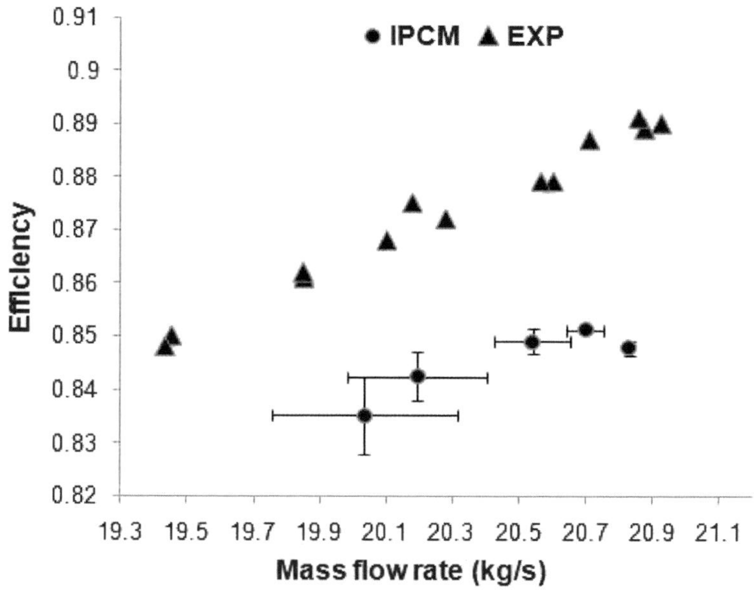


**Fig. 24** NASA Rotor 37: map of efficiency versus mass flow indicating the five simulated running points with error bars as well as experimental data. $p = 2$

**Table 3** NASA Rotor 37: comparison between mean and deterministic values of the mass flow rate (operating regime 2)

| Mean value (kg/s) IPCM [14] | | |
|---|---|---|
| PC order = 2 | PC order = 3 | Deterministic value |
| 20.698019 | 20.743347 | 20.74104 |

**Table 4** NASA Rotor 37: comparison between mean and deterministic values of the efficiency (operating regime 2)

| Mean value (—) IPCM [14] | | |
|---|---|---|
| PC order = 2 | PC order = 3 | Deterministic value |
| 0.8511899 | 0.8510831 | 0.85240 |

**Table 5** NASA Rotor 37: comparison between mean and deterministic values of the total pressure ratio (operating regime 2)

| Mean value (—) IPCM [14] | | |
|---|---|---|
| PC order = 2 | PC order = 3 | Deterministic value |
| 2.00845 | 2.01631 | 2.0083 |

Aforementioned considerations show that working in the paradigm of CFD non-deterministic simulations asks major changes in the whole process of validation, as validation should be based on non-deterministic results, see Hirsch and Dinescu [40].

## References

1. N. Wiener. The Homogeneous Chaos. *Am. J. Math.*, 60:897–936, 1938.
2. D. Xiu and G.E. Karniadakis. Modeling uncertainty in flow simulations via generalized polynomial chaos. *JCP*, 187:137–167, 2003.
3. D. Lucor, D. Xiu, C.-H. Su, and G.E. Karniadakis. Predictability and uncertainty in CFD. *Int. J. Numer. Meth. Fluids*, 43:483–505, 2003.
4. O. Le Maitre, O. Knio, H. Najm, and R. Ghanem. A stochastic projection method for fluid flow i. basic formulation. *J. Comput. Phys.*, 173:481–511, 2001.
5. L. Mathelin, M. Hussaini, and T. Zang. Stochastic approaches to uncertainty quantification in CFD simulations. *Numer. Algorithms*, 38:209–236, 2005.
6. R.W. Walters and L. Huyse. Uncertainty analysis for fluid mechanics with applications. ICASE Rep. no. 2002-1, 2002.
7. Witteveen J.A.S. and Bijl H. Efficient Quantification Of The Effect Of Uncertainties In Advection-Diffusion Problems Using Polynomial Chaos. *Numerical Heat Transfer, Part B*, 53:437–465, 2008.
8. D. Xiu and G.E. Karniadakis. The Wiener-Askey polynomial chaos for stochastic differential equations. *SIAM J. Sci. Comput.*, 24:619–644, 2002.
9. K. Karhunen. Zur spektraltheorie stochastischer prozesse. *Ann. Acad. Sci. Fennicae*, 34:1–7, 1946.
10. M. Loève. Fonctions aleatoires du seconde ordre. Processus Stochastiques et Movement Brownien, 1948. Ed. P. Levy, Paris.
11. Xiu D. FastNumericalMethods for Stochastic Computations: A Review. *Commun. Comput. Phys.*, 5:242–272, 2009.
12. C. Lacor and S. Smirnov. Uncertainty Propagation in the Solution of Compressible Navier-Stokes Equations using Polynomial Chaos Decomposition. RTO-MP-AVT-147 Computational Uncertainty in Military Vehicle Design, 2007. Athens.
13. C. Lacor and S. Smirnov. Non-Deterministic Compressible Navier-Stokes Simulations using Polynomial Chaos. Proc. ECCOMAS Conf., 2008. Venice.
14. C. Dinescu, S. Smirnov, Ch. Hirsch, and C. Lacor. Assessment of intrusive and non-intrusive non-deterministic cfd methodologies based on polynomial chaos expansions. *Int. J. Engineering Systems Modelling and Simulation*, 2:87–98, 2010.
15. G. Onorato, G.J.A. Loeven, G. Ghorbaniasl, H. Bijl, and C. Lacor. Comparison of intrusive and non-intrusive polynomial chaos methods for CFD applications in aeronautics. Proc. ECCOMAS CFD Conf., 2010. Lisbon, June 2010.
16. O. Le Maitre, M. Reagan, H. Najm, R. Ghanem, and O. Knio. A stochastic projection method for fluid flow ii. random process. *J. Comput. Phys.*, 181:9–44, 2002.
17. F. Nobile, R. Tempone, and C.G. Webster. A Sparse Grid Stochastic Collocation Method for Partial Differential Equations with Random Input Data . *SIAM J. Numer. Analysis*, 46: 2309–2345, 2009.
18. M. Berveiller, B. Sudret, and M. Lemaire. Stochastic finite element: a non-intrusive approach by regression. *Rev. Europ enne M canique Num rique*, 15:81–92, 2006.
19. S. Hosder, R.W. Walters, and R. Perez. A Non-Intrusive Polynomial Chaos Method For Uncertainty Propagation in CFD Simulations. 44th AIAA Aerospace Sciences Meeting and Exhibit, 2006.
20. S. Hosder, R.W. Walters, and M. Balch. Efficient uncertainty quantification applied to the aeroelastic analysis of a transonic wing. 46th AIAA Aerospace Sciences Meeting and Exhibit, 2008.
21. C. Hu and B.D. Youn. Adaptive-sparse polynomial chaos expansion for reliability analysis and design of complex engineering systems. *Struct Multidisc Optim*, 43:419–442, 2011.
22. R. Ghanem. Hybrid stochastic finite elements and generalized monte carlo simulation. *ASME J. Appl. Mech.*, 65:1004–1009, 1998.

23. Field R.V. Numerical methods to estimate the coefficients of the polynomial chaos expansion. In *Proc. 15th ASCE Engineering mechanics Conference*, 2002.
24. D. Ghiocel and R. Ghanem. Stochastic finite-element analysis of seismic soil-structure interaction. *J. Eng. mech.*, 128:66–77, 2002.
25. S.A. Smolyak. Quadrature and interpolation formulas for tensor products of certain classes of functions. *Sov Math Dokl*, 4:240–243, 1963.
26. Gerstner T. and Griebel M. Dimension-adaptive tensor-product quadrature. *Computing*, 71(7):65–87, 2003.
27. Ma X. and Zabaras N. An adaptive hierarchical sparse grid collocation algorithm for the solution of stochastic differential equations. *J. Comput. Phys.*, 228:3084–3113, 2009.
28. Rahman S. and Xu H. A univariate dimension-reduction method for multi-dimensional integration in stochastic mechanics . *Probabilistic Engineering Mechanics*, 19:393-408, 2004.
29. Xu H. and Rahman S. A generalized dimension-reduction method for multidimensional integration in stochastic mechanics. *Int. J. Numer. Meth. Engng.*, 61:1992-2019, 2004.
30. Mathelin L., Hussaini M.Y., Zang T.A., and Bataille F. Uncertainty propagation for turbulent, compressible nozzle flow using stochastic methods, . *AIAA J*, 42:1669–1676, 2004.
31. Alekseev A. K., Navon I. M., and Zelentsov M. E. The estimation of functional uncertainty using polynomial chaos and adjoint equations. *Int. J. Numer. Meth. Fluids*, 67:328–341, 2011.
32. W.L. Oberkampf, T.G. Trucano, and Ch. Hirsch. Verification, Validation, and Predictive Capability in Computational Engineering and Physics. Paper presented at FOUNDATIONS 02, Foundations for Verification and Validation in the 21st Century Workshop, USA, 2002.
33. D. Xiu and D.M. Tartakovsky. Numerical Methods for Differential Equations in Random Domains. *SIAM Journal on Scientific Computation*, 28:1167–1185, 2006.
34. Parussini L. and Pediroda V. Investigation of Multi Geometric Uncertainties by Different Polynomial Chaos Methodologies Using a Fictitious Domain Solver. *Computer Modeling in Engineering and Sciences*, 23:29-52, 2008.
35. Parussini L. Fictitious Domain Approach Via Lagrange Multipliers with Least Squares Spectral Element Method. *Journal of Scientific Computing*, 37:316-335, 2008.
36. Perez R. and Walters R. An Implicit Polynomial Chaos Formulation for the Euler Equations. AIAA 2005–1406, 2005.
37. Perez R. Uncertainty Analysis of Computational Fluid Dynamics Via Polynomial Chaos. PhD Thesis, Virginia Polytechnic Institute and State University, 2008.
38. A. Doostan and H. Owhadi. A non-adapted sparse approximation of PDEs with stochastic inputs. *Journal of Computational Physics*, 230:3015–3034, 2011.
39. M. Raets. Application of Polynomial Chaos to CFD. Master Thesis, Research Group Fluid Mechanics and Thermodynamics, VUB, 2007.
40. Hirsch Ch. and Dinescu C. Uncertainty Quantification and Non-deterministic Methodologies for CFD based Design - the NODESIM-CFD Project. Proceedings 7th European Symposium on Aerothermodynamics, SP-692, 2011.
41. J. Dunham. CFD validation for propulsion system components. AGARD-AR-355, 1998.
42. B. Tartinville and Ch. Hirsch. Rotor37 in FLOMANIA - A European Initiative on Flow Physics Modelling. Notes in Numerical Fluid Mechanics and Multidisciplinary Design, Springer, 2006. Vol. 94, pp. 193–202.
43. G.J.A. Loeven, J.A.S. Witteveen, and H. Bijl. Probabilistic Collocation: An Efficient Non-Intrusive Apporach For Arbitrarily Distributed Parametric Uncertainties. AIAA paper 2007–317, 2007. 48th AIAA/ASME/ASCE/AHS/ASC Structures, Structural Dynamics, and Materials Conference.
44. X. Wan and G.E. Karniadakis. An Adaptive Multi-Element Generalized Polynomial Chaos Method for Stochastic Differential Equations. *JCP*, 209:617–642, 2005.
45. O. Le Maitre, O.M. Knio, H.N. Najm, and R.G. Ghanem. Uncertainty propagation using wiener-haar expansions. *J. Comput. Phys.*, 197:28–57, 2004.

# Multi-level Monte Carlo Finite Volume Methods for Uncertainty Quantification in Nonlinear Systems of Balance Laws

**Siddhartha Mishra, Christoph Schwab, and Jonas Šukys**

**Abstract** A mathematical formulation of conservation and of balance laws with random input data, specifically with random initial conditions, random source terms and random flux functions, is reviewed. The concept of random entropy solution is specified. For scalar conservation laws in multi-dimensions, recent results on the existence and on the uniqueness of random entropy solutions with finite variances are presented. The combination of Monte Carlo sampling with Finite Volume Method discretization in space and time for the numerical approximation of the statistics of random entropy solutions is proposed.

The finite variance of random entropy solutions is used to prove asymptotic error estimates for combined Monte Carlo Finite Volume Method discretizations of scalar conservation laws with random inputs. A Multi-Level extension of combined Monte Carlo Finite Volume Method (MC-FVM) discretizations is proposed and asymptotic error bounds are presented in the case of scalar, nonlinear hyperbolic conservation laws. Sparse tensor constructions for the computation of compressed approximations of two- and $k$-point space-time correlation functions of random entropy solutions are introduced.

Asymptotic error versus work estimates indicate superiority of Multi-Level versions of MC-FVM over the plain MC-FVM, under comparable assumptions on

S. Mishra (✉)
SAM, ETH Zürich, HG G 57.2, Rämistrasse 101, Zürich, Switzerland
e-mail: smishra@sam.math.ethz.ch
Supported by ERC StG No. 306279 SPARCCLE.

C. Schwab
SAM, ETH Zürich, HG G 57.1, Rämistrasse 101, Zürich, Switzerland
e-mail: christoph.schwab@sam.math.ethz.ch
Supported by ERC AdG No. 247277 STAHDPDE.

J. Šukys
SAM, ETH Zürich, HG E 62.1, Rämistrasse 101, Zürich, Switzerland
e-mail: jonas.sukys@sam.math.ethz.ch
Supported in part by ETH CHIRP1-03 10-1 and CSCS production project ID S366.

H. Bijl et al. (eds.), *Uncertainty Quantification in Computational Fluid Dynamics*, Lecture Notes in Computational Science and Engineering 92,
DOI 10.1007/978-3-319-00885-1_6, © Springer International Publishing Switzerland 2013

the random input data. In particular, it is shown that these compressed sparse tensor approximations converge essentially at the same rate as the MLMC-FVM estimators for the mean solutions.

Extensions of the proposed algorithms to nonlinear, hyperbolic systems of balance laws are outlined. Multiresolution discretizations of random source terms which are exactly bias-free are indicated.

Implementational aspects of these Multi-Level Monte Carlo Finite Volume methods, in particular results on large scale random number generation, scalability and resilience on emerging massively parallel computing platforms, are discussed.

# 1 Introduction

## 1.1 Weak Solutions of Systems of Balance Laws

Systems of balance laws are nonlinear systems of partial differential equations (PDEs) of the form:

$$
\begin{aligned}
\mathbf{U}_t + \sum_{j=1}^{d} \frac{\partial}{\partial x_j}(\mathbf{F}_j(\mathbf{U})) = \mathbf{S}, \quad x = (x_1, \dots, x_d) \in \mathbb{R}^d, t > 0, \\
\mathbf{U}(x,0) = \mathbf{U}_0(x), \quad x \in \mathbb{R}^d.
\end{aligned}
\tag{1}
$$

Here, $\mathbf{U} : \mathbb{R}^d \mapsto \mathbb{R}^m$ is the vector of unknowns and $\mathbf{F}_j : \mathbb{R}^m \mapsto \mathbb{R}^m$, $j = 1, \dots, d$ denotes the flux vector for the $j$-th direction with the positive integer $m$ denoting the dimension of the state space, and $\mathbf{S} : \mathbb{R}^d \mapsto \mathbb{R}^m$ denotes the so-called *source term*. If $\mathbf{S} = \mathbf{0} \in \mathbb{R}^m$, (1) is termed a *conservation law*.

Examples of balance laws include the Euler equations of gas dynamics, the equations of MagnetoHydroDynamics, the shallow water equations of oceanography and the Buckley-Leverett equations modeling flow of two phases in a porous medium.

It is well known that solutions of (1) develop discontinuities in finite time, even when the initial data is smooth [10]. Hence, solutions of (1) are sought (and computed) in the weak sense: a weak solution $\mathbf{U} \in (L^1_{\text{loc}}(\mathbb{R}^d \times \mathbb{R}_+))^m$ is required to satisfy the integral identity

$$
\int_{\mathbb{R}_+} \int_{\mathbb{R}^d} \left( \mathbf{U}(\varphi)_t + \sum_{j=1}^{d} \mathbf{F}_j(\mathbf{U}) \varphi_{x_j} + \mathbf{S}(\mathbf{U}) \varphi \right) dx dt + \int_{\mathbb{R}^d} \mathbf{U}_0(x) \varphi(x,0) dx = 0 \, , \tag{2}
$$

for all test functions $\varphi \in (C^1_0([0,\infty) \times \mathbb{R}^d))^m$. It is classical that weak solutions are not necessarily unique [10]. Additional admissibility criteria such as *entropy conditions* are necessary to obtain uniqueness. In space dimension $d > 1$, rigorous existence and uniqueness results for deterministic, nonlinear conservation laws and for generic initial data are available only for the scalar case, i.e., in the case $m = 1$.

## *1.2 Numerical Methods*

Numerical schemes have assumed the role of being the main tools for the study of systems of balance (conservation) laws. Many efficient numerical schemes for approximating systems of conservation laws are currently available. They include the Finite Volume, conservative Finite Difference and Discontinuous Galerkin methods, see [19, 25]. For simplicity of exposition, we present the standard Finite Volume Method, following [25].

We consider here (again for the simplicity of exposition) a fixed, positive time step $\Delta t > 0$ and a triangulation $\mathscr{T}$ of the bounded physical domain $\mathbf{D} \subset \mathbb{R}^d$ of interest. Here, the term triangulation $\mathscr{T}$ will be understood as a partition of the physical domain $\mathbf{D}$ into a finite set of disjoint open convex polyhedra $K \subset \mathbb{R}^d$ with boundary $\partial K$ being a finite union of closed plane faces (which are, in these notes, polyhedra contained in $d-1$ dimensional hyperplanes, understood as points in the case $d = 1$). Let $\Delta x_K := \operatorname{diam} K = \sup\{|x-y| : x, y \in K\}$ and by $\Delta x(\mathscr{T}) := \max\{\Delta x_K : K \in \mathscr{T}\}$ denote the *mesh width* of $\mathscr{T}$. For any volume $K \in \mathscr{T}$, we define the *set* $\mathscr{N}(K)$ *of neighbouring volumes*

$$\mathscr{N}(K) := \{K' \in \mathscr{T} : K' \neq K \wedge \operatorname{meas}_{d-1}(\overline{K} \cap \overline{K'}) > 0\}. \tag{3}$$

Note that volumes $K' \in \mathscr{T}$ whose closure shares a set of $d-1$ measure zero with $\overline{K}$ is *not* a neighboring volume. For every $K \in \mathscr{T}$ and $K' \in \mathscr{N}(K)$ denote $\nu_{K,K'}$ to be the exterior unit normal vector, i.e. pointing outward from the volume $K$ at the face $\overline{K} \cap \overline{K'}$. We set:

$$\lambda = \Delta t / \min\{\Delta x_K : K \in \mathscr{T}\} \tag{4}$$

by assuming a uniform discretization in time with constant time step $\Delta t$. The constant $\lambda$ is determined by a standard CFL condition (see [19]) based on the maximum wave speed.

Then, an explicit first-order finite volume [19] for approximating (1) is given by

$$\mathbf{U}_K^{n+1} = \mathbf{U}_K^n - \frac{\Delta t}{\operatorname{meas}(K)} \sum_{K' \in \mathscr{N}(K)} \mathbf{F}(\mathbf{U}_K^n, \mathbf{U}_{K'}^n) + \mathbf{S}_K^n, \tag{5}$$

where

$$\mathbf{U}_K^n \approx \frac{1}{\operatorname{meas}(K)} \int_K \mathbf{U}(\mathbf{x}, t^n) d\mathbf{x}$$

is an approximation to the cell average of the solution and $\mathbf{F}(\cdot,\cdot)$ is a numerical flux that is consistent with $\mathbf{F} \cdot \nu_{K,K'}$. Numerical fluxes are usually derived by (approximately) solving Riemann problems at each cell interface resulting in the Godunov, Roe and HLL fluxes, see e.g. [25]. The discrete source $\mathbf{S}$ in (5) can be a straight-forward evaluation,

$$\mathbf{S}_K^n = \frac{1}{\text{meas}(K)} \int_K \mathbf{S}(\mathbf{x}, \mathbf{U}_K^n) d\mathbf{x}$$

or something more sophisticated, for instance the well-balanced version of the bottom topography source term [15] in shallow-water simulations.

Higher order spatial accuracy is obtained by reconstructing $\mathbf{U}$ from $\mathbf{U}_K^n$ in non-oscillatory piecewise polynomial functions in terms of the TVD [25], ENO [21] and WENO [40] procedures or by the Discontinuous Galerkin method (see, e.g. [9]). Higher order temporal accuracy is achieved by employing strong stability preserving Runge-Kutta methods [20]. Space-time DG-discretizations can also be employed for uniformly high-order spatio-temporal accuracy [6].

## *1.3 Uncertainty Quantification (UQ)*

Any numerical scheme approximating (1) requires the initial data $\mathbf{U}_0$, the source term $\mathbf{S}$ and the flux functions $\mathbf{F}_j$ as inputs. However, in practice, these inputs cannot be measured precisely. As a first example, consider the modeling of propagation of tsunami waves with the shallow water equations. It is not possible to measure the initial water displacement (at tsunami source) with any precision in real time (cp. eg. [26]). Similarly, the bottom topography is measured with sonar equipment and this data collection is prone to uncertainty. Thus, the inputs (initial data and source terms) to the underlying shallow water equations are uncertain. As a second example, consider the modeling of an oil and gas reservoir. Water flooding is modeled by the equations of two phase flow. However, the rock permeability as well as the relative permeabilities of each phase with respect to the other, need to measured. Again, the measurement process is characterized by uncertainty. Consequently, the inputs (the fluxes) to the underlying two-phase flow equations are uncertain. This uncertainty in the inputs for (1) results in the propagation of uncertainty into the solution. The modeling and approximation of the propagation of uncertainty in the solution due to uncertainty in inputs constitutes the theme of uncertainty quantification (UQ).

Uncertainty in inputs and solutions of PDEs is frequently modeled in a probabilistic manner. The inputs are random fields with prescribed probability laws. Then, the solution is also realized as a random field and its law and the (deterministic!) statistical moments of the solutions like the expectation and variance are the quantities of engineering interest.

It is a non-trivial matter to develop efficient algorithms for quantifying uncertainty in solutions of balance (conservation) laws with random inputs. The biggest challenge lies in the fact that singularities in physical space (which inevitably arise in solutions of nonlinear hyperbolic conservation laws) may propagate into parametric representations of the probability densities (laws) of the random solutions. A robust numerical method should be able to deal with this phenomenon. Another challenge lies in dealing with the fact that the number of random sources driving the uncertainty may be very large (possibly countably infinite in the case of random field inputs parametrized by Karhunen–Loève expansions).

The design of efficient numerical schemes for quantifying uncertainty in solutions of partial differential equations has seen a lot of activity in recent years.

Among the most popular methods (particularly for elliptic and parabolic PDEs) are the stochastic Galerkin methods based on generalized Polynomial Chaos (gPC for short). An incomplete list of references on gPC methods for uncertainty quantification in hyperbolic conservation laws includes [1, 8, 27, 35, 43, 45] and other references therein. Although these *deterministic* methods show some promise, they suffer from the disadvantage that they are highly *intrusive*: existing codes for computing deterministic solutions of balance (conservation) laws need to be completely reconfigured for implementation of the gPC based stochastic Galerkin methods. An alternative class of methods for quantifying uncertainty in PDEs are the stochastic collocation methods, see [48] for a general review and [28, 47] for modifications of these methods near discontinuities. Stochastic collocation methods are non-intrusive and easier to parallelize than the gPC based stochastic Galerkin methods. However, the lack of regularity of the solution with respect to the stochastic variables (the solution can be discontinuous in the stochastic variables) impedes efficient performance of both the stochastic Galerkin as well as the stochastic collocation methods (see, however, [31]).

Another class of methods for computational uncertainty quantification in numerical solutions of PDEs are *statistical sampling methods*, most notably Monte Carlo (MC) sampling. In a MC method, the probability space is *sampled* and the underlying deterministic PDE is solved for each sample. The MC samples of numerical solutions of the PDE are combined into statistical estimates of expectation and other statistical moments of the random solution which are necessary to quantify uncertainty. In uncertainty quantification for hyperbolic scalar conservation laws with random initial data, MC type methods together with Finite Volume (FV) spatio-temporal discretizations of the PDE were proposed in a recent paper [30]. The MC-FVM methods were analyzed in the context of a scalar conservation law with random initial data and corresponding estimates of the combined discretization and statistical sampling errors were obtained. MC methods are non-intrusive; they can, therefore, be based on existing, deterministic CFD solvers. As it was shown in [30], MC methods converge at rate $1/2$ as the number $M$ of MC samples increases with each "sample" corresponding to a full, deterministic flow simulation. The asymptotic convergence rate $M^{-1/2}$ in terms of the number $M$ of MC samples is non-improvable by the central limit theorem. To achieve a sampling error which is of the order of the discretization error, MC Finite Volume Methods therefore require a large number of "samples", with each sample consisting of the numerical solution of (1) with a given draw of initial data (and/or random flux and random source term). This slow convergence entails high computational costs for MC type UQ methods in CFD. In particular, accurate quantification of uncertainty by direct MC methods combined with available solvers for hyperbolic systems of conservation or balance laws in several space dimensions becomes very costly, even with a moderately large number of random inputs.

One is therefore led to explore alternative approaches. In recent years, adaptive deterministic discretization methods of *polynomial chaos type* have received substantial attention. These methods have been, in connection with elliptic and parabolic problems, found to be able to facilitate convergence rates which are higher than the (mean square) rate $1/2$ afforded by MC sampling, under appropriate conditions on the input data. While their implementation is intrusive and therefore more involved than that of MC methods (see, e.g. [43]), potentially higher convergence rates than MC-FVM can be achieved by these methods since they approximate directly certain statistical moments of random solutions (in the form of polynomial chaos expansions of random solutions) which recently have been found to exhibit additional smoothness as compared to "pathwise" solutions [39] which, typically, feature discontinuities. In general, however, the lack of regularity of solutions in nonlinear hyperbolic conservation laws and the nonstandard nature of the strongly coupled, large hyperbolic systems (i.e. the dimension $m$ of the state space in (1) is a discretization parameter) which result from the so-called "stochastic Galerkin projection" (i.e. a mean – square projection of the conservation law onto a $m$-term truncated polynomial chaos expansion) indicates at present for this approach only a limited range of applicability (see, however, [39] for evidence of a *mechanism for smoothing through ensemble averaging* in random entropy solutions of hyperbolic conservation laws).

In order to address the slow convergence of MC methods, we proposed in [30] a novel *Multi-Level Monte Carlo* Finite Volume (MLMC-FVM) algorithm for *scalar* conservation laws in [30]. Multi-Level MC methods were introduced by S. Heinrich for numerical quadrature [22] and developed by M. Giles to enhance the efficiency of path simulations for Itô stochastic ordinary differential equations in [17, 18]. More recently, MLMC Finite Element Methods for elliptic problems with stochastic coefficients were introduced by Barth, Schwab and Zollinger in [5]. The analysis in these references, in particular in [5, 30], reveals that the *MLMC is able to deliver converged numerical approximations to statistics of uncertain solutions of partial differential equations in computational complexity comparable to that of one numerical solve of a single "path", i.e. a single realization of the random input data*, under in a sense minimal regularity on the solution. Specifically, only finiteness of second moments of the random solution is needed, when the size of solution is measured in terms of a slightly stronger norm than the norm appearing in energy bounds.

We note that the Multi-Level Monte Carlo method is not the only way to enhance the standard MC method. For instance, variance reduction techniques (such as importance sampling, control variates, stratified sampling, correlated sampling, and conditional Monte Carlo) [14] or quasi Monte Carlo methods (see, e.g., [11] and the references there for a survey) are also available. Efficient variance reduction, however, requires additional a-priori analysis and knowledge about the second moments of the a-priori unknown solution of the random PDE; such knowledge is rarely available. Quasi Monte Carlo methods on the other hand require a parametrization and smoothness assumptions of the unknown solution field which may not hold for nonlinear problems with discontinuous solutions considered here. If these assumptions are not verified, an adhoc application of QMC integration in problems with many random sources (such as the ones considered in Sect. 6.3 of these notes) may lead to simulation methods with a lack of robustness with respect to the curse of the stochastic dimension (i.e. cases when many random variables are involved) as compared to standard Monte Carlo methods.

### 1.4 Objectives of These Notes

The present paper has several objectives. First, we will outline the concept of random entropy solutions for *scalar, multi-dimensional* conservation laws with random inputs. We present a mathematical framework of well-posedness of such problems and provide, in particular, precise statements on the existence and the uniqueness of random entropy solutions for *scalar, multi-dimensional* conservation laws with random inputs.

To this end, we recapitulate results of our recent paper [30] on random entropy solutions for scalar conservation laws with uncertain initial data. Furthermore, we outline extensions of the results on wellposedness and the existence and uniqueness of random entropy solutions for a scalar conservation law with random flux. Further details and complete mathematical developments of these results are available in [31]. The corresponding theory will provide a rigorous basis for the design and analysis of Multi-Level Monte Carlo Finite Volume Methods for the efficient computational quantification of uncertainty in a scalar, hyperbolic conservation law with random input data.

The second objective of this paper is to outline essentials on statistical sampling methods of the Monte Carlo (MC) and Multi-Level Monte Carlo (MLMC) type, with particular attention to their use in computational fluid dynamics. We summarize recent results from [30–33], describe the algorithms, outline the convergence and complexity analysis and present several numerical experiments to demonstrate the efficiency of the proposed algorithms. Systems of conservation laws with uncertain initial data, uncertain source terms and uncertain flux functions are considered in our numerical examples.

The remainder of the paper is organized as follows: in Sect. 2, the mathematical theory of random entropy solutions of scalar conservation laws with uncertain initial data and uncertain flux functions is outlined. The MC algorithms and MLMC algorithms are presented in Sects. 3 and 4 respectively. Details of implementation are provided in Sect. 5 and numerical experiments are presented in Sect. 6. Sparse tensor methods to efficiently compute higher statistical moments of the random entropy solutions are also discussed within Sect. 6. The paper concludes with a description and demonstration of MLMC-FVM approximation of random event probabilities in Sect. 7.

## 2 Random Entropy Solutions

In this section, we introduce the notion of random entropy solutions for conservation laws with random initial data and with random flux functions. We show that scalar conservation laws are well-posed in the sense that we have existence and uniqueness of random entropy solutions for scalar conservation laws with, in particular, continuous dependence of random entropy solutions on the statistical input data of the scalar conservation laws. Since, even in the deterministic case, rigorous results are available only for the scalar problem, in this section we will restrict the mathematical developments to the scalar case ($m = 1$ in (1)). We start with some mathematical preliminaries from probability (cp. eg. [36]).

### *2.1 Random Fields*

Let $(\Omega, \mathscr{F})$ be a measurable space, with $\Omega$ denoting the set of all elementary events, and $\mathscr{F}$ a $\sigma$-algebra of all possible events in our probability model. If $(E, \mathscr{G})$ denotes a second measurable space, then an *E-valued random variable* (or random variable taking values in $E$) is any mapping $X : \Omega \to E$ such that the set $\{\omega \in \Omega \colon X(\omega) \in A\} = \{X \in A\} \in \mathscr{F}$ for any $A \in \mathscr{G}$, i.e. such that $X$ is a $\mathscr{G}$-measurable mapping from $\Omega$ into $E$.

Assume now that $E$ is a metric space; with the Borel $\sigma$-field $\mathscr{B}(E)$, $(E, \mathscr{B}(E))$ is a measurable space and we shall always assume that $E$-valued random variables $X : \Omega \to E$ will be $(\mathscr{F}, \mathscr{B}(E))$ measurable. If $E$ is a separable Banach space with norm $\| \circ \|_E$ and (topological) dual $E^*$, then $\mathscr{B}(E)$ is the smallest $\sigma$-field of subsets of $E$ containing all sets

$$\{x \in E : \varphi(x) \leq \alpha\}, \ \varphi \in E^*, \ \alpha \in \mathbb{R} \,. \tag{6}$$

Hence if $E$ is a separable Banach space, $X : \Omega \to E$ is an $E$-valued random variable if and only if for every $\varphi \in E^*$, $\omega \longmapsto \varphi(X(\omega)) \in \mathbb{R}^1$ is a $\mathbb{R}^1$-valued random variable.

The random variable $X : \Omega \to E$ is called *Bochner integrable* if, for any probability measure $\mathbb{P}$ on the measurable space $(\Omega, \mathscr{F})$,

$$\int_\Omega \|X(\omega)\|_E \, \mathbb{P}(d\omega) < \infty \,. \tag{7}$$

A probability measure $\mathbb{P}$ on $(\Omega, \mathscr{F})$ is any $\sigma$-additive set function from $\Omega$ into $[0, 1]$ such that $\mathbb{P}(\Omega) = 1$, and the measure space $(\Omega, \mathscr{F}, \mathbb{P})$ is called probability space. We shall always assume, unless explicitly stated, that $(\Omega, \mathscr{F}, \mathbb{P})$ is complete.

If $X : (\Omega, \mathscr{F}) \to (E, \mathscr{E})$ is a random variable, $\mathscr{L}(X)$ denotes the *law of X under* $\mathbb{P}$, i.e.

$$\mathscr{L}(X)(A) = \mathbb{P}(\{\omega \in \Omega : X(\omega) \in A\}) \quad \forall A \in \mathscr{E} . \tag{8}$$

The image measure $\mu_X = \mathscr{L}(X)$ on $(E, \mathscr{E})$ is called law or distribution of $X$.

We shall require for $1 \leq p \leq \infty$ Bochner spaces of $p$-summable random variables $X$ taking values in the Banach-space $E$. By $L^1(\Omega, \mathscr{F}, \mathbb{P}; E)$ we denote the set of all (equivalence classes of) integrable, $E$-valued random variables $X$. We equip it with the norm

$$\|X\|_{L^1(\Omega;E)} = \int_\Omega \|X(\omega)\|_E \, \mathbb{P}(d\omega) = \mathbb{E}(\|X\|_E) . \tag{9}$$

More generally, for $1 \leq p < \infty$, we define $L^p(\Omega, \mathscr{F}, \mathbb{P}; E)$ as the set of $p$-summable random variables taking values $E$ and equip it with norm

$$\|X\|_{L^p(\Omega;E)} := (\mathbb{E}(\|X\|_E^p))^{1/p}, \ 1 \leq p < \infty . \tag{10}$$

For $p = \infty$, we denote by $L^\infty(\Omega, \mathscr{F}, \mathbb{P}; E)$ the set of all $E$-valued random variables which are $\mathbb{P}$-almost surely bounded. This set is a Banach space equipped with the norm

$$\|X\|_{L^\infty(\Omega;E)} := \operatorname{ess}\sup_{\omega \in \Omega} \|X(\omega)\|_E . \tag{11}$$

If $T < \infty$ and $\Omega = [0, T]$, $\mathscr{F} = \mathscr{B}([0, T])$, we write $L^p([0, T]; E)$. Note that for any separable Banach-space $E$, and for any $r \geq p \geq 1$,

$$L^r(0, T; E), \ C^0([0, T]; E) \in \mathscr{B}(L^p(0, T; E)) . \tag{12}$$

## 2.2 k-th Moments

For $k \in \mathbb{N}$ and separable Banach space $X$, we denote by $X^{(k)} = \underbrace{X \otimes \cdots \otimes X}_{k \text{ times}}$ the $k$-fold tensor product of $k$ copies of $X$. Throughout the following, we shall assume the $k$-fold tensor product of the Banach-space $X$ with itself, i.e. $X^{(k)}$, to be equipped with a cross norm $\| \circ \|_{X^{(k)}}$ which satisfies

$$\|u_1 \otimes \cdots \otimes u_k\|_{X^{(k)}} = \|u_1\|_X \ldots \|u_k\|_X . \tag{13}$$

We refer to [30, Sect. 3.4] and to the references of [30] for more information on $k$-fold tensor products $X^{(k)}$ of a Banach space $X$ and for norms on $X^{(k)}$.

In particular, for $X = L^p(\mathbb{R}^d)$, $1 \le p < \infty$, we get from Fubini's theorem the isomorphism

$$L^p(\mathbb{R}^d)^{(k)} \cong L^p(\mathbb{R}^{kd}). \tag{14}$$

For $k \in \mathbb{N}$ and for $u \in L^k(\Omega; X)$, we consider the random field $(u)^{(k)}$ defined by $\underbrace{u(\omega) \otimes \cdots \otimes u(\omega)}_{k-\text{times}}$. Then

$$(u)^{(k)} = u \otimes \cdots \otimes u \in L^1(\Omega; X^{(k)}) \tag{15}$$

and, by (13), we have

$$\|(u)^{(k)}\|_{L^1(\Omega;X^{(k)})} = \int_\Omega \|u(\cdot,\omega)\|_X^k \,\mathbb{P}(d\omega) = \|u\|_{L^k(\Omega,X)}^k < \infty. \tag{16}$$

Therefore, $(u)^{(k)} \in L^1(\Omega, X^{(k)})$ and *the* $k$*-th moment* (or $k$-point correlation function of $u$)

$$\mathscr{M}^k(u) := \mathbb{E}[(u)^{(k)}] \in X^{(k)} \tag{17}$$

is well-defined as a (deterministic) element of $X^{(k)}$ for $u \in L^k(\Omega; X)$.

## 2.3 Random Initial Data

Equipped with the above notions, we first model uncertain initial data by assuming $(\Omega, \mathscr{F}, \mathbb{P})$ as the underlying probability space and realizing the uncertain initial data as a random field $u_0$, i.e. a $L^1(\mathbb{R}^d)$-valued random variable which is a $L^1(\mathbb{R}^d)$ measurable map

$$u_0 : (\Omega, \mathscr{F}) \longmapsto \big(L^1(\mathbb{R}^d),\ \mathscr{B}(L^1(\mathbb{R}^d))\big). \tag{18}$$

We assume further that

$$u_0(\cdot,\omega) \in L^\infty(\mathbb{R}^d) \cap BV(\mathbb{R}^d) \quad \mathbb{P}\text{-a.s.}, \tag{19}$$

which is to say that

$$\mathbb{P}(\{\omega \in \Omega : u_0(\cdot,\omega) \in (L^\infty \cap BV)(\mathbb{R}^d)\}) = 1. \tag{20}$$

Since $L^1(\mathbb{R}^d)$ and $C^1(\mathbb{R}^d;\mathbb{R}^d)$ are separable, (18) is well defined and we may impose for $k \in \mathbb{N}$ the *k-th moment condition*

$$\|u_0\|_{L^k(\Omega;L^1(\mathbb{R}^d))} < \infty\,, \tag{21}$$

where the Bochner spaces with respect to the probability measure are defined in (9) and (10) above.

## 2.4 Random Flux Functions for SCL

Noting that the space $E = C^1(\mathbb{R};\mathbb{R}^d)$ is separable, we concentrate on the case of spatially homogeneous random flux functions and follow [31]. The definition of random flux for scalar conservation laws (i.e. for the case $m = 1$ in (6)) that we shall work with is

**Definition 1.** A *(spatially homogeneous) random flux* for a scalar conservation law is a random field taking values in the separable Banach space $E = C^1(\mathbb{R}^1;\mathbb{R}^d)$, i.e. a measurable mapping from $(\Omega,\mathscr{F})$ to $(C^1(\mathbb{R}^1;\mathbb{R}^d);\mathscr{B}(C^1(\mathbb{R}^1;\mathbb{R}^d)))$. A *bounded random flux* is a random flux whose $C^1(\mathbb{R}^1;\mathbb{R}^d)$-norm is bounded $\mathbb{P}$-a.s., i.e.

$$\exists 0 < B(f) < \infty : \quad \|f(\omega;\cdot)\|_{C^1(\mathbb{R}^1;\mathbb{R}^d)} \le B(f) \quad \mathbb{P}-\text{a.s.} \tag{22}$$

We observe that a bounded random flux has finite statistical moments of any order. Of particular interest will be the *second moment of a bounded random flux* (i.e. its "two-point correlation in state-space"). The existence of such a state-space correlation function is addressed in the following lemma from [31], to which we refer for the proof.

**Lemma 1.** *Let $f$ be a bounded random flux as in Definition 1 which belongs to $L^2(\Omega, d\mathbb{P}; C^1(\mathbb{R};\mathbb{R}^d))$. Then its covariance function, i.e. its centered second moment defined by*

$$\text{Cov}[f](v,v') := \mathbb{E}\big[(f(\cdot;v) - \mathbb{E}[f(\cdot;v)]) \otimes (f(\cdot;v') - \mathbb{E}[f(\cdot;v')])\big] \tag{23}$$

*is well-defined for all $v, v' \in \mathbb{R}$ and there holds*

$$\text{Cov}[f] \in C^1(\mathbb{R}\times\mathbb{R};\mathbb{R}^{d\times d}_{sym}). \tag{24}$$

The two point correlation function of a bounded random flux allows, as is well-known in statistics, for spectral decompositions of the random flux in terms of eigenpairs of its covariance operator, which is a compact and self-adjoint integral operator on square-integrable flux functions with kernel function $\text{Cov}[f](v,v')$ defined in (23). We remark in passing that our assumption of continuous differentiability of (realizations of) random flux functions entails linear growth of such fluxes as the state variables tend to infinity, i.e. as $|v| \to \infty$ which, at first sight, precludes considering the covariance operator on the space of square integrable flux functions. In [31], we circumvent these integrability issues for scalar conservation laws by truncating the state space to a bounded interval $[-R, R]$ with sufficiently large $R > 0$. By classical $L^\infty(\mathbb{R}^d)$ bounds on entropy solutions of scalar conservation laws, for sufficiently large values of the flux cutoff $R$, any realization of the random scalar conservation law will "see" only the flux function for states which (in absolute value) are below the threshold values $R$; accordingly, it suffices to consider the flux covariance operator only as integral operator on $L^2(-R, R)$ which is the view taken in [31].

As a concrete example for a random flux, we have the following representation using the Karhunen-Loeve (KL) expansion.

**Example: Karhunen–Loève expansion of bounded random flux.** Consider a bounded random flux $f(\omega; u)$ in the sense of Definition 1. By Lemma 1, its covariance function $\mathrm{Cov}[f]$ is well-defined; for $0 < R < \infty$ we denote by $\mathscr{C}_f^R$ the integral operator with continuously differentiable kernel $\mathrm{Cov}[f](u, v)$, defined on $L^2(-R, R)$ by

$$\mathscr{C}_f^R[\Phi](u) := \int_{|v|\le R} \mathrm{Cov}[f](u, v)\Phi(v)dv \,. \tag{25}$$

As explained above, the covariance operator $\mathscr{C}_f^R$ describes the covariance structure of the random flux on the set $[-R, R]$ of states. Given initial data $u_0 \in L^\infty(\mathbb{R}^d)$ with a-priori bound $\|u_0\|_{L^\infty(\mathbb{R}^d)} \le R$, the unique entropy solution $S(t)u_0$ of the *deterministic* SCL will, for all $t > 0$, take values in $[-\|u_0\|_{L^\infty(\mathbb{R}^d)}, \|u_0\|_{L^\infty(\mathbb{R}^d)}]$. For random flux and random initial data, therefore, we continue under the assumption

$$R > \operatorname*{ess\,sup}_{\omega\in\Omega} \|u_0(\omega, \cdot)\|_{L^\infty(\mathbb{R}^d)}. \tag{26}$$

This ensures that $\mathscr{C}_f^R$ will "capture" all possible states.

By (24), for every $0 < R < \infty$, the integral operator $\mathscr{C}_f$ is a compact, self-adjoint operator on $L^2(-R, R)$. By the spectral theorem, it admits for every fixed value $0 < R < \infty$ a sequence $(\lambda_j^R, \Phi_j^R)_{j\ge1}$ of real eigenvalues $\lambda_j^R$ (accumulate only at zero), which are assumed to be enumerated in decreasing magnitude and repeated according to multiplicity, and a corresponding set of eigenfunctions $\Phi_j^R$; to exclude trivial degeneracies, we shall assume throughout that *the sequence* $(\Phi_j^R)_{j\ge1}$ *is a complete, orthonormal base of* $L^2(-R, R)$.

It follows from the continuous differentiability (24) of $\mathrm{Cov}[f]$ and from the eigenvalue equation

$$(\mathscr{C}_f^R \Phi_j^R)(u) = \lambda_j^R \Phi_j^R(u) \,, \qquad |u| \le R \,, \tag{27}$$

that $\Phi_j^R \in C^1([-R,R];\mathbb{R}^d)$: for $u, u' \in [-R,R]$, there holds by Lemma 1 and by the eigenvalue equation (27)

$$
\begin{aligned}
|\Phi_j^R(u) - \Phi_j^R(u')| &= \frac{1}{\lambda_j^R}\left|\int_{|v|<R} \left(\mathrm{Cov}[f](u,v) - \mathrm{Cov}[f](u',v)\right) \Phi_j^R(v)dv\right| \\
&\leq \frac{8RB(f)}{\lambda_j^R}|u-u'| \sup_{|v|<R}\left(\int_\Omega \|f(\omega;v) - \bar{f}(v)\|_2^2\, d\mathbb{P}(\omega)\right)^{1/2} \left\|\Phi_j^R\right\|_{L^1(-R,R)} \\
&\leq \frac{8R^{3/2}B(f)}{\lambda_j^R}|u-u'| \sup_{|v|<R}\left(\int_\Omega \|f(\omega;v) - \bar{f}(v)\|_2^2\, d\mathbb{P}(\omega)\right)^{1/2} .
\end{aligned}
$$

Any bounded random flux $f(\omega;u)$ therefore admits, for every fixed $0 < R < \infty$, a Karhunen–Loève expansion

$$
f(\omega;u) = \bar{f}(u) + \sum_{j\geq 1} Y_j^R(\omega)\Psi_j^R(u), \quad |u| \leq R , \tag{28}
$$

which converges in $L^2(\Omega, d\mathbb{P}; L^2(-R,R)^d)$. In (28), the nominal flux $\bar{f}(u) = \mathbb{E}[f(\cdot;u)]$ and the sequence $(Y_j^R)_{j\geq 1}$ is a sequence of pairwise uncorrelated random variables given by

$$
\forall j \in \mathbb{N}: \quad Y_j^R(\omega) := \sqrt{\lambda_j^R}\int_{|v|<R} f(\omega;v)\Phi_j^R(v)dv . \tag{29}
$$

and the principal components of the random flux are given by

$$
\forall j \in \mathbb{N}: \quad \Psi_j^R(u) := \frac{1}{\sqrt{\lambda_j^R}}\Phi_j^R(u) .
$$

We remark that under suitable smoothness conditions on two-point correlation function $\mathrm{Cov}[f]$ of the random flux the convergence of the expansion (28) is (a) pointwise with respect to $u$ and (b) the convergence rates increase with increasing smoothness of $\mathrm{Cov}[f]$ (see, e.g. [31]).

## 2.5 Random Entropy Solutions of Scalar Conservation Laws

Equipped with the above notions of random initial data and random fluxes, we consider a random scalar conservation law (RSCL):

$$
\begin{aligned}
\partial_t u(x,t;\omega) + \sum_{j=1}^d \frac{\partial}{\partial x_j} f_j(\omega; u(x,t;\omega)) = 0 , \\
u(x,0;\omega) = u_0(x;\omega) , \quad x \in \mathbb{R}^d .
\end{aligned} \tag{30}
$$

and define,

**Definition 2.** A random field $u : \Omega \ni \omega \to u(x,t;\omega)$, i.e. a measurable mapping from $(\Omega,\mathscr{F})$ to $C([0,T];L^1(\mathbb{R}^d))$, is a *random entropy solution* of the SCL (30) with random initial data $u_0$ satisfying (18)–(21) for some $k \geq 2$ and with a *spatially homogeneous* random flux $f(\omega;u)$ as in Definition 1 that is statistically independent of $u_0$, if it satisfies the following conditions:

(i) Weak solution: for $\mathbb{P}$-a.e $\omega \in \Omega$, $u(\cdot,\cdot;\omega)$ satisfies the following integral identity, for all test functions $\varphi \in C_0^1(\mathbb{R}^d \times [0,\infty))$:

$$\int_0^\infty \int_{\mathbb{R}^d} \left( u(x,t;\omega)\varphi_t(x,t) + \sum_{j=1}^d f_j(\omega;u(x,t;\omega)) \frac{\partial}{\partial x_j}\varphi(x,t) \right) dxdt + \int_{\mathbb{R}^d} u_0(x,\omega)\varphi(x,0)dx = 0 \,. \tag{31}$$

(ii) Entropy condition:

For any pair of (deterministic) entropy $\eta$ and (stochastic) entropy flux $Q(\omega;\cdot)$ i.e. $\eta, Q_j$ with $j = 1,2,\dots,d$ are functions such that $\eta$ is convex and such that $Q_j'(\omega;\cdot) = \eta' f_j'(\omega;\cdot)$ for all $j$, and for $\mathbb{P}$-a.e $\omega \in \Omega$, $u$ satisfies the following integral identity,

$$\int_0^\infty \int_{\mathbb{R}^d} \left( \eta(u(x,t;\omega))\varphi_t(x,t) + \sum_{j=1}^d Q_j(\omega;u(x,t;\omega)) \frac{\partial}{\partial x_j}\varphi(x,t) \right) dxdt \geq 0 \,, \tag{32}$$

for all deterministic test functions $0 \leq \varphi \in C_0^1(\mathbb{R}^d \times (0,\infty))$, $\mathbb{P}$-a.s.

Throughout what follows, we assume that the deterministic entropy function $\eta(\cdot)$ in (32) is a *Kružkov entropy*, i.e. $\eta(u) = |u-k|$ for some $k \in \mathbb{R}$.

In a forthcoming paper [31], we show the following well-posedness result for random entropy solutions of (30).

**Theorem 1.** *Consider the RSCL* (30) *with spatially homogeneous, bounded random flux* $f : \Omega \to C^1(\mathbb{R};\mathbb{R}^d)$ *as in Definition 1 and with (independent of* $f$*) random initial data* $u_0 : \Omega \to L^1(\mathbb{R}^d)$ *satisfying* (19), (20) *and the* $k$*-th moment condition* (21) *for some integer* $k \geq 2$*. In particular, then, there exists a constant* $\bar{R} < \infty$ *such that*

$$\|u_0(\omega,\cdot)\|_{L^\infty(\mathbb{R}^d)} \leq \bar{R} \qquad \mathbb{P}-a.e.\ \omega \in \Omega \,. \tag{33}$$

*Assume moreover that the random flux admits the representation* (28) *with* (29) *where the continuously differentiable scaled flux components* $\overline{\Psi_j^R}$ *have Lipschitz constants* $B_j^R$ *such that* $\boldsymbol{B}^R := (B_j^R)_{j\geq 1} \in \ell^1(\mathbb{N})$ *with some* $R \geq \bar{R}$ *as in* (33)*.*

*Then there exists a unique random entropy solution* $u : \Omega \ni \omega \to C_b(0,T;L^1(\mathbb{R}^d))$ *which is "pathwise", i.e. for* $\mathbb{P}-a.e.\ \omega \in \Omega$*, described in terms of a nonlinear mapping* $S(\omega;t)$ *depending only on the random flux, such that*

$$u(\cdot,t;\omega) = S(\omega;t)u_0(\cdot,\omega)\,, \quad t > 0,\ \mathbb{P}-a.e.\omega \in \Omega \tag{34}$$

*such that for every* $k \geq m \geq 1$ *and for every* $0 \leq t \leq T < \infty$ *holds*

$$\|u\|_{L^k(\Omega;C(0,T;L^1(\mathbb{R}^d)))} \leq \|u_0\|_{L^k(\Omega;L^1(\mathbb{R}^d))}, \tag{35}$$

$$\|S(\omega;t)\, u_0(\cdot,\omega)\|_{(L^1\cap L^\infty)(\mathbb{R}^d)} \leq \|u_0(\cdot,\omega)\|_{(L^1\cap L^\infty)(\mathbb{R}^d)} \tag{36}$$

*and such that we have* $\mathbb{P}$*-a.s.*

$$TV(S(\omega;t)u_0(\cdot,\omega)) \leq TV(u_0(\cdot,\omega)). \tag{37}$$

*and, with* $\bar{R}$ *as in* (33)*,*

$$\sup_{0\leq t\leq T} \|u(\cdot,t;\omega)\|_{L^\infty(\mathbb{R}^d)} \leq \bar{R} \quad \mathbb{P}-a.e.\ \omega \in \Omega. \tag{38}$$

*Remark 1.* The above theorem establishes that RSCLs are well-posed (in several space dimensions) for uncertain initial data as well as for random fluxes. An extension of these definitions and results to include random source terms $S(u,\omega)$ for bounded sources as well as to spatially inhomogeneous flux functions $f(\omega;x,u)$ is possible, provided their dependence on the spatial coordinate is continuously differentiable: they are measurable mappings from $(\Omega,\mathscr{F})$ into $(E,\mathscr{B}(E))$ where $E = C^1(\mathbb{R}^{d+1};\mathbb{R}^d)$.

*Remark 2.* So far, we considered the Karhunen–Loève expansion only for RSCLs (30), i.e. for the case $m = 1$ in (1). It is straightforward to extend the principal component representation and the notion of covariance operator to flux functions for hyperbolic systems (1): in this case, the covariance operator is to be interpreted as abstract, symmetric bilinear form on the space $C^1(\mathbb{R}^m;\mathbb{R}^{m\times d})$ whose kernel coincides with a symmetric, fourth order tensor function on the state space $\mathbb{R}^m$. Spectral decompositions analogous to (28) for $\mathbb{R}^{m\times d}$ matrix-valued random flux functions which arise in (1) can then be defined in an analogous fashion. However, due to the lack of bounds like (38), the approach of [31] can not be directly applied for the mathematical investigation of random hyperbolic systems (1) at present. Nevertheless, the spectral expansion (28) for (1) may be a useful tool to achieve a parsimonious parametric representation of a general, given random flux also in the numerical treatment of random hyperbolic systems (1).

The notions of random entropy solutions for a system of balance laws (1) with uncertain initial data, fluxes and sources can be analogously defined. Currently, there are no global well-posedness results for systems of balance laws. Hence, we are unable to extend the well-posedness results of Theorem 1 to the case of systems of balance laws such as (1).

# 3 Monte Carlo Finite Volume Method

Our aim is to approximate the random balance law (1). The spatio-temporal discretization can be performed by any standard Finite Volume or DG scheme, for instance (5). The probability space will be discretized using a statistical sampling technique. The simplest sampling method is the Monte Carlo (MC) algorithm consisting of the following three steps:

1. **Sample:** We draw $M$ independent identically distributed (i.i.d.) initial data, flux and source samples $\{\mathbf{U}_0^k, \mathbf{F}_j^k, \mathbf{S}^k\}$ with $j = 1, \ldots, d$ and $k = 1, \ldots, M$ from the random fields $\{\mathbf{U}_0, \mathbf{F}_j, \mathbf{S}\}$ and approximate these by piece-wise constant cell averages.
2. **Solve:** For each realization $\{\mathbf{U}_0^k, \mathbf{F}_j^k, \mathbf{S}^k\}$, the underlying balance law (1) is solved numerically by the Finite Volume Method (5). We denote the FVM solutions by $\mathbf{U}_{\mathscr{T}}^{k,n}$, i.e. by cell averages $\{\mathbf{U}_C^{k,n} : C \in \mathscr{T}\}$ at time $t^n$,

$$\mathbf{U}_{\mathscr{T}}^{k,n}(\mathbf{x}) = \mathbf{U}_C^{k,n}, \quad \forall \mathbf{x} \in C, \ C \in \mathscr{T}, \ k = 1, \ldots, M.$$

3. **Estimate Statistics:** We estimate the expectation of the random solution field with the sample mean (sample average) of the approximate solution:

$$E_M[\mathbf{U}_{\mathscr{T}}^n] := \frac{1}{M} \sum_{k=1}^{M} \mathbf{U}_{\mathscr{T}}^{k,n}. \tag{39}$$

   Higher statistical moments can be approximated analogously under suitable statistical regularity of the underlying random entropy solutions [30].

The above algorithm is, at first sight, straightforward to implement. We remark that step 1 requires a (pseudo) random number generator. Here, care must be exercised in ensuring good statistical properties in massively parallel implementations (see Sect. 5 for details). In step 2, any standard (high-order) finite volume or DG scheme can be used. Hence, existing ("legacy") code for FVM (or DG) can be (re)used and there is no need to rewrite FVM (or DG) code. However, in doing so, particular care must be exercised that the "forward solver" thus employed is particularly robust: MC sampling will generate equistributed samples which cover the entire "senario space" of data, and include, in particular, also data instances which might appear "unlikely" or "unphysical" to experts. Nevertheless, such extremal scenarios do contribute to the approximate ensemble averages and must be resolved by the forward solver with accuracy versus CPU that is comparable to the accuracy of so-called "benchmark problems" commonly accepted as test for code performance in various CSE communities. "Legacy codes" whose performance might have been optimized on "benchmark problems" may lack such robustness for MC-generated scenarios which fall outside sets of commonly accepted benchmarks.

Furthermore, the only (data) interaction between different samples is in step 3 when ensemble averages are computed. Thus, the MC-FVM algorithms for UQ are non-intrusive as well as easily parallelizable.

Although a rigorous error estimate for the MCFVM approximating systems of balance laws appears to be currently out of reach, we rely on the analysis for a scalar conservation law in [30,31] and on our computational experience with MLMC-FVM solution of systems of balance laws with random inputs in [32, 33] to postulate that the following estimate holds:

$$\|\mathbb{E}[\mathbf{U}(\cdot,t^n)] - E_M[\mathbf{U}^n_{\mathcal{T}}]\|_{L^2(\Omega;L^1(\mathbb{R}^d))} \leq C_{\text{stat}} M^{-\frac{1}{2}} + C_{\text{st}} \Delta x^s \,. \tag{40}$$

The positive constants $C_{\text{stat}}, C_{\text{st}}$ depend only on the second moments (must be *finite*) of the random initial data and of the source term in (1). In the above, we have assumed that the underlying Finite Volume (or DG) scheme converges to the solutions of the deterministic balance law (1) at rate $s > 0$. One rationale for adopting MLMC-FVM for the numerical solution of (1) with random data and fluxes lies in the fact that, in practice, even second or higher order schemes are known to realize only convergence rates $0 < s < 1$, due to the lack or regularity of the exact solutions. Therefore, the use of deterministic sampling strategies which promise a convergence rate that is higher than 1/2 will, in general, not substantially improve the work versus accuracy of FVM for these problems.

Based on the error analysis of [30], we equilibrate the discretization and the sampling errors in the a-priori estimate (40) and choose [30, 32]

$$M = \mathcal{O}(\Delta x^{-2s}). \tag{41}$$

Next, we assume that the computational work (e.g. FLOPs, i.e. the number of required floating point operations) of the FVM solver for *one* complete run is given (asymptotically) by

$$\text{Work}_{\text{FVM}}(\Delta x) = C_s \Delta x^{-(d+1)}, \tag{42}$$

where $C_s > 0$ is independent of $\Delta x$ and $\Delta t$, but depends on the order $s$ of the FVM scheme.

Then, with the choice (41) and definition (42), it is straightforward to deduce that the asymptotic error vs. work estimate of the MC estimator (40) is given by

$$\|\mathbb{E}[u(\cdot,t^n)] - E_M[u^n_{\mathcal{T}}]\|_{L^2(\Omega;L^1(\mathbb{R}^d))} \lesssim (\text{Work})^{-s/(d+1+2s)}. \tag{43}$$

The above error vs. work estimate is considerably more expensive when compared to the FVM discretization error for the corresponding deterministic problem [30]:

$$\|u(\cdot,t^n) - u^n_{\mathcal{T}}\|_{L^1(\mathbb{R}^d)} \leq C_{\text{Err}} C_s^{s/(d+1)} \text{Work}^{-s/(d+1)} =: C_{\text{FVM}} \text{Work}^{-s/(d+1)}. \tag{44}$$

# 4 Multi-level Monte Carlo Finite Volume Method

The low convergence rate (43) of MC-FVM motivates the use Multi-Level Monte Carlo Finite Volume Method (MLMC-FVM). The key idea behind the MLMC-FVM is to simultaneously sample a *hierarchy of discretizations of the PDE with random inputs with a level-dependent number of samples*. In the present setting, this amounts to running a deterministic FV solver on a sequence of nested grids in space with correspondingly adapted time step sizes, so as to ensure the validity of a CFL condition uniformly over all space-discretizations of the hierarchy [30].

## 4.1 MLMC-FVM Error Analysis

### 4.1.1 MLMC-FVM Algorithm

The Multi Level Monte Carlo Finite Volume algorithm (MLMC-FVM for short) consists of the following four steps:

1. **Hierarchy of space-time discretizations:** Consider *nested* triangulations $\{\mathscr{T}_\ell\}_{\ell=0}^{\infty}$ of the spatial domain $\mathbf{D}$ with corresponding mesh widths $\Delta x_\ell$ that satisfy:

$$\Delta x_\ell = \Delta x(\mathscr{T}_\ell) = \sup\{\mathrm{diam}(K) : K \in \mathscr{T}_\ell\} = O(2^{-\ell}\Delta x_0), \quad \ell \in \mathbb{N}_0, \qquad (45)$$

   where $\Delta x_0$ is the mesh width for the coarsest resolution and corresponds to the lowest level $\ell = 0$.
2. **Sample:** For each level of resolution $\ell \in \mathbb{N}_0$, we draw a *level-dependent number* $M_\ell$ of independent, identically distributed (i.i.d) samples from the input random fields $\{\mathbf{U}_0(\omega), \mathbf{F}_j(\omega), \mathbf{S}(\omega)\}$. Importantly, these random field inputs are only sampled on $\mathscr{T}_\ell$ in spatially discrete form, i.e. as realizations of cell-averages on triangulation $\mathscr{T}_\ell$, to yield *vectors* $\{\mathbf{U}_{0,\ell}(\omega), \mathbf{F}_{j,\ell}(\omega), \mathbf{S}_\ell(\omega)\}$. on mesh $\mathscr{T}_\ell$. We index the level-dependent number $M_\ell$ of samples of these vectors by $k$, i.e. we write for $\ell = 0, 1, \ldots$

$$\{\mathbf{U}_{0,\ell}^k, \mathbf{F}_\ell^k, \mathbf{S}_\ell^k\}_{k=1}^{M_\ell} = \{\mathbf{U}_{0,\ell}(\omega^k), \mathbf{F}_{j,\ell}(\omega^k), \mathbf{S}_\ell(\omega^k) : k = 1, \ldots, M_\ell\}. \qquad (46)$$

3. **Solve:** For each resolution level $\ell \in \mathbb{N}_0$ and for each realization of the random input data $\{\mathbf{U}_{0,\ell}^k, \mathbf{F}_{j,\ell}^k, \mathbf{S}_\ell^k\}$, $k = 1, \ldots, M_\ell$, the resulting *deterministic balance law* (1) (for this particular realization) is solved numerically by the Finite Volume Method (5) with mesh width $\Delta x_\ell$. We denote the resulting ensemble of Finite Volume solutions by $\mathbf{U}_{\mathscr{T}_\ell}^{k,n}$, $k = 1, \ldots, M_\ell$. These constitute vectors of approximate cell averages, i.e. $\mathbf{U}_{\mathscr{T}_\ell}^{k,n} = \{\mathbf{U}_C^{k,n} : C \in \mathscr{T}_\ell\}$ of the corresponding realization of the random balance law at time level $t^n$ and at spatial resolution level $\ell$.

4. **Estimate solution statistics:** Fix some positive integer $L < \infty$ corresponding to the highest level. We estimate the expectation of the random solution field with the following estimator:

$$E^L[\mathbf{U}(\cdot, t^n)] := \sum_{\ell=0}^{L} E_{M_\ell}[\mathbf{U}^n_{\mathcal{T}_\ell} - \mathbf{U}^n_{\mathcal{T}_{\ell-1}}], \tag{47}$$

with $E_{M_\ell}$ being the MC estimator defined in (39) for the mesh level $\ell$.

*Remark 3.* In the present article, we assume (for ease of exposition only) the sequence of triangulations $\{\mathcal{T}_\ell\}_{\ell=0}^{\infty}$ to be nested. This assumption was also made in the proofs of [30]. We emphasize here that an inspection of the arguments in [30] reveals that *the nestedness assumption on the meshes is not essential for the error bounds to hold.* However, in order to execute Step 4. (estimate solution statistics), in the case that the grid hierarchy is non-nested, an efficient intergrid transfer resp. prolongation must be available. This is often the case, when *multilevel discretizations* have been employed in the deterministic solver which is used for the discrete solutions.

*Remark 4.* Higher statistical moments (17) of the random entropy solution can be approximated analogously (see, e.g., the sparse tensor discretization of [30]). An additional, new issue arises in the efficient numerical computation of *space-time correlation functions* due to the high-dimensionality of such correlation functions.

The MLMC-FVM is *non-intrusive* as any standard FVM (or DG) code can be used in step 3. Furthermore, MLMC-FVM is amenable to *efficient parallelization* as data from different grid resolutions and different samples only interacts in step 4. We refer to [42] and to Sect. 5 for details.

### 4.1.2 MLMC-FVM Error Bounds

Again, based on the rigorous estimate for scalar conservation laws in [30, 31] and on our experience for systems of balance laws [32, 33], we postulate the following error estimate [12]:

$$\|\mathbb{E}[u(\cdot,t^n)] - E^L[u(\cdot,t^n)]\|_{L^2(\Omega;L^1(\mathbb{R}^d))} \le C_1 \Delta x_L^s + C_2\Big\{\sum_{\ell=0}^{L} M_\ell^{-\frac{1}{2}} \Delta x_\ell^s\Big\} + C_3 M_0^{-\frac{1}{2}}. \tag{48}$$

Here, the parameter $s > 0$ refers to the convergence rate of the finite volume scheme for the deterministic problem and $C_{1,2,3}$ are positive constants depending only on the second moments of the initial data and the source term.

From the error estimate (48), we obtain that the number of samples to equilibrate the statistical and spatio-temporal discretization errors in (47) is given by

$$M_\ell = \mathcal{O}(2^{2(L-\ell)s}), \quad \ell = 0, 1, \dots, L. \tag{49}$$

Notice that (49) implies that the largest number of MC samples is required on the coarsest mesh level $\ell = 0$, whereas only a small fixed number $M_L \in \mathbb{N}$ of MC samples are needed on the finest discretization levels. To achieve this, we choose

$$M_\ell = M_L 2^{2(L-\ell)s}. \tag{50}$$

Using choice (49) for $M_\ell$, the error bound (48) reduces to

$$\|\mathbb{E}[u(\cdot, t^n)] - E^L[u(\cdot, t^n)]\|_{L^2(\Omega;L^1(\mathbb{R}^d))} \leq C_{\mathrm{Err}} \Delta x_L^s (L+1), \tag{51}$$

with positive constant $C_{\mathrm{Err}}$ independent of $\Delta x$, $\Delta t$ and $L$.

### 4.1.3 MLMC-FVM Work Estimates

Total amount of work required for the complete MLMC-FVM estimator (47) can be estimated using (50) and (42):

$$\begin{aligned} \mathrm{Work}_{\mathrm{MLMC}}(\Delta x_L, L) &= \sum_{\ell=0}^{L} M_\ell \mathrm{Work}_{\mathrm{FVM}}(\Delta x_\ell) \\ &= \sum_{\ell=0}^{L} M_L 2^{2s(L-\ell)} C_s \Delta x_\ell^{-(d+1)} \\ &= \sum_{\ell=0}^{L} M_L 2^{2s(L-\ell)} C_s 2^{-(d+1)(L-\ell)} \Delta x_L^{-(d+1)} \\ &= M_L C_s \Delta x_L^{-(d+1)} 2^{\alpha L} \sum_{\ell=0}^{L} 2^{\alpha \ell}, \end{aligned} \tag{52}$$

where $\alpha := 2s - (d+1)$. Next, we bound the sum in (52) using geometric series,

$$\sum_{\ell=0}^{L} 2^{-\alpha\ell} \leq \begin{cases} \frac{2^{-\alpha L}}{1-2^{\alpha}} & \alpha < 0, \\ L+1 & \alpha = 0, \\ \frac{1}{1-2^{-\alpha}} & \alpha > 0\,. \end{cases} \tag{53}$$

### 4.1.4 MLMC-FVM Error Versus Work Bounds

Combining error estimate (51) with work estimates (52)–(53), we obtain (see [32]) the error vs. work estimate for (47),

$$\left\| \mathbb{E}[u] - E^L[u] \right\|_{L^2(\Omega;L^1(\mathbb{R}^d))} \leq \begin{cases} C_{\text{MLMC}} \cdot \text{Work}^{-s/(d+1)} \cdot \log(\text{Work}) & s < (d+1)/2, \\ C_{\text{MLMC}} \cdot \text{Work}^{-1/2} \cdot \log(\text{Work})^{3/2} & s = (d+1)/2, \\ C_{\text{MLMC}} \cdot \text{Work}^{-1/2} \cdot \log(\text{Work}) & s > (d+1)/2. \end{cases} \tag{54}$$

with constant $C_{\text{MLMC}}$ given by (disregarding constants from the logarithm base)

$$C_{\text{MLMC}} = C_{\text{Err}}(M_L C_s)^{s/(d+1)} \cdot \begin{cases} (1-2^{\alpha})^{-s/(d+1)} & s < (d+1)/2, \\ 1 & s = (d+1)/2, \\ (1-2^{-\alpha})^{-s/(d+1)} & s > (d+1)/2. \end{cases} \tag{55}$$

For usual convergence rates, i.e. $s \geq 1/2$, $s \in \mathbb{N}/2$, constant $C_{\text{MLMC}}$ is bounded by

$$C_{\text{MLMC}} \leq (2M_L)^{s/(d+1)} \cdot C_{\text{Err}} C_s^{s/(d+1)}. \tag{56}$$

In the numerical experiments of Sect. 6.7, constants $C_{\text{Err}}$ in (44) and (51) have almost equal numerical value. Hence constant $C_{\text{MLMC}}$ can be bounded in terms of $C_{\text{FVM}}$,

$$C_{\text{MLMC}} \leq (2M_L)^{s/(d+1)} \cdot C_{\text{FVM}} \leq \sqrt{2M_L} \cdot C_{\text{FVM}}, \quad \text{for } s \leq (d+1)/2. \tag{57}$$

The error estimates in (54) show that the MLMC-FVM is superior to the MC-FVM as the asymptotic computational cost for MLMC-FVM scales as $\text{Work}^{-s/(d+1)}$ (disregarding logarithmic term, see [32]); compare to $\text{Work}^{-s/(d+1+2s)}$ for the MC-FVM scheme as in (43). Hence, the MLMC-FVM is expected to be (asymptotically) considerably faster than the MC-FVM for the same magnitude of error.

Notice, that for $s > (d+1)/2$, the asymptotic error convergence rate in (43) no longer increases as it is limited by the convergence rate of Monte Carlo sampling which equals $1/2$. Moreover, the constant $C_{\text{MLMC}}$ also *increases* with $s$, see (55). However, such high ($s > 2$) convergence rates of the *deterministic* solver can only be observed in very special cases, where the pathwise regularity of the stochastic solution is high (i.e. in the absence of shocks such as, for example, the Ringleb flow).

Furthermore, if $s \leq (d+1)/2$ then the error vs. work estimate (54) is almost (i.e. up to a logarithmic term) of the same order as the error vs. work of the deterministic finite volume scheme which implies that the total amount of work to achieve a certain error level $\varepsilon$, say, in approximation of the random entropy solution's mean field will, asymptotically, be equal to that of approximating the entropy solution of one deterministic balance law at the same level $L$ of resolution.

In addition to *optimal* asymptotic convergence rates of the MLMC-FVM solver, bounds (57) provide estimates on how much more work is required for *stochastic* version of the simulation compared to its *deterministic* version. The difference in constants is comparably small, i.e. the *stochastic* simulation is *at most* $\sqrt{2M_L}$ times less accurate (disregarding logarithmic terms) compared to its *deterministic* version, where the free parameter $M_L$ is usually chosen to be small, i.e. $\mathcal{O}(1) - \mathcal{O}(10)$.

*Remark 5.* Notice, that the standard MC-FVM estimator (39) is monotonic, i.e. mean estimator of *positive* quantities (such as pressure, density) is also *positive*. However, the MLMC-FVM estimator (47) is not monotonic, i.e. negative undershoots might occur, which are *not* related to numerical instability, bias of the estimators, inappropriate FVM schemes or other implementation errors. The awareness of this *non-monotonicity* of the MLMC-FVM estimator is essential for the interpretation of the numerical results. More detailed explanation and examples are provided in Remark 7.

## 4.2 Sparse Tensor Approximations of k-Point Correlations

We now consider the efficient MLMC-FVM approximation of two and of $k$-point correlation functions of random entropy solutions of the system (1). Throughout, we consider only the scalar case, i.e. $m = 1$ in (1), in order to simplify the notation. All concepts and methods have (tensor) analogues in the case of hyperbolic systems.

Throughout this section, we assume that there exists a unique random entropy solution of (1) which satisfies, for a given order $k \geq 1$ of correlation of interest,

$$u(\cdot, t; \omega) \in L^{2k}(\Omega; C^0([0, T]; W^{s,1}(D))) , \quad \text{for some} \quad 0 < s \leq 1 . \tag{58}$$

I.e., we assume that the function is integrable to the power $2k$ if $k \geq 1$ is the order of the moment of interest, and admits, as a function of the spatial variable, a fractional derivative of order $s$ which belongs to the space $L^1(D)$ where $D \subset \mathbb{R}^d$ denotes the computational domain.

For a mesh hierarchy $\{\mathcal{T}_\ell\}_{\ell=0}^{\infty}$ in $D$, we define the space $S_\ell$ of simple, i.e. piecewise constant functions on $\mathcal{T}_\ell$, and the associated projector by

$$S_\ell := S(\mathcal{T}_\ell), \quad P_\ell := P_{\mathcal{T}_\ell} : L^1(\boldsymbol{D}) \mapsto S_\ell, \qquad \ell \geq 0 . \tag{59}$$

Here, for a given triangulation $\mathcal{T}$, $P_{\mathcal{T}}$ denotes the operator which associates to a function $v \in L^1(D)$ the piecewise constant function of cell-averages $P_{\mathcal{T}} v \in S(\mathcal{T})$.

Then the MC-FVM approximations of $\mathcal{M}^k(u(\cdot, t))$ are defined as statistical estimates from the ensemble

$$\{\hat{v}^i_{\mathcal{T}}(\cdot, t)\}_{i=1}^{M} \tag{60}$$

obtained by from samples of the RSCL: specifically, the first moment of the random solution $u(\cdot, t; \omega)$ at time $t > 0$, is estimated as

$$\mathcal{M}^1(u(\cdot, t)) \approx E_M[v_{\mathcal{T}}(\cdot, t)] := \frac{1}{M} \sum_{i=1}^{M} \hat{v}^i_{\mathcal{T}}(\cdot, t) , \tag{61}$$

and, for $k > 1$, the $k$th moment (or $k$ point correlation function) $\mathscr{M}^k(u(\cdot,t)) = \mathbb{E}[(u(\cdot,t))^{(k)}]$ defined in (17) is estimated by

$$E_M^{(k)}[v_{\mathscr{T}}(\cdot,t)] := \frac{1}{M}\sum_{i=1}^{M} \underbrace{(\hat{v}^i_{\mathscr{T}} \otimes \cdots \otimes \hat{v}^i_{\mathscr{T}})}_{k-\text{times}}(\cdot,t)\,. \tag{62}$$

More generally, for $k > 1$, we consider time instances $t_1,\ldots,t_k \in (0,T]$, $T < \infty$, and define the statistical FVM estimate of $\mathscr{M}^k(u)(t_1,\ldots,t_k)$ by

$$E_M^{(k)}[v_{\mathscr{T}}](t_1,\ldots,t_k) := \frac{1}{M}\sum_{i=1}^{M} \underbrace{(\hat{v}^i_{\mathscr{T}}(\cdot,t_1) \otimes \cdots \otimes \hat{v}^i_{\mathscr{T}}(\cdot,t_k))}_{k-\text{times}}\,. \tag{63}$$

The work to form a single tensor product in the ensemble average (63) over a finite spatial domain $D \subset \mathbb{R}^d$ grows as $O(\Delta x^{-kd})$ which, in general, entails a computational effort that is, for moment orders $k \geq 2$, prohibitive. To reduce the complexity of $k$-th moment estimation, we introduce in the following a *compressive approximation* of two- and of $k$-point correlation functions that is similar to the strategy for high order moment approximation in elliptic problems with random data in [5, 44].

### 4.2.1 Sparse Tensorization of FV Solutions

Since the cell-average projections $P_\ell : L^1(\mathbb{R}^d) \to S_\ell$ in (59) are onto, we may define the linear space of *increment* or *details* of FV functions between successive meshes in the grid hierarchy $\{\mathscr{T}_\ell\}_{\ell=0}^{\infty}$ by

$$W_\ell := (P_\ell - P_{\ell-1})S_\ell, \qquad \ell \geq 0, \tag{64}$$

where $P_{-1} := 0$ so that $W_0 = S_0$. Typically, this increment space is spanned by the so-called *Haar-wavelets* on $\mathscr{T}_\ell$; however, in what follows we aim at retaining the nonintrusive nature of the MLMC-FVM and therefore we will *never explicitly assume redesign of the FV solvers in terms of multiresolution analysis on the triangulations* $\mathscr{T}$. We rather propose an algorithm which is based only on the piecewise constant approximations at each timestep, and on recursively identifying the projections of the FV solution onto the increment or detail space $W_\ell$ in (64) "on the fly", by the so-called pyramid scheme. Such schemes are available for approximations in structured meshes, but moreover can also be developed for unstructured grids (we refer to [37, Chap. 2] and [38] for details).

With (64), for any $L \in \mathbb{N}_0$, we have the multilevel decomposition

$$S_L = W_0 \oplus \ldots \oplus W_L = \bigoplus_{\ell=0}^{L} W_\ell \tag{65}$$

and the $k$-point correlation functions $(v_L(\cdot,t))^{(k)}$ of the FV solutions on mesh $\mathscr{T}_L$ at time $t > 0$ take values in the tensor product space

$$(S_L)^{(k)} := \underbrace{S_L \otimes \ldots \otimes S_L}_{k \text{ times}} = \sum_{|\boldsymbol{\ell}|_\infty \le L} S_{\ell_1} \otimes \ldots \otimes S_{\ell_k} = \bigoplus_{|\boldsymbol{\ell}|_\infty \le L} \bigotimes_{j=1}^{k} W_{\ell_j} \,. \tag{66}$$

Then, the full tensor projections

$$P_L^{(k)} v := \underbrace{P_L \otimes \ldots \otimes P_L}_{k \text{ times}} : L^1(\mathbb{R}^{kd}) \to (S_L)^{(k)} \tag{67}$$

are bounded, linear and onto. Here, $|\boldsymbol{\ell}|_\infty := \max\{\ell_1, \ldots, \ell_k\}$ and the last sum in (66) is a direct one. Obviously, if $N_L := \dim S_L < \infty$ (as is the case when, e.g., the spaces $S_\ell$ are only defined on a bounded domain $D \subset \mathbb{R}^d$) then $\dim((S_L)^{(k)}) = N_L^k$ which is prohibitive. Sparse Tensor approximations of $k$-point correlations $(v(\cdot,t))^{(k)}$ will be approximations in tensor products of spaces of piecewise constant functions on meshes on coarser levels which are defined similar to (66) by

$$\widehat{(S_L)}^{(k)} := \bigoplus_{|\boldsymbol{\ell}|_1 \le L} \bigotimes_{j=1}^{k} W_{\ell_j} \tag{68}$$

where now $|\boldsymbol{\ell}|_1 := \ell_1 + \ldots + \ell_k$. If the mesh family $\{\mathscr{T}_\ell\}_{\ell=0}^\infty$ is generated by recursive dyadic refinements of the initial triangulation $\mathscr{T}_0$, when $N_L = \dim S_L < \infty$ (as is the case e.g. on bounded domains $D \subset \mathbb{R}^d$) it holds

$$\dim \widehat{(S_L)}^{(k)} = O(N_L (\log_2 N_L)^{k-1}) \,. \tag{69}$$

With $\widehat{(S_L)}^{(k)}$ in (68), we also define the *sparse tensor projection*

$$\widehat{(P_L)}^{(k)} := \bigoplus_{|\boldsymbol{\ell}|_1 \le L} \bigotimes_{j=1}^{k} (P_{\ell_j} - P_{\ell_j - 1}) : L^1(\mathbb{R}^{kd}) \to \widehat{(S_L)}^{(k)} \,. \tag{70}$$

The approximation properties of the sparse tensor projection are as follows (cf. the Appendix of [30]): for any function $U(x_1, \ldots, x_k)$ which belongs to $(W^{s,1}(\mathbb{R}^d))^{(k)}$, it holds

$$\|U - \widehat{(P_L)}^{(k)} U\|_{L^1(\mathbb{R}^{kd})} \le C (\Delta x_L)^s |\log \Delta x_L|^{k-1} \|U\|_{(W^{s,1}(\mathbb{R}^d))^{(k)}}, \tag{71}$$

where $C > 0$ depends only on $k$, $d$ and on the shape regularity of the family $\{\mathscr{T}_\ell\}_{\ell \ge 0}$ of triangulations, but is independent of $\Delta x$.

### 4.2.2 Definition of the Sparse Tensor MLMC-FVM Estimator

With the above notions in hand, we proceed to the *definition of the sparse tensor MLMC-FVM estimator of* $\mathscr{M}^{(k)}(u(\cdot,t))$. To this end, we modify the full tensor product MLMC-FV estimator which is based on forming the $k$-fold tensor products of the FV solution in the spaces $S_L^{(k)}$ in (66), as follows (recall from (61) that $E_M[\cdot]$ denotes the MC estimate based on $M$ samples): for a given sequence $\{M_\ell\}_{\ell=0}^L$ of MC samples at level $\ell$, the sparse tensor MLMC estimate of $\mathscr{M}^k[u(\cdot,t)]$ is, for $0 < t < \infty$, defined by

$$\widehat{E^{L,(k)}}[u(\cdot,t)] := \sum_{\ell=0}^{L} E_{M_\ell}[\widehat{P_\ell}^{(k)}(v_\ell(\cdot,t))^{(k)} - \widehat{P_{\ell-1}}^{(k)}(v_{\ell-1}(\cdot,t))^{(k)}] \,. \tag{72}$$

We observe that (72) is identical to the full tensor product formation of the Finite Volume solution if the sparse projectors $\widehat{P_\ell^{(k)}}$ in (72) are replaced by the full tensor projections $P_\ell^{(k)}$, except for the sparse formation of the $k$-point correlation functions of the FV solutions corresponding to the initial data samples $\hat{u}_0^i$. In bounded domains, this reduces the work for the formation of the $k$-point correlation function from $N_L^k$ to $O(N_L(\log_2 N_L)^{k-1})$ per sample at mesh level $L$. As our convergence analysis ahead will show, use of sparse rather than full tensor products will not entail any reduction in the order of convergence of the $k$-th moment estimates.

### 4.2.3 Error and Complexity Analysis of the Sparse Tensor MLMC-FVM

The following, basic result on the complexity of the MLMC-FVM proved in [30] shows that *MLMC-FVM estimates of two- and of k-point correlations of the random entropy solutions are possible in log-linear complexity of one single, deterministic solve on the finest mesh level L*.

**Theorem 2.** *Assume the regularity (58). Assume further that we are given a FVM such that* (4) *holds and such that the deterministic FVM scheme converges at rate* $s > 0$ *in* $L^\infty(0,\infty; L^1(\mathbb{R}^d))$. *Then the MLMC-FVM estimate* $\widehat{E^{L,(k)}}[u(\cdot,t)]$ *defined in (72) satisfies, for every sequence* $\{M_\ell\}_{\ell=0}^L$ *of MC samples, the error bound*

$$\|\mathscr{M}^k u(\cdot,t) - \widehat{E^{L,(k)}}[u(\cdot,t;\omega)]\|_{L^2(\Omega;L^1(\mathbb{R}^{kd}))}$$

$$\lesssim (1 \vee t)\Delta x_L^s |\log \Delta x_L|^{k-1} \left\{ \|\mathrm{TV}(u_0(\cdot,\omega))\|_{L^k(\Omega;d\mathbb{P})}^k + \|u_0(\cdot;\omega)\|_{L^\infty(\Omega;W^{s,1}(\mathbb{R}^d))}^k \right\}$$

$$+ \left\{ \sum_{\ell=0}^{L} \frac{\Delta x_\ell^s |\log \Delta x_\ell|^{k-1}}{M_\ell^{1/2}} \right\} \left\{ \|u_0(\cdot;\omega)\|_{L^{2k}(\Omega;W^{s,1}(\mathbb{R}^d))}^k + t\|\mathrm{TV}(u_0(\cdot;\omega))\|_{L^{2k}(\Omega;d\mathbb{P})}^k \right\} .$$

*The total work to compute the MLMC estimates $\widehat{E^{L,(k)}}[u(\cdot\,;t)]$ on compact domains $D \subset \mathbb{R}^d$ is therefore (with $O(\cdot)$ depending on the size of $D$)*

$$\widehat{\mathrm{Work}}_L^{MLMC} = O\left(\sum_{\ell=0}^{L} M_\ell \Delta x_\ell^{-(d+1)} |\log \Delta x|^{k-1}\right). \tag{73}$$

Based on Theorem 2, we infer that the choice of sample sizes $M_\ell$ at level $\ell$ should also be used in the MLMC-FVM estimation of $k$-point correlation functions of order $k > 1$ of the random entropy solution, provided the order $s$ of the underlying deterministic FVM scheme is at most 1 (see [30] for details). Due to the linear complexity of the pyramid scheme, the conversion of the FVM approximations of the draws $\hat{u}^i(\cdot\,,t;\omega)$ of the random entropy solution at time $t > 0$ into a multilevel representation and the sparse tensor product formation in the MLMC estimator (72) increases the work bounds for the first moments only by a logarithmic factor, so that, in terms of the computational work, we have with the choices (49) of MC samples $M_\ell$, the following error bound in terms of work in a bounded computational domain $D \subset \mathbb{R}^d$:

$$\|\mathcal{M}^k u(\cdot,t) - \widehat{E^{L,(k)}}[u(\cdot,t;\omega)]\|_{L^2(\Omega;L^1(D^k))} \leq C(\widehat{\mathrm{Work}}_L^{MLMC})^{-s'/(d+1)} \tag{74}$$

for any $0 < s' < s$ with the constant depending on $D$ and growing as $0 < s' \to s \leq 1$.

## 5 Efficient Implementation of MLMC-FVM

As stated in the previous section, the MLMC-FVM algorithm has four stages. We discuss implementation issues that arise in each stage below.

### *5.1 Step 1: Hierarchy of Nested Grids*

We will solve systems of balance laws (1) in one and two space dimensions. In the numerical results which are reported in Sect. 6 ahead, in two space dimensions, we will choose Cartesian meshes for simplicity. It is relatively straightforward to choose any hierarchy of nested grids consisting of either triangles/tetrahedra or quad/hexahedral volume in either one, two or three space dimensions.

### *5.2 Step 2: Sample*

In this step, we have to draw $M_\ell$ i.i.d. samples for the initial data and source random fields $\mathbf{U}_0, \mathbf{F}_j, \mathbf{S}$ corresponding to the underlying probability distribution. Standard random number generators (RNG) can be readily used to draw such samples. For the serial implementation, any reasonable RNG works well in practice. However, random number generation becomes a very sensitive part of Monte Carlo type algorithms on massively parallel architectures. Inconsistent seeding and insufficient period length of the RNG might cause correlations in presumably i.i.d. draws which might potentially lead to biased solutions, see [32]. We used the `WELL`-series of pseudo random number generators from [23, 24]. These generators have been designed with particular attention towards large periods and good equidistribution. To deal with the seeding issues, we *injectively* map the *unique* rank of each core (in a parallel algorithm) to some corresponding element in the hardcoded array of prime numbers (henceforth, the array of seeds), see [42] for detailed explanations. In this way statistical independence is preserved. For all numerical experiments reported here, the RNG `WELL512a` was used. We found `WELL512a` to have a sufficiently large period $2^{512} - 1$ and to be reasonably efficient (33 CPU sec for $10^9$ draws). We emphasize that there are plenty of alternatives to `WELL512a` with even longer periods (which, however, use more memory than `WELL512a`). To name a few: `WELL1024a` with period $2^{1,024} - 1$, takes 34 s and `WELLRNG44497` with period $2^{44,497} - 1$ which takes 41 s to generate $10^9$ draws; yet another alternative could be "Marsenne Twister" [29] – it has the same period of $2^{19,937} - 1$ as its counterpart `WELLRNG19937`. Within our numerical experiments, we did not observe any *qualitative* difference between the described RNGs.

## 5.3 *Step 3: Solve*

For each realization of the random input fields, we need to solve (1) for each realization of the random input with a finite volume or DG scheme.

### 5.3.1 General Consideration

As the solve step in the MLMC-FVM algorithm will be repeated for a large number of data samples on different space-time resolution levels, a robust and efficient FVM code for systems of balance laws. We recall that the MC and MLMC methods rely on *equidistribution of samples in the data space*. Therefore, in MC-FVM and MLMC-FVM, also extreme (i.e. "improbable" or physically practically impossible in the eyes of experts) data scenarios are generated, and care must be taken that the numerical solver for these scenarios is of comparable efficiency as for the more standard, "benchmark" cases. *Robustness of the numerical solver* is therefore a key issue in the development and deployment of MLMC techniques for nonlinear PDEs. In our large scale numerical experiments, we choose the code named `ALSVID` [2] that was designed by researchers at CMA, University of Oslo and SAM, ETH Zürich. Based on this platform, we developed a version called `ALSVID-UQ` which is specifically tailored to UQ for hyperbolic systems of conservation and balance laws, and the complete source code of which is publicly available for download under [3]. As `ALSVID` is extensively used in the examples of this paper, we describe it briefly below.

### 5.3.2 ALSVID

This finite volume code approximates the Shallow-water, Euler equations and MHD equations in one, two and (in the last two cases) in three space dimensions. It is based on the following ingredients:

1. **Approximate Riemann solver:** The numerical fluxes in the Finite Volume Scheme (5) used in `ALSVID` are based on approximate Riemann solvers of the HLL type for the Euler and MHD equations [16]. For shallow-water equations with bottom topography, the energy stable well-balanced schemes of [15] have been implemented in `ALSVID-UQ`.
2. **Divergence constraint.** The divergence constraint in the MHD equations [16] is handled in `ALSVID` by adding the Godunov-Powell source term to the MHD equations. This source term is proportional to the divergence and allows divergence errors to be swept out of the domain. Numerical stability can only be ensured by a careful *upwinding* of the source term, see [16].
3. **Non-oscillatory reconstructions.** `ALSVID` employs a variety of piecewise polynomial non-oscillatory reconstruction procedures for attaining high order of spatial accuracy. In particular, second order ENO and WENO procedures are employed, see Sect. 2 of [16]. However, these procedures need to be modified in order to preserve positivity of the density and pressure. Such modifications are described in Sect. 2 of [16] (cp. also [41]).
4. **Time stepping.** High-order accurate time stepping procedures of the SSP Runge-Kutta [20] are employed in `ALSVID`.

Fluxes on the boundary of the computational domain are defined using so-called ghost cells, see, for example, Chap. 10 in [25]. The FV solver library `ALSVID` uses a modular structure in `C++` with a `Python` front end for pre- and post-processing. One and two dimensional visualizations are performed with `MatPlotLib` and three dimensional data sets are visualized using `MayaVi2`. Extensive testing of `ALSVID` has been performed and reported in [16].

A massively parallel version of `ALSVID` had already been developed for deterministic problems; we refer to [2] for further details. The parallelization paradigm for `ALSVID` is based on domain decomposition using Message Passing Interface (`MPI`) standard [51] and its particular implementation `OpenMPI` [52].

## 5.4 *Stable Computation of Sample Statistics*

For both MC-FVM and MLMC-FVM algorithms, we need to combine ensembles of individual realizations of numerical solutions for the statistical estimation of ensemble averages.

### 5.4.1 Discussion of Round-Off Effects

It is straightforward to evaluate the sample mean for the MC-FVM and the estimator (47) for MLMC-FVM. A straightforward algorithm to compute an *unbiased* estimate of the variance for scalar $u = u(x,t)$ with fixed $x,t$ is the following statistical estimator $\mathrm{Var}_M[u]$ of the variance $\mathrm{Var}[u] := \mathbb{E}[u^2] - \mathbb{E}[u]^2$:

$$\mathrm{Var}_M[u] := \frac{1}{M-1}\sum_{i=1}^{M}(u^i)^2 - \frac{1}{M(M-1)}\left(\sum_{i=1}^{M}u^i\right)^2 , \tag{75}$$

where $u^i$ are MC-FVM samples. This way, it suffices to loop over all samples only once; unfortunately, both quantities are almost equal in regions of nearly vanishing variance. This is typically the case in smooth regions of the flow, i.e. outside of shocks and of viscous boundary layers. We observed in our numerical experiments that the straightforward use of the estimator (75) is prone to *subtractive cancellation* and *loss of accuracy* in finite (IEEE double) precision floating point arithmetic. This problem is well-known, and in [46], the authors propose an alternative *stable* "on-line" variance computation algorithm:
Set $\bar{u}^0 = 0$ and $\Phi^0 = 0$; then proceed iteratively:

$$\bar{u}^i = \sum_{j=1}^{i} u^j / i, \tag{76}$$

$$\Phi^i := \sum_{j=1}^{i}(u^j - \bar{u}^i)^2 = \Phi^{i-1} + (u^i - \bar{u}^i)(u^i - \bar{u}^{i-1}) . \tag{77}$$

Then, unbiased mean and variance estimates are given by:

$$E_M[u] = \bar{u}^M, \quad \mathrm{Var}_M[u] = \Phi^M/(M-1). \tag{78}$$

Although identical in exact arithmetic, the above algorithm can deal with small cancellation errors. Note, however, that the estimators (75) and (78) will only converge under the provision that *fourth statistical moments of the random solution, evaluated pointwise in space and time, are finite*; an alternative approach which does not require pointwise finite fourth moments proceeds via estimating the space-time two-point correlation functions of the random entropy solutions, see [30]. For this approach, finite fourth moments of $u$ are also required, albeit with values in $(C_b([0,\ T], L^1(\mathbb{R}^d)))^m$.

### 5.4.2 Efficient Parallelization

The key issue in the parallel implementation of the whole algorithm (the solve steps) is to distribute computational work evenly among the cores. Without going into the details, we refer the reader to the novel static load balancing strategy on homogeneous parallel architectures (i.e. all cores are assumed to have identical CPUs and RAM per node, and equal bandwidth and latency to all other cores) in recent papers [32, 42].

## 6 Performance Studies of the MLMC-FVM for Conservation Laws

In this section, we will test the MLMC-FVM algorithm, presented in the previous section, and demonstrate its robustness and efficiency. We run numerical tests for five different problem sets: two of them will consider a multi-dimensional system of conservation laws with uncertain initial data, one will consider a system of balance laws with random bottom topography (source term) and the remaining three numerical experiments will address the performance of our Multi-Level MC-FVM for conservation laws with random fluxes.

Recalling that the discretization of the random conservation law involves discretizing in space-time with a standard Finite Volume Method and the discretizing the probability space with a statistical sampling method, we tabulate various combinations of methods that are to be tested:

| | | |
|---|---|---|
| **MC** | Monte Carlo with 1st order FVM scheme | $M = \mathcal{O}(\Delta x^{-1})$, |
| **MC2** | Monte Carlo with 2nd order FVM scheme | $M = \mathcal{O}(\Delta x^{-2})$, |
| **MLMC** | Multilevel MC with 1st order FVM scheme | $M_\ell = M_L 2^{(L-\ell)}$, |
| **MLMC2** | Multilevel MC with 2nd order FVM scheme | $M_\ell = M_L 4^{(L-\ell)}$. |

Furthermore, we need to following parameters, which will be specified for every simulation in the form of a table below the corresponding figure:

| **Parameter** | **Description** |
|---|---|
| $L$ | Number of hierarchical mesh levels |
| $M_L$ | Number of samples at the finest mesh level |
| Grid size | Number of cells in X and in Y directions |
| CFL | CFL number based on the fastest wave |
| Cores | Total number of cores used in the simulation |
| Runtime | Clock-time (serial runs) or wall-time (parallel runs); hrs:min:sec |
| Efficiency | MPI efficiency, as defined in [42]. |

As we will present numerical convergence analysis results, we need to specify the following error estimator.

*Error estimator.* Since the solution is a random field, the discretization error is a random quantity as well. For our computational convergence analysis we therefore compute a statistical estimator by averaging estimated discretization errors from several independent runs. We will compute the error in (40) by approximating $L^2(\Omega; L^1(\mathbb{R}^d))$ norm with MC quadrature. Let $U_{\mathrm{ref}}$ denote the reference solution and $\{U_k\}_{k=1,\dots,K}$ be a sequence of independent approximate solutions obtained by running MC-FVM or MLMC-FVM solver $K$ times corresponding to $K$ realizations of the stochastic space. Then the $L^2(\Omega; L^1(\mathbb{R}^d))$-based relative error estimator is defined as in [30],

$$\mathscr{R}E = \sqrt{\sum_{k=1}^{K} (\mathscr{R}E_k)^2 / K}, \tag{79}$$

where:

$$\mathscr{R}E_k = 100 \times \frac{\|U_{\mathrm{ref}} - U_k\|_{\ell^1(\mathscr{T})}}{\|U_{\mathrm{ref}}\|_{\ell^1(\mathscr{T})}}. \tag{80}$$

The extensive analysis for the appropriate choice of $K$ is conducted in [30]; unless indicated otherwise, we choose $K = 30$ which proved to be sufficient in our numerical experiments to remove statistical fluctuations in the convergence plots.

Equipped with the above notation and concepts, we present the following six numerical experiments.

## 6.1 Euler Equations with Uncertain Initial Data

The Euler equations of gas dynamics are

$$\begin{cases} \rho_t + \operatorname{div}(\rho\mathbf{u}) = 0, \\ (\rho\mathbf{u})_t + \operatorname{div}(\rho\mathbf{u} \otimes \mathbf{u} + p\mathbf{ID}) = 0, \\ E_t + \operatorname{div}((E + p)\mathbf{u}) = 0. \end{cases} \tag{81}$$

Here, $\rho$ is the density and $\mathbf{u}$ is the velocity field. The pressure $p$ and total energy $E$ are related by the ideal gas equation of state:

$$E := \frac{p}{\gamma - 1} + \frac{1}{2}\rho|\mathbf{u}|^2, \tag{82}$$

with $\gamma$ being the ratio of specific heats.

The MLMC-FVM algorithm is tested on a problem with a large number of sources of uncertainty. We consider the so-called *cloud-shock interaction problem*. The computational domain is taken to be $\mathbf{D} = [0, 1] \times [0, 1]$. Let $Y \sim \frac{1}{25} + \mathscr{U}(0, \frac{1}{50})$ and let $Y_1, \ldots, Y_7 \sim \mathscr{U}(0, 1)$ denote i.i.d. random variables independent of $Y$.

The initial data consists of an initial shock with uncertain amplitude and uncertain location given by:

$$\{\rho_0(\mathbf{x}, \omega),\ \mathbf{u}_0(\mathbf{x}, \omega),\ p_0(\mathbf{x}, \omega)\} =$$

$$= \begin{cases} \left\{3.86859 + \frac{1}{10} Y_6(\omega),\ (11.2536,\ 0)^\top,\ 167.345 + Y_7(\omega)\right\} & \text{if} \quad \mathbf{x}_1 < Y(\omega), \\ \{1,\ (0,\ 0)^\top,\ 1\} & \text{if} \quad \mathbf{x}_1 > Y(\omega). \end{cases} \tag{83}$$

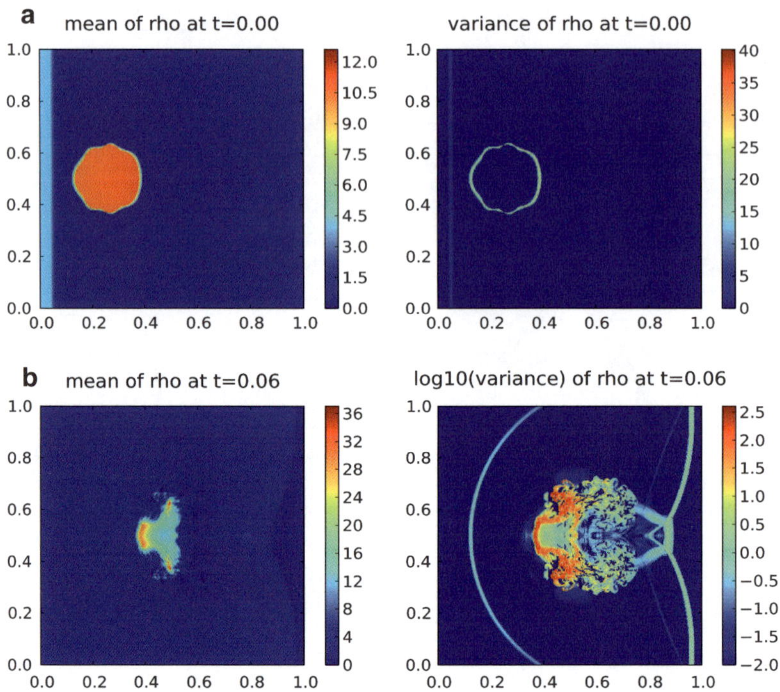


| $L$ | $M_L$ | Grid size | CFL | Cores | Runtime | Efficiency |
|---|---|---|---|---|---|---|
| 9 | 8 | 4,096 × 4,096 | 0.4 | 1,023 | 5:38:17 | 96.9 % |

**Fig. 1** Cloud shock at $t = 0$ and $t = 0.06$ using MLMC-FVM. (**a**) Initial data. (**b**) Solution at $t = 0.06$

Furthermore, a high density cloud or bubble with uncertain amplitude and uncertain shape of the form

$$\boldsymbol{\rho_0}(\mathbf{x},\omega) = 10 + \frac{1}{2}Y_1(\omega) + Y_2(\omega)\sin(4(\mathbf{x}_1 - \frac{1}{4})) + \frac{1}{2}Y_3(\omega)\cos(8(\mathbf{x}_2 - \frac{1}{2})) \quad \text{if} \quad r \leq 0.13 + \frac{1}{50}Y_4(\omega)\sin\theta + \frac{1}{100}Y_5(\omega)\sin(10\theta), \tag{84}$$

where

$$r = \sqrt{(\mathbf{x}_1 - 0.25)^2 + (\mathbf{x}_2 - 0.5)^2}, \quad \theta = \frac{\mathbf{x}_1 - 0.25}{r}, \tag{85}$$

lies to the right of the shock. The mean and the variance of the initial data are depicted in Fig. 1a. Note that there are eight sources of uncertainty in the above problem. A parametric representation of the initial data results in a 11 dimensional problem consisting of 2 space, 1 time and 8 stochastic dimensions. The mean and variance of the solution at time $t = 0.06$ is shown in Fig. 1b. The results are from a MLMC-WENO run with 10 nested levels of resolution ($L = 9$) and the finest resolution is set to 4,096 × 4,096 mesh. The number $M_L$ of MC samples at the finest resolution is 8 and number of cores for this run is 1,023.

The physics of the flow in this case consists of the supersonic initial shock moving to the right, interacting with the high density bubble and leading to a complex flow pattern that consists of a leading bow shock, trailing tail shocks and a very complex region (near the center) possessing sharp gradients as well as turbulent like smooth features. The mean flow (for the density) consists of the bow shock, tail shocks and a complex region with sharp gradients as well as smooth regions. The variance is concentrated in the smooth region at the center; it is significantly smaller at the tail shocks and almost vanishing at the bow shock. The initial uncertainty in the shape of the bubble seems to lead to a more complex distribution of the variance.

## 6.2 MHD Equations of Plasma Physics

Next, we consider the MHD equations:

$$\begin{cases} \rho_t + \operatorname{div}(\rho \mathbf{u}) = 0, \\ (\rho \mathbf{u})_t + \operatorname{div}(\rho \mathbf{u} \otimes \mathbf{u} + (p + \frac{1}{2}|\mathbf{B}|^2)I - \mathbf{B} \otimes \mathbf{B}) = 0, \\ \mathbf{B}_t + \operatorname{div}(\mathbf{u} \otimes \mathbf{B} - \mathbf{B} \otimes \mathbf{u}) = 0, \\ E_t + \operatorname{div}((E + p + \frac{1}{2}|\mathbf{B}|^2)\mathbf{u} - (\mathbf{u} \cdot \mathbf{B})\mathbf{B}) = 0, \\ \operatorname{div}(\mathbf{B}) = 0. \end{cases} \tag{86}$$

Here, $\mathbf{B}$ denotes the magnetic field and the total energy is given by the equation of state (82). In this example, the random initial data is a parametric version of the celebrated *Orszag-Tang* vortex which is *randomly* perturbed in two different ways:

1. **Two sources of uncertainty.** Let $Y_1, Y_2 \sim \mathcal{U}(0, 1)$. The phases of the velocities are uncertain and depend on the scaled random variables $Y_1, Y_2$:

$$\begin{aligned} &\{\rho_0(\mathbf{x}, \omega),\ p_0(\mathbf{x}, \omega)\} = \{\gamma^2,\ \gamma\}, \\ &\mathbf{u}_0(\mathbf{x}, \omega) = \left(-\sin\left(\pi \mathbf{x}_2 + \frac{1}{20}Y_1(\omega)\right),\ \sin\left(\pi \mathbf{x}_1 + \frac{1}{10}Y_2(\omega)\right)\right)^\top, \\ &\mathbf{B}_0(\mathbf{x}, \omega) = (-\sin(\pi \mathbf{x}_2),\ \sin(2\pi \mathbf{x}_1))^\top. \end{aligned} \tag{87}$$

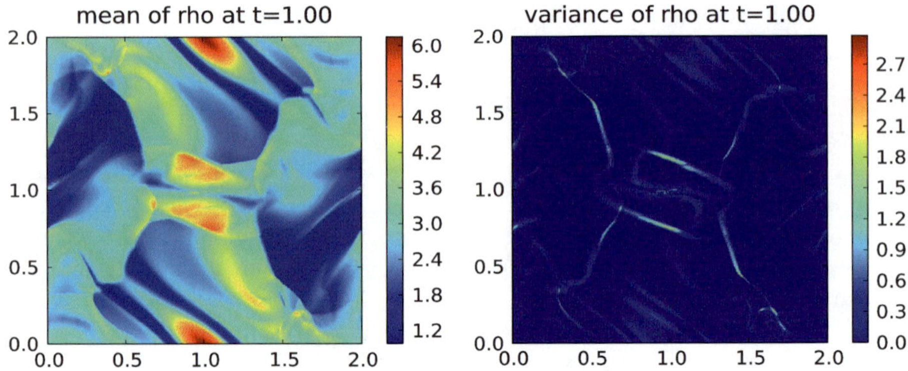


| $L$ | $M_L$ | Grid size | CFL | Cores | Runtime | Efficiency |
|---|---|---|---|---|---|---|
| 7 | 4 | 2,048 × 2,048 | 0.475 | 128 | 5:02:14 | 98.4 % |

**Fig. 2** Uncertain Orszag-Tang vortex solution at $t = 1.0$ using MLMC-FVM (two sources of uncertainty). Variance is very large near discontinuities of the path-wise solutions

2. **Eight sources of uncertainty.** Let $Y_i \sim \mathcal{U}(-1,1)$, $i = 1,\ldots,8$. The amplitudes of the initial density and pressure are uncertain

$$\begin{aligned}\rho_0(\mathbf{x},\omega) &= \gamma^2\left(1+\frac{1}{20}Y_1(\omega)\right),\\ p_0(\mathbf{x},\omega) &= \gamma\left(1+\frac{1}{20}Y_4(\omega)\right),\end{aligned} \tag{88}$$

and, additionally, the phases of the initial velocities and the phases with the amplitudes of the initial magnetic fields are also uncertain,

$$\begin{aligned}\mathbf{u}_0(\mathbf{x},\omega) &= \left(-\sin\left(\pi\mathbf{x}_2+\frac{1}{20}Y_2(\omega)\right),\ \sin\left(\pi\mathbf{x}_1+\frac{1}{10}Y_3(\omega)\right)\right)^\top,\\ \mathbf{B}_1(\mathbf{x},\omega) &= -\left(1+\frac{1}{20}Y_6(\omega)\right)\sin\left(\pi\mathbf{x}_2+\frac{1}{25}Y_5(\omega)\right),\\ \mathbf{B}_2(\mathbf{x},\omega) &= \left(1+\frac{1}{20}Y_8(\omega)\right)\sin\left(2\pi\mathbf{x}_1+\frac{1}{20}Y_7(\omega)\right).\end{aligned} \tag{89}$$

Here, as in the setup for cloud-shock interaction problem in Fig. 1a, a parametric representation of the initial data results in a 11 dimensional problem consisting of two space, one time and eight stochastic dimensions.

The MLMC-FVM solution is then considered for both versions of the initial data, i.e. with two sources (87) and with eight sources (88) of uncertainty. The mean field and the variance (for the plasma density) of the solutions are shown in Figs. 2 and 3, respectively.

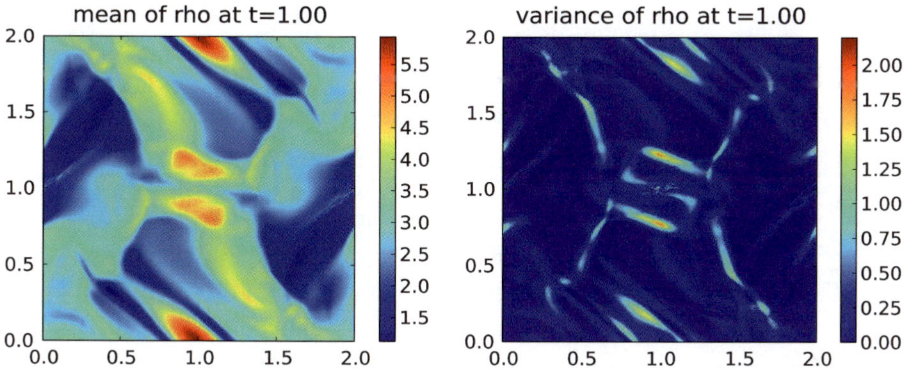


| $L$ | $M_L$ | Grid size | CFL | Cores | Runtime | Efficiency |
|---|---|---|---|---|---|---|
| 8 | 4 | 4,096 × 4,096 | 0.475 | 2,044 | 3:17:18 | 97.0 % |

**Fig. 3** Uncertain Orszag-Tang vortex solution at $t = 1.0$ using MLMC-FVM (eight sources of uncertainty). Again, the largest variances appear near discontinuities of the path-wise solutions

The computation is performed using the MLMC-FVM scheme with second-order WENO reconstruction, and with the HLL three wave solver of [16]. The code uses an upwind discretization of the Godunov-Powell source term. The results shown in these figures are from a computation with 8 levels of refinement ($L = 7$) and the finest mesh resolutions of 2,048 × 2,048 mesh cells and 4,096 × 4,096 mesh cells for the problem with two sources of uncertainty (87) and with eight sources of uncertainty (88), respectively. The number of MC samples at the finest resolution for both problems is 4. The problems have more than $10^9$ degrees of freedom per time step and the total number of time steps is about $10^4$ amounting to an overall computational cost of this simulation between $10^{12}$ and $10^{13}$ FLOPS. These numbers show that the simulations are extremely challenging and requires massively parallel architectures. In fact, the problem with two sources of uncertainty took about 6 h (wall-clock) on 128 cores (simulated on ETH's parallel cluster Brutus [49]) and the problem with eight sources of uncertainty took about 3.5 h (wall-clock) on 2,040 cores (simulated on Palu, CSCS [50]). We also observe that the variance for the problem with eight sources of uncertainty is more diffused than the variance for the problem with two sources of uncertainty.

It is well known (see [16]) that stable computation of numerical solutions of the Orszag-Tang problem on highly refined meshes (which, by the CFL condition (4), entails a correspondingly large number of timesteps) is quite challenging. Since our spatial resolution at mesh level $L = 7$ is very fine, we need an *extremely* robust code like `ALSVID` for the solve step in MLMC-FVM in order to resolve this problem.

The mean density is quite complicated with shocks along the diagonals of the domain as well a (smooth) current sheet at the center of the domain. The solution consists of discontinuities interspersed within interesting smooth features. Our simulations show that the variance is concentrated at shocks as well as at the current

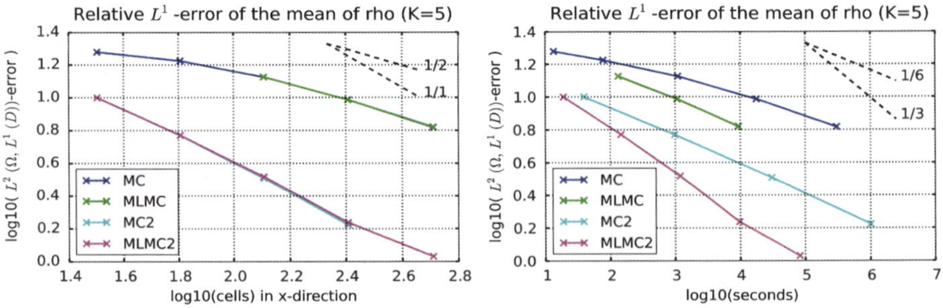


**Fig. 4** Convergence of mean in the uncertain Orszag-Tang vortex simulation (two sources of uncertainty)

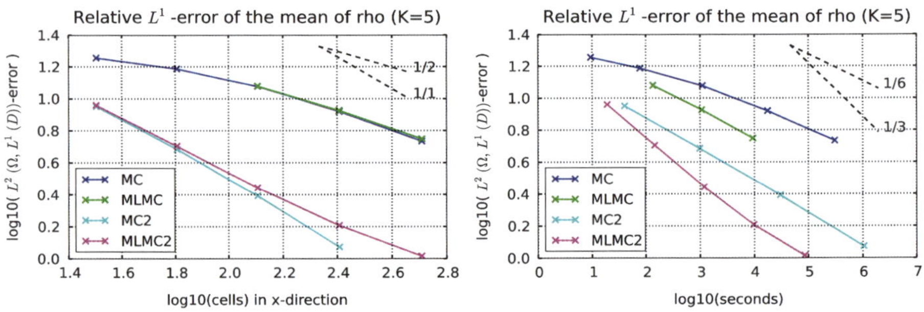


**Fig. 5** Convergence of mean in the uncertain Orszag-Tang vortex simulation (eight sources of uncertainty)

sheets and other interesting smooth regions. From this problem as well as the results of the previous section, we observe that the variance is a very good indicator of where the discontinuities and sharp gradients of the solution are concentrated and would serve as a good a posteriori error indicator for adaptive mesh refinement.

### 6.2.1 Numerical Convergence Analysis

We analyze these particular two dimensional numerical experiments (Orszag-Tang vortex with two and eight sources of uncertainty) in greater detail. Again, we use the high-resolution MLMC-FVM simulations from Figs. 2 and 3 as the reference solutions, respectively. We investigate convergence of error vs. work in Figs. 4 and 6 for two sources of uncertainty and in Figs. 5 and 7 for eight sources of uncertainty. The error in the mean field converges at expected rates. At comparable numerical resolution and accuracy, the MLMC(2) is about two orders of magnitude faster than the MC(2) method for both problems. We observe a slight deterioration in the estimated convergence rates for the variance. This could well be a pre-asymptotic

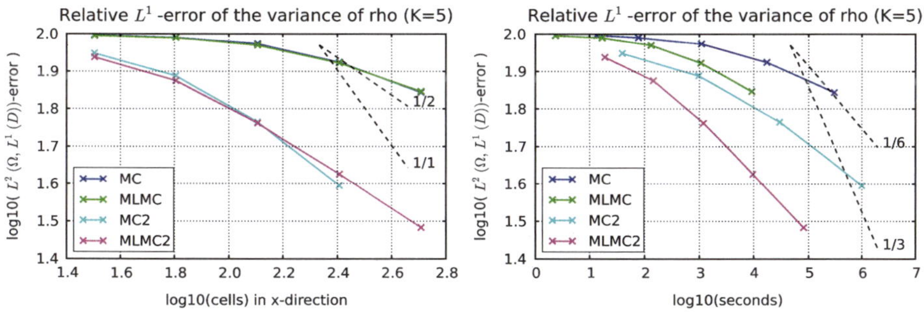


**Fig. 6** Convergence of variance in the uncertain Orszag-Tang vortex simulation (two sources of uncertainty)

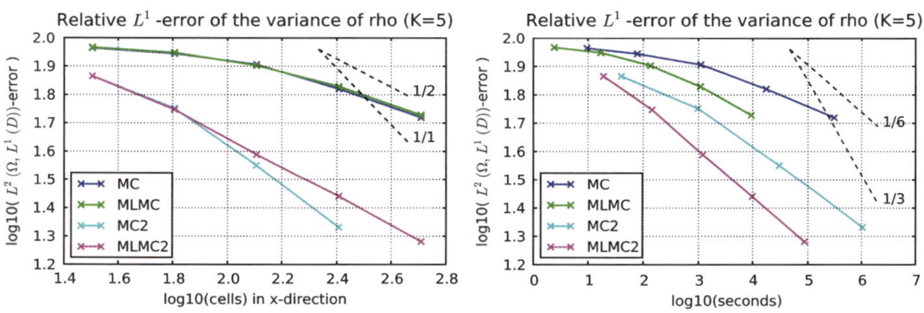


**Fig. 7** Convergence of variance in the uncertain Orszag-Tang vortex simulation (eight sources of uncertainty)

effect. As seen in Figs. 6 and 7, the curves are steepening which seems to indicate better rates with further refinement. Again, the MLMC(2) appears considerably faster than the corresponding MC(2) method in delivering variance estimates of comparable numerical accuracy.

*Remark 6.* Our aim in computing the Orszag-Tang vortex with two and with eight sources of uncertainty in the initial data is to compare the robustness of the MLMC method with respect of an increase in the number of sources of uncertainty. To this end, we plot the error vs. resolution and the error vs. runtime for the MLMC(2) FVM with both two and with eight sources of uncertainty in Fig. 8. The results in this figure show that the runtime for a fixed level of error is nearly identical whether there are two or eight sources of uncertainty in the initial data. This shows that the MLMC method is quite robust with respect to large number of sources of uncertainty in the data: an increase in the number of sources of uncertainty does not appear to lead to a deterioration in the computational efficiency. Thus, the MLMC method can be used for computing uncertainty in problems with a very large number of sources of randomness.

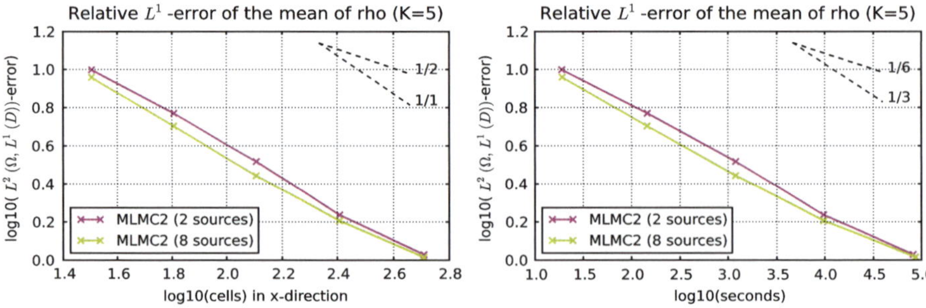


**Fig. 8** Error convergence sensitivity to the number of sources of uncertainty. For both two and eight sources of uncertainty, the convergence of error vs. computational work is almost identical

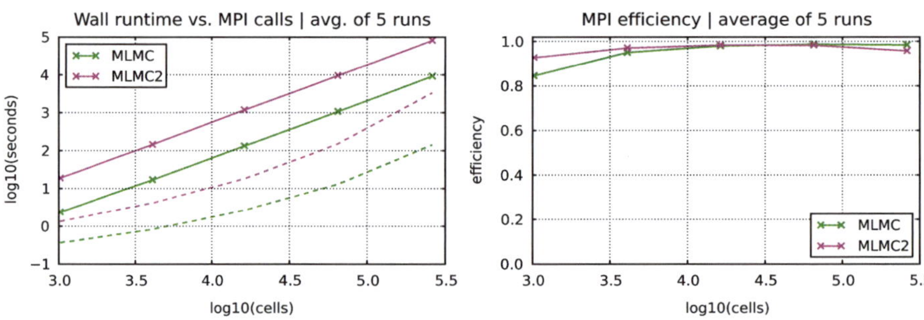


**Fig. 9** MPI overhead

### 6.2.2 Efficiency of Parallelization

We test the efficiency of static load balancing for the parallelization procedure described in [42] in this two-dimensional example. Here, the parallelization efficiency is defined as

$$\text{efficiency} := \frac{(\text{cumulative wall-time}) - (\text{cumulative wall-time of MPI calls})}{\text{cumulative wall-time}}. \tag{90}$$

It separates the amount of time spent in computing from the amount of time spent in communicating (the latter is indicated with *dashed* lines in runtime plots). In Fig. 9, we show the parallelization efficiency of the MLMC-FVM and see that the algorithm is quite efficient and most of the time is spent computing rather than communicating or waiting.

The strong scaling (fixed discretization and sampling parameters while increasing #cores) for this problem is shown in Fig. 10. We see that the algorithm scales linearly up to around 4,000 cores. Similarly, Fig. 11 shows a weak scaling (problem size is equivalent to #cores) up to a similar number of processors. We have tested the algorithm on many different other problems and also verified

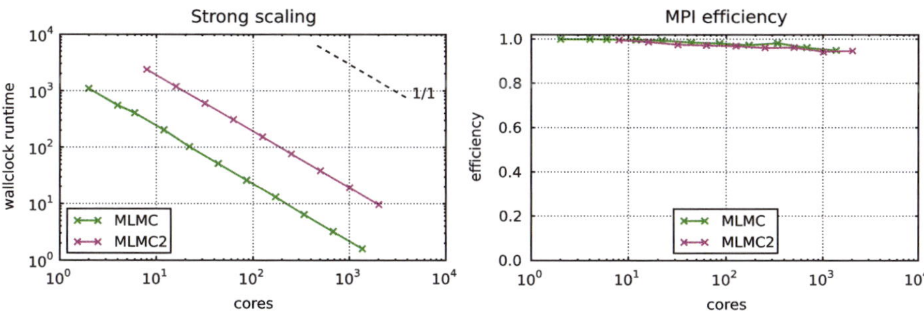


**Fig. 10** Strong scaling. The inferior scalability of DDM has no significant influence on the overall strong scaling for $d = 2$

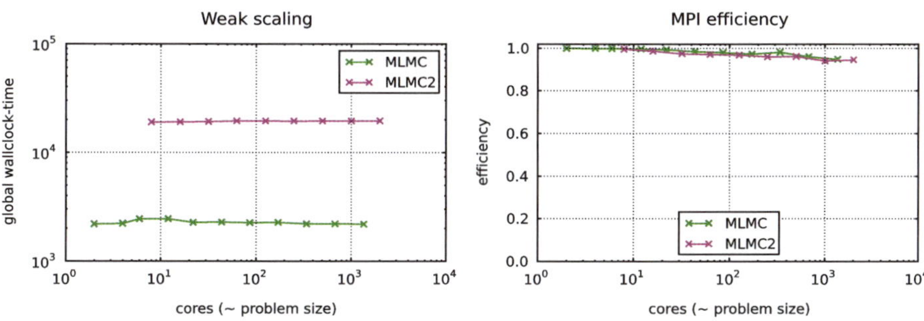


**Fig. 11** Weak scaling. The inferior scalability of DDM has no significant influence on the overall weak scaling for $d = 2$

strong (linear) scaling up to 40,000 cores, which is almost the limit size of the machine [50]; we expect it to scale up to a much larger number of cores. The results in both one and two space dimensions indicate that our static load balancing algorithm is quite efficient.

## *6.3 Shallow-Water Equations with Uncertain Bottom Topography*

In this section, we consider the shallow-water equations (in two space dimensions):

$$\begin{cases} h_t + (hu)_x + (hv)_y = 0, \\ (hu)_t + \left(hu^2 + \dfrac{1}{2}gh^2\right)_x + (huv)_y = -ghb_x, \\ (hv)_t + (huv)_x + \left(hv^2 + \dfrac{1}{2}gh^2\right)_y = -ghb_y. \end{cases} \tag{91}$$

Here, $h$ is the height of the fluid column above the bottom topography $b = b(x, y)$ over which the fluid flows and $(u, v)$ is the vertically averaged horizontal fluid velocity field. The constant $g$ denotes the acceleration due to gravity.

### 6.3.1 Multi-level Representation of the Bottom Topography

An approximation to the exact bottom topography $b(\mathbf{x}) \in W^{1,\infty}(\mathbf{D})$ is often obtained from the measurements. For instance [7, 13], in the two-dimensional case, nodal measurements $b_{i+\frac{1}{2},j+\frac{1}{2}} := b(\mathbf{x}_{i+\frac{1}{2},j+\frac{1}{2}})$ are obtained at locations $\mathbf{x}_{i+\frac{1}{2},j+\frac{1}{2}} = (x_{i+\frac{1}{2}}, y_{j+\frac{1}{2}})$, i.e. at vertices of an axiparallel quadrilateral *topography* mesh $\bar{\mathscr{T}}$ (possibly different from FVM mesh $\mathscr{T}$) on the rectangular two-dimensional domain $\mathbf{D}$. Since each measurement $b_{i+\frac{1}{2},j+\frac{1}{2}}$ is prone to uncertainty [13], all measured values are treated as random variables with some prescribed probability distribution; we choose

$$b_{i+\frac{1}{2},j+\frac{1}{2}}(\omega) := b(x_{i+\frac{1}{2},j+\frac{1}{2}}) + Y_{i,j}(\omega), \quad Y_{i,j} \sim \mathscr{U}(-\varepsilon_{i,j}, \varepsilon_{i,j}),\ \varepsilon_{i,j} > 0, \tag{92}$$

i.e. $b_{i+\frac{1}{2},j+\frac{1}{2}}(\omega) \in L^2(\Omega, \mathbb{R})$ are random variables (*not* necessarily independent), which deviate from the measurements $b_{i+\frac{1}{2},j+\frac{1}{2}}$ by $\pm\varepsilon_{i+\frac{1}{2},j+\frac{1}{2}}$. Thus, (92) provides an approximation to the *uncertain* topography $b(\mathbf{x}, \omega) \in L^2(\Omega, W^{1,\infty}(\mathbf{D}))$.

It is shown in a recent paper [33] that the uncertain bottom topography needs to be represented in a multi-level framework in-order to accelerate the MLMC-FVM algorithm. To introduce the multi-level topography representation, we recall some notation: levels $\ell = 0, \dots, L$ enumerate nested grids $\mathscr{T}_0, \dots, \mathscr{T}_L$ that are used in the MLMC-FVM *solver*. Apart from $\mathscr{T}_0, \dots, \mathscr{T}_L$, we consider an additional hierarchical structure, that will be used in the multi-level representation of the *bottom topography*. More precisely, assume a nested sequence $\{\bar{\mathscr{T}}_{\bar{\ell}} = \bar{\mathscr{T}}^1_{\bar{\ell}} \times \cdots \times \bar{\mathscr{T}}^d_{\bar{\ell}},\ \bar{\ell} = 0, \dots, \bar{L}\}$ of isotropic regular $d$-dimensional axiparallel quadrilateral meshes for the physical *bounded* domain $\mathbf{D} = I_1 \times \cdots \times I_d \subset \mathbb{R}^d,\ I_r \subset \mathbb{R},\ d = 1, 2$, each of them obtained by $\bar{\ell}$ uniform refinements of some initial, regular mesh $\bar{\mathscr{T}}_0$ (of domain $\mathbf{D}$) consisting of the cells $C^0_k,\ k = 1, \dots, \#\bar{\mathscr{T}}_0$. Note, that a-priori we do *not* assume any relation between $\bar{L}$ and $L$. However, for the sake of consistency, we assume

$$\bar{\mathscr{T}}_{\bar{\ell}} = \mathscr{T}_\ell, \quad \text{provided } \bar{\ell} = \ell.$$

For $p \in \mathbb{N}_0$, define $\mathscr{Q}^p(\mathbf{D}, \bar{\mathscr{T}})$ to be the space of piece-wise multivariate tensor product polynomials of degree $p$ on a mesh $\bar{\mathscr{T}}$ of a bounded domain $\mathbf{D}$ having essentially bounded weak derivatives up to order $p$, i.e.

$$\mathscr{Q}^p(\mathbf{D}, \bar{\mathscr{T}}) := \{f \in W^{p,\infty}(\mathbf{D}) : f|_C \in \mathscr{Q}_p(C),\ \forall C = C_1 \times \cdots \times C_d \in \bar{\mathscr{T}}\}, \tag{93}$$

where $\mathscr{Q}_p(C)$ is the space of multivariate tensor product polynomials on cell $C$,

$$\mathscr{Q}_p(C) := \{\mathbf{x} \mapsto p_1(\mathbf{x}_1) \cdot \dots \cdot p_d(\mathbf{x}_d) : p_r \in \mathscr{P}_d(C_r),\ \forall r = 1, \dots, d\}.$$

We assume that *uncertain* measurements $b_{i+\frac{1}{2},j+\frac{1}{2}}(\omega) := b(x_{i+\frac{1}{2}}, y_{j+\frac{1}{2}}, \omega)$ of the exact bottom topography $b(x, y)$ are available, as in (92). Then $b_{i+\frac{1}{2},j+\frac{1}{2}}(\omega)$ are treated as nodal values and are linearly interpolated in each dimension using the bilinear *hierarchical* interpolation operator,

$$\begin{aligned} \mathscr{I}^{\bar{L}} b(x, y, \omega) &= \sum_{\bar{\ell}=0}^{\bar{L}} \sum_{\bar{\ell}'=0}^{\bar{L}} b^{\bar{\ell},\bar{\ell}'}(x, y, \omega), \\ b^{\bar{\ell},\bar{\ell}'} &:= \mathscr{I}_{\bar{\ell},\bar{\ell}'} b - \mathscr{I}_{\bar{\ell}-1,\bar{\ell}'} b - \mathscr{I}_{\bar{\ell},\bar{\ell}'-1} b, \quad \mathscr{I}_{-1,\cdot} \equiv \mathscr{I}_{\cdot,-1} \equiv 0, \end{aligned} \tag{94}$$

where $\mathscr{I}_{\bar{\ell},\bar{\ell}'}$ denotes bilinear *nodal* interpolation operator on the quadrilateral mesh $\bar{\mathscr{T}}^1_{\bar{\ell}} \times \bar{\mathscr{T}}^2_{\bar{\ell}'}$.

Each $b^{\bar{\ell},\bar{\ell}'}(x, y, \omega) \in L^2(\Omega, \mathscr{Q}^1(I_1 \times I_2, \bar{\mathscr{T}}^1_{\bar{\ell}} \times \bar{\mathscr{T}}^2_{\bar{\ell}'}))$ is a linear combination of the multivariate tensor products of two *hierarchical* "hat" ("Schauder") basis functions,

$$b^{\bar{\ell},\bar{\ell}'}(x, y, \omega) = \sum_{k=1}^{\hat{N}_{\bar{\ell}}} \sum_{k'=1}^{\hat{N}_{\bar{\ell}'}} b^{\bar{\ell},\bar{\ell}'}_{k,k'}(\omega) \varphi^{\bar{\ell}}_k(x) \varphi^{\bar{\ell}'}_{k'}(y), \quad b^{\bar{\ell},\bar{\ell}'}_{k,k'} \in L^2(\Omega, \mathbb{R}). \tag{95}$$

The interpolated bottom topography belongs to the space

$$\mathscr{I}_{\bar{L}} b(\mathbf{x}, \omega) \in L^2(\Omega, \mathscr{Q}^1(I_1 \times I_2, \bar{\mathscr{T}}_{\bar{L}})).$$

### 6.3.2 2-D Numerical Experiments: Random Perturbation of Lake at Rest

We consider the shallow-water equations in the computational domain $\mathbf{D} = [0, 2] \times [0, 2]$, and investigate the evolution of a random perturbation of the lake at rest coupled with outflow boundary conditions.

The uncertain bottom topography $b(x, \omega)$ is represented in terms of the nodal, bivariate hierarchical basis (94)–(95) with random amplitudes. Notice that, formally, this bilinear basis can be obtained by tensorizing the univariate Schauder basis of $C^0([0, 2])$. Notice also that we used in the present study only *isotropically supported* product functions. The bottom topography was resolved to 6 levels (i.e. $\bar{L} = 5$, $\bar{\ell}, \bar{\ell}' = 0, \dots, 5$) where coefficients $b^{\bar{\ell},\bar{\ell}'}_{k,k'}(\omega)$ are given by mean values $\bar{b}^{\bar{\ell},\bar{\ell}'}_{k,k'}$ that are perturbed by independent uniformly distributed centered random variables with decaying variances,

$$b^{\bar{\ell},\bar{\ell}'}_{k,k'}(\omega) = \bar{b}^{\bar{\ell},\bar{\ell}'}_{k,k'} + Y^{\bar{\ell},\bar{\ell}'}_{k,k'}(\omega) \sim \frac{2}{5} \mathscr{U}(-\varepsilon_{\bar{\ell},\bar{\ell}'}, \varepsilon_{\bar{\ell},\bar{\ell}'}), \tag{96}$$

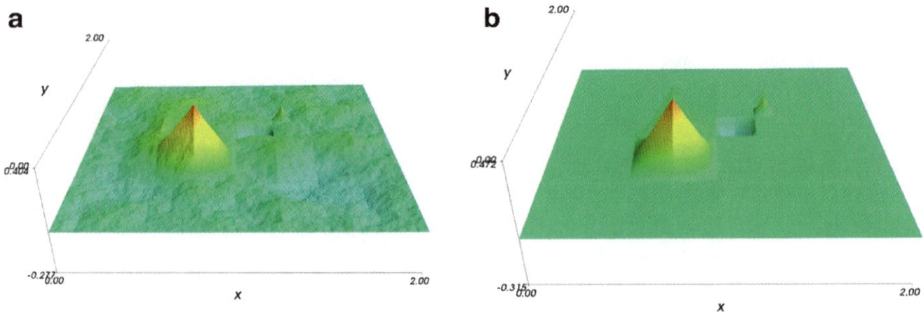


**Fig. 12** Uncertain bottom topography (96) with 9 hierarchical levels ($\bar{L} = 8$). (**a**) one realization for some **fixed** $\omega \in \Omega$. (**b**) Mean and variance of $b(\mathbf{x}, \omega)$

where all coefficients $\bar{b}_{k,k'}^{\bar{\ell},\bar{\ell}'}$ are zero except

$$\bar{b}_{2,2}^{3,3} = 0.4, \quad \bar{b}_{6,6}^{4,4} = -0.32, \quad \bar{b}_{11,11}^{5,5} = 0.12, \tag{97}$$

and

$$\varepsilon_{0,\cdot} = \varepsilon_{\cdot,0} = 0, \quad \varepsilon_{\bar{\ell},\bar{\ell}'} = 2^{-\max\{\bar{\ell},\bar{\ell}'\}}, \ \forall \bar{\ell} \geq 1. \tag{98}$$

A realization of the uncertain bottom topography and the corresponding mean and variance are shown in Fig. 12.

Next, we consider the initial data $\mathbf{U}_0$ to be a *random* perturbation of a lake-at-rest. Let $Y \sim \frac{1}{50} + \frac{1}{100}\mathscr{U}(-1, 1)$ be a random variable independent of $\{Y_{k,k'}^{\bar{\ell},\bar{\ell}'}\}$. An initial perturbation around $\mathbf{x}_0 = (x_0, y_0) = (1.0, 0.7)$ with a radius $r = \frac{1}{10}$ reads

$$h_0(x, y, \omega) = \begin{cases} 1.0 + Y(\omega) - b(x, y, \omega) & \text{if} \quad |\mathbf{x} - \mathbf{x}_0| < r, \\ 1.0 - b(x, y, \omega) & \text{if} \quad |\mathbf{x} - \mathbf{x}_0| > r, \end{cases} \tag{99}$$

with $b(x, \omega)$ as defined in (96) and the initial layer velocities set to zero, i.e.

$$\{u_0(x, y, \omega),\ v_0(x, y, \omega)\} = \{0.0,\ 0.0\}. \tag{100}$$

Note, that here we have a very large number of sources of uncertainty ($(2^5 - 1)^2 - 1 = 962$).

Reference solutions, computed with the second-order entropy stable scheme [15] at time $T = 0.1$ is depicted in Fig. 13. The results are computed on 9 nested levels of resolution ($L = 8$) with the finest resolution being on a 4,096 × 4,096 mesh and with time steps reduced accordingly in order to maintain the same CFL constant over all discretization levels. The simulation is run on 2,044 cores and 16 samples are taken for the finest mesh resolution.

The above problem is quite involved due to large number of sources of uncertainty as well as the underlying difficulty of simulating small perturbations

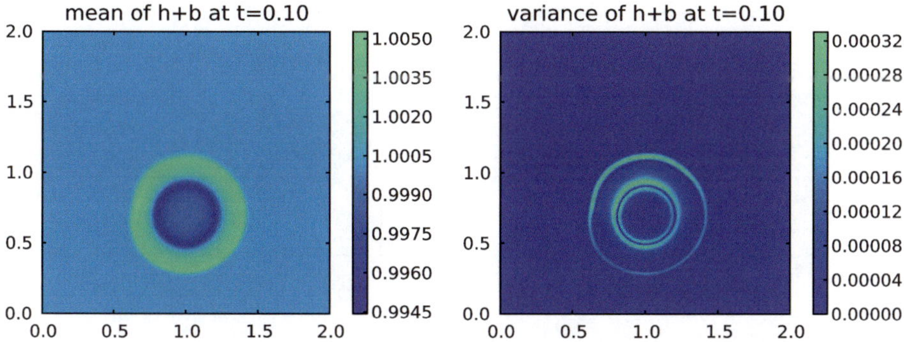


| $L$ | $M_L$ | Grid size | CFL | Cores | Runtime | Efficiency |
|---|---|---|---|---|---|---|
| 8 | 16 | 4,096 × 4,096 | 0.45 | 2,044 | 2:13:51 | 97.5 % |

**Fig. 13** Reference solution for perturbed steadystate (99) using MLMC-FVM. Initial perturbation evolves into asymmetric ribbon wave with uncertain amplitude

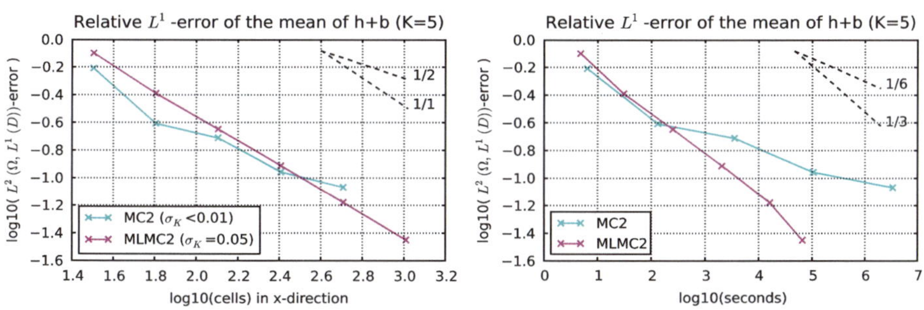


**Fig. 14** Convergence of estimated mean in the 2-D simulation (99). MLMC methods are three orders of magnitude faster than MC

of steady states. The reference solution show that the wave (in mean) spreads out of the initial source. The variance is distributed in a non-linear and complicated manner with large amount of variance corresponding to the uncertainties in the bottom topography.

### 6.3.3 Numerical Convergence Analysis

We investigate convergence of error vs. work in Figs. 14 and 15. Here we use the MLMC-FVM simulation from Fig. 13 with 9 levels of resolution with the finest resolution being on a 2,048 × 2,048 mesh as the reference solution $\mathbf{U}_{\text{ref}}$. The error in the mean field converges at expected rates. At comparable numerical resolution and accuracy, the MLMC2 is about two orders of magnitude faster than the MC2 method for this problem. We observe a slight deterioration in the

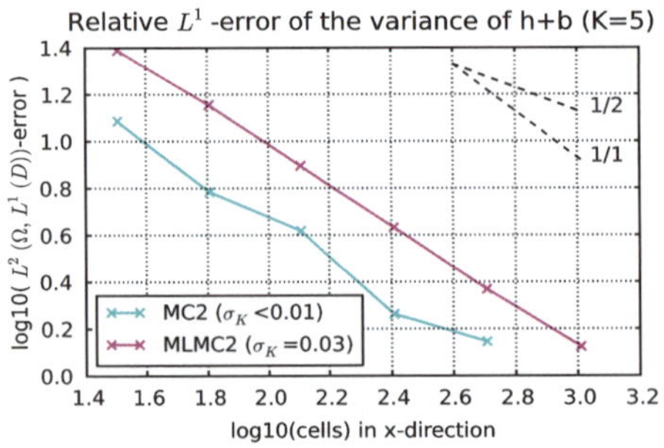


**Fig. 15** Convergence of estimated variance in the 2-D simulation (99). MLMC methods are asymptotically faster than MC

estimated convergence rates for the variance. This could well be a pre-asymptotic effect. Again, the MLMC2 appears to be slightly faster than the corresponding MC2 method in delivering variance estimates of comparable numerical accuracy.

### 6.3.4 Speed Up Due to Hierarchical Topology Representation

We test the gain in efficiency due to the multi-level hierarchical representation of the uncertain bottom topography (94) by comparing with a simulation that uses the standard MLMC algorithm. In other words, the MLMC2 (full) simulation uses the underlying bottom topography (at the resolution of the underlying topography mesh) for all shallow water samples. In particular, simulations at the coarsest level of the FVM mesh use the topography at the finest level of the underlying topography mesh. We compare MLMC2 (full) with MLMC2 (truncated) which uses the representation (94) on the perturbations of lake at rest steady state problem in Fig. 16. As suggested by the theory of [33], the two methods should lead to an identical order of the error for a given space-time resolution. We verify this in Fig. 16. On the other hand, the MLMC2 (truncated) is at least an order of magnitude faster than the MLMC2 (full) showing that the multi-level representation of the uncertain bottom topography really provides a significant gain in efficiency.

## *6.4 Burgers' Equation with Random Flux*

The deterministic Burgers' equation is the simplest example of the *non-linear* scalar conservation law. It is given by

$$u_t + f(u)_x = 0, \qquad f(u) = \frac{u^2}{2}. \tag{101}$$

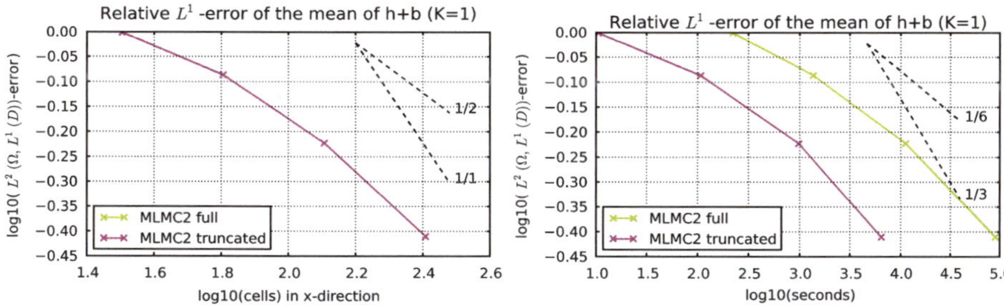


**Fig. 16** Convergence of estimated mean for 2-D steadystate (99) with full ($\bar{L} = 8$) and truncated ($\ell + 1$) number of levels in the hierarchical representation (96) of bottom topography. For a given mesh resolution, both estimators coincide. The implementation with the truncated number of levels is more than 10 times faster on a mesh of $256 \times 256$ cells

The solutions to (101) are well posed provided initial data $u_0(\cdot) = u(\cdot, 0)$ is given. We consider deterministic initial data of the form

$$u_0(x) = \sin(\pi x). \tag{102}$$

Notice that $\|u_0\|_{L^\infty(\mathbb{R})} = 1$, hence one can choose $\bar{R} = 1$ in (33) of Theorem 1.

### 6.4.1 Uniformly Perturbed Flux

We consider a *random* version of Burgers' equation (101) with the *random* flux

$$f(u, \omega) = \frac{u^{p(\omega)}}{p(\omega)}, \qquad p \sim \mathscr{U}(1.5, 2.5). \tag{103}$$

It is straightforward to verify that the random flux $f$ defined above is a *bounded random flux* with each realization $f(\omega) \in C^1([-R, R])$ with $R = \bar{R}$.

The initial data (102) and the reference solution (obtained by MLMC-FVM) at time $t = 4$ are depicted in Fig. 17. There are 13 levels ($L = 12$) of FVM mesh resolution with the finest resolution (at the finest level $\ell = L$) being 32,768 cells. The Rusanov numerical flux with a second order accurate WENO reconstruction was used. At every point $x \in [0, 2]$ the solid line represents the mean and the dashed lines represent the mean $\pm$ standard deviation of the (random) solution. For each *sample* (realization) of the *random* flux (103), the smooth initial data evolves into (as expected) discontinuity in the physical space and a shock forms at $\mathbf{x}_1 = 1.0$. Given the fact that the flux function is random, the variance is high over the entire physical domain and is not just concentrated at the discontinuity.

Next, we use this high-resolution MLMC-FVM simulation from Fig. 17 as the reference solution. We investigate the convergence of error vs. work in Fig. 18. The error in the mean field converges at expected rates; furthermore, the MLMC2

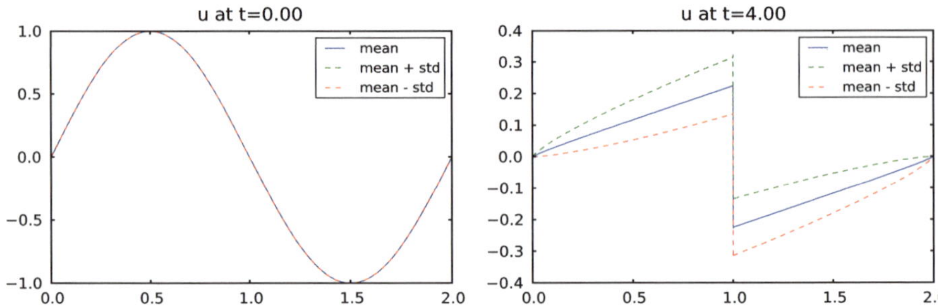


| $L$ | $M_L$ | Grid size | CFL | Cores | Runtime | Efficiency |
|---|---|---|---|---|---|---|
| 12 | 16 | 32,768 | 0.475 | 104 | 6:31:48 | 99.3 % |

**Fig. 17** MLMC-FVM solution of Burgers' equation with uniformly perturbed flux (103)

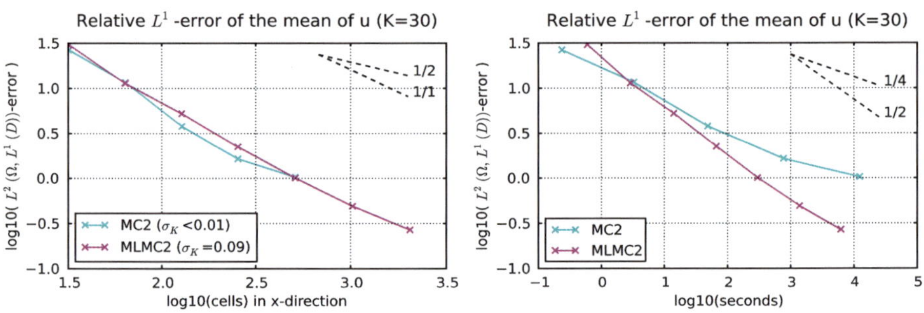


**Fig. 18** Convergence of mean for MLMC-FVM solution of Burgers' equation with uniformly perturbed flux (103)

method is almost two orders of magnitude faster than the MC2 method (for the same numerical accuracy).

## 6.5 Two Phase Flows in a Porous Medium with Uncertain Permeabilities

Many interesting phenomena (such as water flooding) in an oil and gas reservoir can be modeled by using two phase flows in a porous medium. For simplicity, we consider the flow of two phases (oil and water) in a one dimensional reservoir [4]. The model reduces to a one dimensional scalar conservation law (with the Buckley-Leverett flux):

$$\begin{aligned} S_t + f(S)_x &= 0, \\ f(S) &= \frac{qK\lambda_o(S)}{\lambda_w(S) + \lambda_o(S)}. \end{aligned} \tag{104}$$

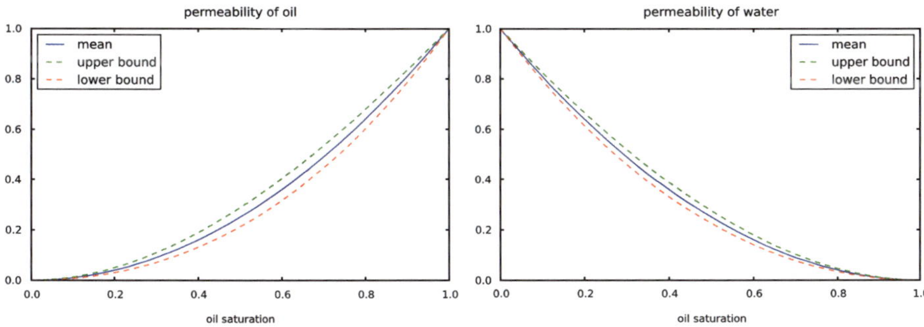


**Fig. 19** The mean and the upper/lower bounds for the *uncertain* relative oil and water permeabilities as defined in (105)

Here, the variable $S$ represents the saturation of oil, $q$ is total flow rate, $K$ the rock permeability and $\lambda_w, \lambda_o : [0, 1] \mapsto \mathbb{R}$ are the relative permeabilities of the water and oil phases, respectively. In practice, the rock permeability needs to be measured and is prone to uncertainty. Similarly, the relative permeabilities are measured in laboratory experiments and are characterized by uncertainty. In this example, we focus on the case of uncertain relative permeabilities, which are frequently taken to be of the form

$$\lambda^o = S^2, \qquad \lambda^w(S) = (1 - S)^2.$$

We add *random perturbations* to $\lambda^o$ and $\lambda^w$, i.e.

$$\begin{aligned} \lambda^o(S) &= S^2 + \varepsilon_o Y_o(\omega) S^2 (1 - S), \\ \lambda^w(S) &= (1 - S)^2 + \varepsilon_w Y_w(\omega)(1 - S)^2 S, \end{aligned} \tag{105}$$

with

$$\varepsilon_o = 0.3, \qquad \varepsilon_w = 0.2, \qquad Y_o, Y_w \sim \mathcal{U}[-1, 1].$$

*Uncertain* relative permeabilities defined in (105) are depicted in Fig. 19.

### 6.5.1 1-D Numerical Experiment: Deterministic Initial Shock

We set $K = 1.0$ and $q = 1.0$. The initial data is given by a *deterministic* shock,

$$S^o = \begin{cases} 0.25 & \text{if} \quad \mathbf{x}_1 < 1.0, \\ 0.85 & \text{if} \quad \mathbf{x}_1 > 1.0. \end{cases} \tag{106}$$

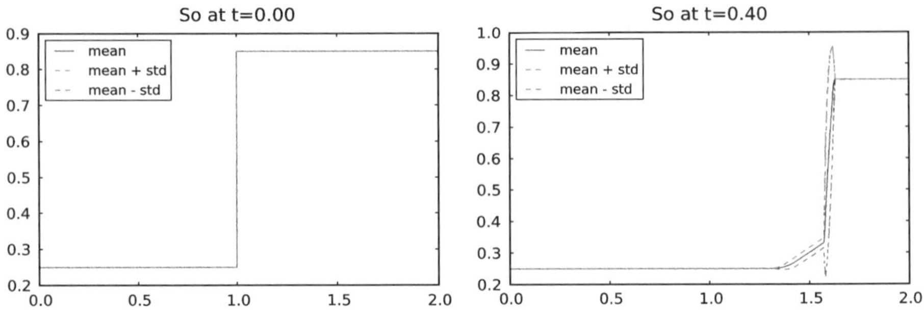


| $L$ | $M_L$ | Grid size | CFL | Cores | Runtime | Efficiency |
|---|---|---|---|---|---|---|
| 12 | 16 | 32,768 | 0.475 | 104 | 1:12:27 | 99.0 % |

**Fig. 20** MLMC-FVM solution of the Buckley-Leverett equation (104) with uncertain permeabilities (105) and deterministic initial shock (106)

Notice that $\|u_0\|_{L^\infty(\mathbb{R})} = 0.85$, hence one can choose $\bar{R} = 0.85$ in (33) of Theorem 1. Furthermore, the random flux defined in (104) with random permeabilities (105) is a *bounded random flux* with each realization $f(\omega) \in C^1([-R, R])$ with $R = \bar{R}$.

The initial data (106) and the reference solution at time $t = 0.4$ are depicted in Fig. 20. There are 13 levels ($L = 12$) of FVM mesh resolution with the finest resolution (at the finest level $\ell = L$) being 32,768 cells. At every point $x \in [0, 2]$, the solid line represents the mean and the dashed lines represent the mean $\pm$ standard deviation of the (random) solution. For each *sample* of the *random* permeabilities $\lambda^o, \lambda^w$, the initial shock splits into a compound shock, consisting of right going rarefaction that is immediately followed by a right moving shock wave. Notice the improvement of the regularity in the stochastic solution: deterministic path-wise solutions for each sample are discontinuous due to formation of the shock; nevertheless, the mean of the solution appears to be continuous. Furthermore, the uncertainty seems to be concentrated on the compound shock in this case.

Next, we use the high-resolution MLMC-FVM simulation from Fig. 20 as the reference solution. We investigate convergence of error vs. work in Figs. 21 and 22. The error in the mean field converges at expected rates. If we compare the MC2 method (for the same numerical accuracy) with the MLMC2, the latter is approximately an order of magnitude faster for the approximation of the mean field and approximately two orders of magnitude faster for the approximation of variance.

### 6.5.2 Exposure of the Non-monotonicity of the MLMC-FVM Estimator

In this section we provide examples of MLMC-FVM estimated mean and variance fields with much smaller discretization parameters (number of mesh cells and mesh

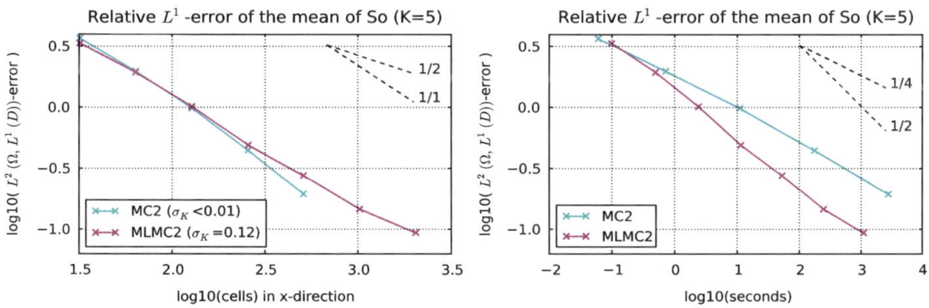


**Fig. 21** Convergence of mean in the Buckley-Leverett equation (104) with uncertain permeabilities (105)

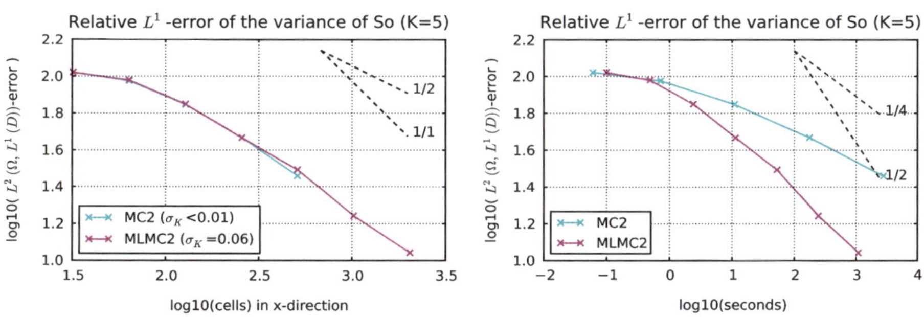


**Fig. 22** Convergence of variance in the Buckley-Leverett equation (104) with uncertain permeabilities (105)

levels) in order to expose the non-monotonicity of the MLMC-FVM estimator mentioned in Remark 5. In particular, we consider the same problem as in Sect. 6.5.1, only with substantially smaller space and time step size.

Two cases are considered: mesh size of 128 with 5 levels and mesh size 1,024 with 7 levels. The results are presented in Figs. 23 and 24, respectively.

For the first case, where the resolution is *coarse*, small overshoots in the approximated mean and a significant negative overshoot to the left of the peak in the approximated variance appear. These phenomena are, indeed, discretization artifacts which are due to the MLMC-FVM estimation. For a detailed explanation, we refer to Remark 7.

For the second case, where the resolution is *fine*, the approximations of the mean and the variance have almost converged, i.e. the numerical overshoots have substantially smaller amplitude as compared to results presented in Fig. 23. In fact, these overshoots are not significant at all and are expected to eventually vanish due to the convergence of the MLMC-FVM.

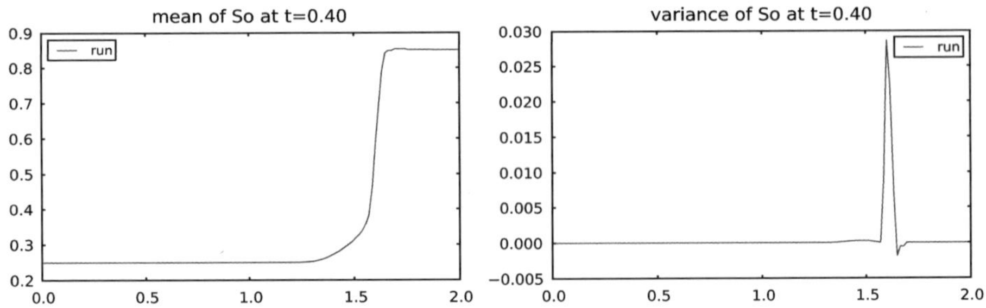


**Fig. 23** MLMC-FVM solution as in Fig. 20, with only 128 cells and 5 levels ($L = 4$). Notice the small overshoot in the approximated mean and a significant negative overshoot to the left of the peak in the approximated variance. Such overshoots are explained by the non-monotonicity of the MLMC-FVM estimator

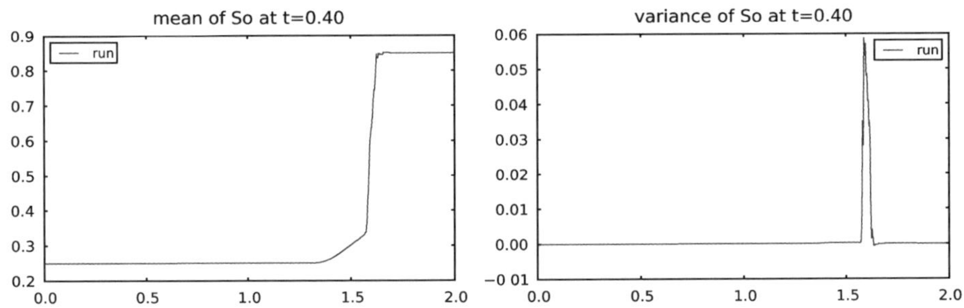


**Fig. 24** MLMC-FVM solution as in Fig. 20, with only 1,024 cells and 7 levels ($L = 6$). The approximations of the mean and the variance have almost converged, i.e. the overshoots are *much* smaller compared to results presented in Fig. 23. In fact, these overshoots are not significant at all and eventually vanish due to the convergence of the MLMC-FVM method

## 6.6 Euler Equations with Uncertain Equation of State

Next we consider a *random* version of the Euler equations (81) from Sect. 6.1 with the *random* constant of specific heats in (82)

$$\gamma = \gamma(\omega), \qquad \gamma \sim \mathscr{U}(5/3 - \varepsilon, 5/3 + \varepsilon), \quad \varepsilon = 0.1. \tag{107}$$

### 6.6.1 Uniformly Perturbed Specific Heats in 1d: Sod Shock Tube

We consider one dimensional version of the Euler equations (81) with (107) in the domain $\mathbf{D} = [0, 2]$. The initial data consists of the shock at $x = 1$:

$$U_0(x,\omega) = \{\rho_0(x,\omega), u_0(x,\omega), p_0(x,\omega)\} = \begin{cases} \{3.0, 0.0, 3.0\} & \text{if} \quad x < 1, \\ \{1.0, 0.0, 1.0\} & \text{if} \quad x > 1. \end{cases} \tag{108}$$

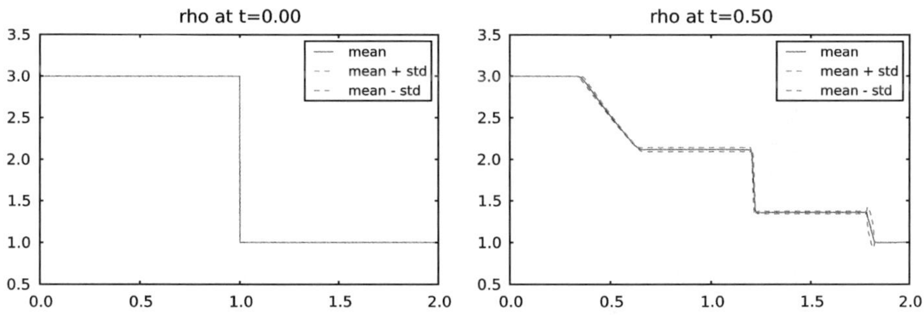


| $L$ | $M_L$ | Grid size | CFL | Cores | Runtime | Efficiency |
|---|---|---|---|---|---|---|
| 8 | 16 | 2,048 | 0.475 | 36 | 0:00:39 | 99.1 % |

**Fig. 25** MLMC-FVM solution of the Sod shock tube (108) with *random* specific heats constant (107)

The initial data (108) and the reference solution at time $t = 0.5$ are depicted in Fig. 25. There are 9 levels ($L = 8$) of FVM mesh resolution with the finest resolution (at the finest level $\ell = L$) being 2,048 cells. Rusanov flux with second order accurate positivity preserving WENO reconstruction was used. At every point $x \in [0, 2]$ the solid line represents the mean and the dashed lines represent the mean $\pm$ standard deviation of the (random) solution. For each *sample* the *random* constant $\gamma$ of specific heats, the initial shock splits into three waves: a left going rarefaction wave, a right going contact discontinuity and a right going shock wave. The variances are relatively small (compared to the mean) with a concentration of the variance near the shock.

### 6.6.2 Uniformly Perturbed Specific Heats in 2d: Cloud-Shock

The MLMC-FVM algorithm is tested on a problem that is analogous to the one presented in Sect. 6.1. The computational domain is again taken to be $\mathbf{D} = [0, 1] \times [0, 1]$. Compared to Sect. 6.1, we assume a *random* constant of specific heats (107) and we consider the *deterministic* initial data for cloud shock problem:

$$\{\rho_0(\mathbf{x}, \omega),\ \mathbf{u}_0(\mathbf{x}, \omega),\ p_0(\mathbf{x}, \omega)\} = \\ = \begin{cases} \{3.86859,\ (11.2536,\ 0)^\top,\ 167.345\} & \text{if} \quad \mathbf{x}_1 < 0.05, \\ \{1,\ (0,\ 0)^\top,\ 1\} & \text{if} \quad \mathbf{x}_1 > 0.05, \end{cases} \tag{109}$$

with a high density cloud (or bubble) lying to the right of the shock

$$\rho_0(\mathbf{x}, \omega) = 10, \quad \text{if} \quad \sqrt{(\mathbf{x}_1 - 0.25)^2 + (\mathbf{x}_2 - 0.5)^2} \leq 0.15. \tag{110}$$

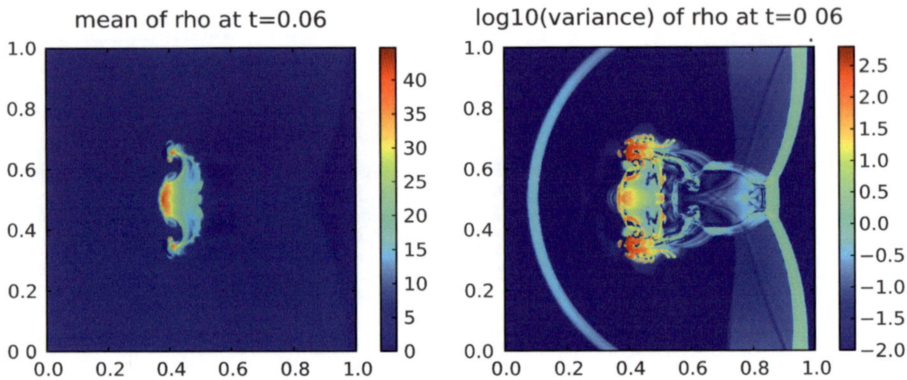


| $L$ | $M_L$ | Grid size | CFL | Cores | Runtime | Efficiency |
|---|---|---|---|---|---|---|
| 8 | 8 | 2,048 × 2,048 | 0.475 | 128 | 6:06:48 | 86.3 % |

**Fig. 26** MLMC-FVM solution of the cloud-shock (109)–(110) with *random* specific heats constant (107)

The mean and variance of the solution at time $t = 0.06$ obtained by the numerical simulation using MLMC-FVM are given in Fig. 26. The results are from a MLMC-WENO run with 9 nested levels of resolution ($L = 8$) and the finest resolution is set to 2,048 × 2,048 mesh. Rusanov flux with second order accurate positivity preserving WENO reconstruction was used. The number $M_L$ of MC samples at the finest resolution is 8 and number of cores for this run is 128. Although the flow and uncertainty appear to be similar to the one discussed in Sect. 6.1, there are important differences. In particular, the variance near the bow and tail shocks appears to be spread over a larger region compared to the case of random initial data. Furthermore, the smooth regions after the bow shock have a very different distribution of uncertainty in this case.

## *6.7 Verification of the Derived Constants in the Asymptotic Error Bounds*

In this section, we would like to verify the results derived in Sect. 4.1, i.e. that there is only a *small* (of order $\sqrt{2M_L}$) difference in constants of the asymptotic error convergence bounds (44) and (54). To this end, consider deterministic and stochastic one-dimensional *Sod shock tube* [32] problems modeled by Euler equations of gas dynamics (81). Four different configurations are considered:

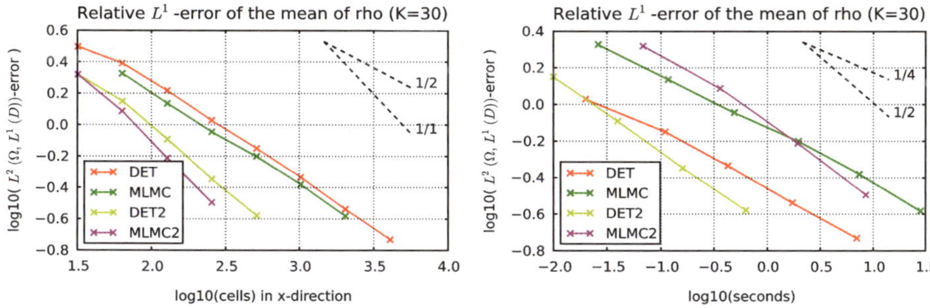


| | $C_{\mathrm{FVM}}$ | $C_{\mathrm{MLMC}}$ | Quotient |
|---|---|---|---|
| DET, MLMC | 0.63 | 0.88 | 1.40 |
| DET2, MLMC2 | 0.52 | 0.91 | 1.75 |

**Fig. 27** Error convergence for deterministic and stochastic *Sod shock tube* problem. For some *fixed* computational time, errors of the *stochastic* (MLMC(2)) runs are only approximately 1.4–1.75 times larger than errors of the *deterministic* (DET(2)) runs

| | Initial data | Solver | Reconstruction | CFL |
|---|---|---|---|---|
| **DET** | Deterministic | HLL 3-wave | – | 0.9 |
| **MLMC** | Stochastic | MLMC + 3-wave HLL | – | 0.9 |
| **DET2** | Deterministic | HLLC 3-wave | WENO | 0.475 |
| **MLMC2** | Stochastic | MLMC + HLL 3-wave | WENO | 0.475 |

The convergence results, i.e. error vs. number of cells and error vs. computational runtime are depicted in Fig. 27. In the table below the figure, the approximated constants $C_{\mathrm{FVM}}$ and $C_{\mathrm{MLMC}}$ from asymptotic error bounds (44) and (54) are provided. The values of the constants were obtained by the least squares fitting of the linear polynomial to the obtained logarithmic error convergence data. The measured differences in the constants are *small* (approximately 1.4 and 1.75 times) and hence support the derived estimate (57) presented in Sect. 4.1.

## 6.8 Sparse Tensor MLMC Estimation of Two-Point Correlations

So far, we presented numerical experiments which addressed the MLMC-FVM estimation of mean fields and of variances of random entropy solutions as functions of space and time. In Sect. 4.2, we also addressed theoretically the efficient computation of two- and of $k$-point correlation functions. Theorem 2 stated that in parallel to any MLMC-FVM simulation, MLMC estimates of two- and $k$-point correlation functions in the entire domain can be obtained by sparse tensorization of Finite Volume solutions in complexity which equals, up to logarithmic terms, that of the MLMC-FVM estimate of the mean field. The computational realization of the sparse tensor product projects is based on a *multilevel splitting* of the space of piecewise constant functions (and, therefore, of a multilevel decomposition of the Finite Volume solutions) on the mesh hierarchy $\{\mathscr{T}_\ell\}_{\ell=0}^{\infty}$.

### 6.8.1 Multiresolution Basis

For the efficient numerical realization of the sparse projectors $\hat{P}_L^{(k)}$ defined in (70), a *Multiresolution* basis of the spaces $S_L$ containing the FV approximations of the (pathwise) entropy solutions must be chosen. We assume here, for simplicity of exposition, that we are given a FV solver which is based on *cell averages*, so that the spaces $S_\ell$ are the space of simple (or step) functions on the triangulation $\mathscr{T}_\ell$.

In the following numerical experiments in one space dimension, which are based on the cartesian grid FV solver ALSVID (see [2, 3]), this will always be the following:

**Definition 3.** The hierarchical basis

$$\hat{\psi}_k^l(x) = \begin{cases} 1, & x \in \left[2k \cdot 2^{-l}, (2k+1)2^{-l}\right) \\ -1, & x \in \left[(2k+1)2^{-l}, (2k+2)2^{-l}\right) \\ 0, & \textit{otherwise} \end{cases} \tag{111}$$

is called *Haar-wavelet basis*.

We remark that Haar (multi-) wavelets are also available on two- and three-dimensional meshes consisting either of triangles/tetrahedra or of quadrilaterals/hexahedra (see, e.g. [5] for constructions of any polynomial degree on triangles in two space dimensions). The MLMC and MC require that sparse tensors can be added and multiplied by scalars. By writing a sparse tensor as a linear combination of its basis

$$\begin{aligned} \widehat{u^{(k)}} &= \sum_{|\mathbf{l}|_1 \le L} \sum_{\mathbf{j}} \alpha_{\mathbf{j}}^{\mathbf{l}} (\psi_{j_1}^{l_1} \otimes \cdots \otimes \psi_{j_k}^{l_k}) \\ &= \sum_{l=0}^{L} \sum_{j} \psi_j^l \otimes \left( \sum_{|\mathbf{l}'|_1 \le L-l} \sum_{\mathbf{j}'} \psi_{j_1'}^{l_1'} \otimes \cdots \otimes \alpha_{\mathbf{j}'}^{\mathbf{l}'} \psi_{j_{k-1}'}^{l_{k-1}'} \right) \\ &=: \sum_{l=0}^{L} \sum_{j} \psi_j^l \otimes u_j^l \end{aligned}$$

it becomes clear that one can store a sparse tensor by mimicking the hierarchical structure, i.e. `u[l][k] =` $u_k^l$, with the 'pointwise' operations

```
    (u+v)[l][k] = u[l][k] + v[l][k]
(alpha*u)[l][k] = alpha*u[l][k].
```

The conversion of a numerical solution based on the (piecewise constant) cell averages to the hierarchic, multiresolution representation can be effected on triangulation $\mathscr{T}_\ell$ in $O(\#\mathscr{T}_\ell)$ operations by the so-called pyramid scheme (see, e.g., [38]) for which efficient implementations are available in image compression, for example.

We emphasize that these concepts are *not* restricted to cartesian grids: for unstructured, triangular or tetrahedral meshes in complex geometries, subspace splittings (65) and $L^2(D)$-stable bases for the detail spaces $W_\ell$ in (65) also exist but are, naturally, triangulation dependent and must be generated be recursive agglomeration of cells and recursive SVD's of the corresponding mass matrices. We refer to [38] and [37, Chap. 2] for details.

### 6.8.2 Numerical Experiments

The number of samples $M_l$ and size of the mesh $N_l$ on level $l$ is given by $M_0, N_0$ and the maximal level $L$ through

$$N_l N_0 2^l \ , \quad M_l M_0 2^{L-l} \qquad l = 1, \ldots, L \tag{112}$$

If the MC-Estimator is described in these terms this means $M = N = N_L$. In the following numerical experiments $M_0 = N_0 = 8, L = 6$ and the PDE is evolved until time $T = 0.4$.

All PDEs are solved with the first order FVM using a Rusanov flux and explicit timestepping with the CFL condition $|f'| \leq C \frac{\Delta x}{\Delta t}$ with CFL number $C = 0.45$.

A *sparse tensor product with reduced sparsity* defined by

$$\widetilde{S_L^{(k)}} = \bigoplus_{\|\mathbf{l}\|_1 \leq L+3} \bigotimes_{j=1}^{k} W_{l_j} \qquad (l_j \in \{0, \ldots, L\}) \tag{113}$$

is also studied along with the full and sparse tensor products defined through (66), (68), respectively. With $N_0 = 2^3$ the reduced sparsity tensor product is, in fact, a full tensor on the coarsest level and realizes the sparse tensor product construction on a higher level of mesh refinement. Evidently, however, its formation out of Finite Volume solutions is of the same asymptotic complexity as forming the standard sparse tensor product.

The same numerical experiments are performed for three different equations, the few differing parameters will be mentioned accordingly. The first experiment compares MC- and MLMC-Estimates of $E[u^{(2)}]$, $E[\hat{u}^{(2)}]$ and $E[\tilde{u}^{(2)}]$. These are maps from $\mathbb{R}^2$ to $\mathbb{R}$. and their graph is plotted as a heat map (i.e. individual values contained in an array are represented as colors).

The second map compares different methods for computing the variance. Four distinct methods are presented. The first is direct estimation. The other three are obtained by observing that $E[u^{(2)}](x,x) - E[u]^2$ is an estimate of the variance. Analogous results hold for sparse and reduced sparsity tensor products instead of a full tensor product.

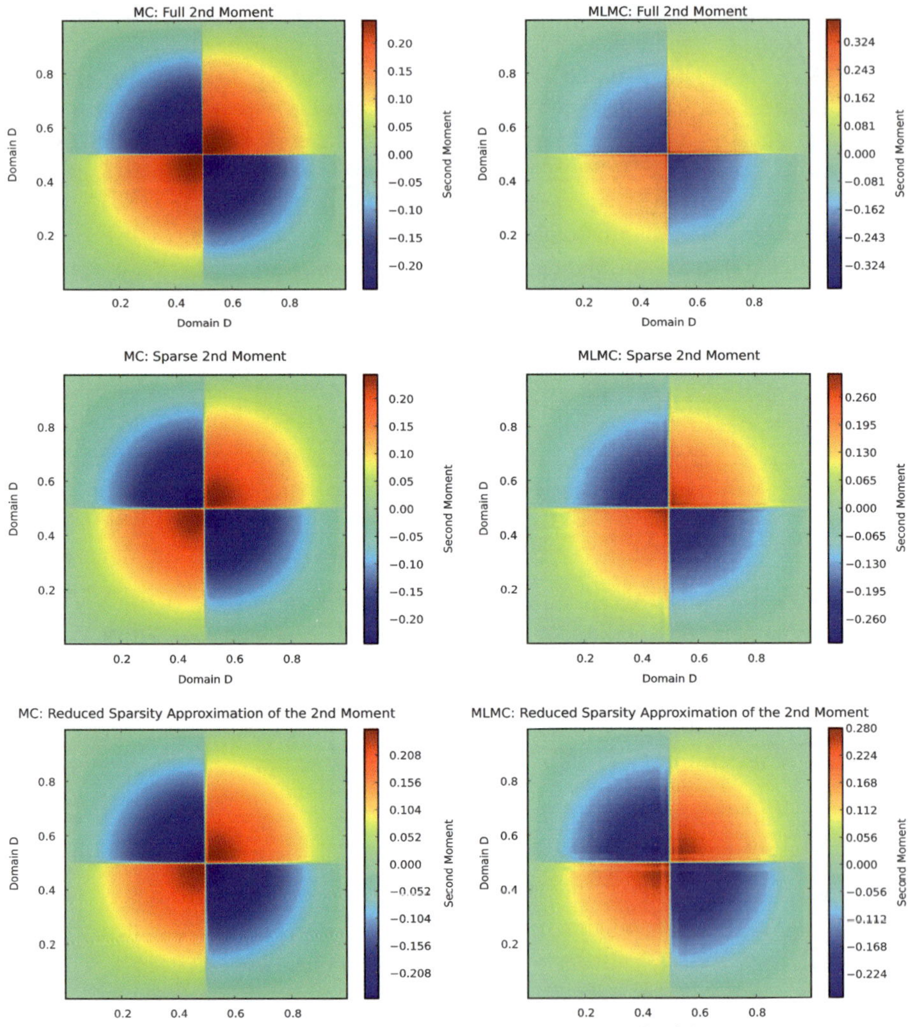


**Fig. 28** Estimated second moment of Burgers' equation (114) for $\alpha = 0$

### 6.8.3 Burgers' Equation: Initial Data with Random Amplitude

The stochastic PDE is Burgers' equation (101) with sinusoidal initial conditions where the amplitude is uncertain. More precisely

$$u_0(x, \omega) = X(\omega) \sin(2\pi(x - \alpha)), \quad X \sim \mathscr{U}(0, 1) \tag{114}$$

where $\alpha$ is some fixed parameter. As seen in Figs. 28–30, the quality of the sparse second moment depends on $\alpha$. In the two cases where the sparse approximation is

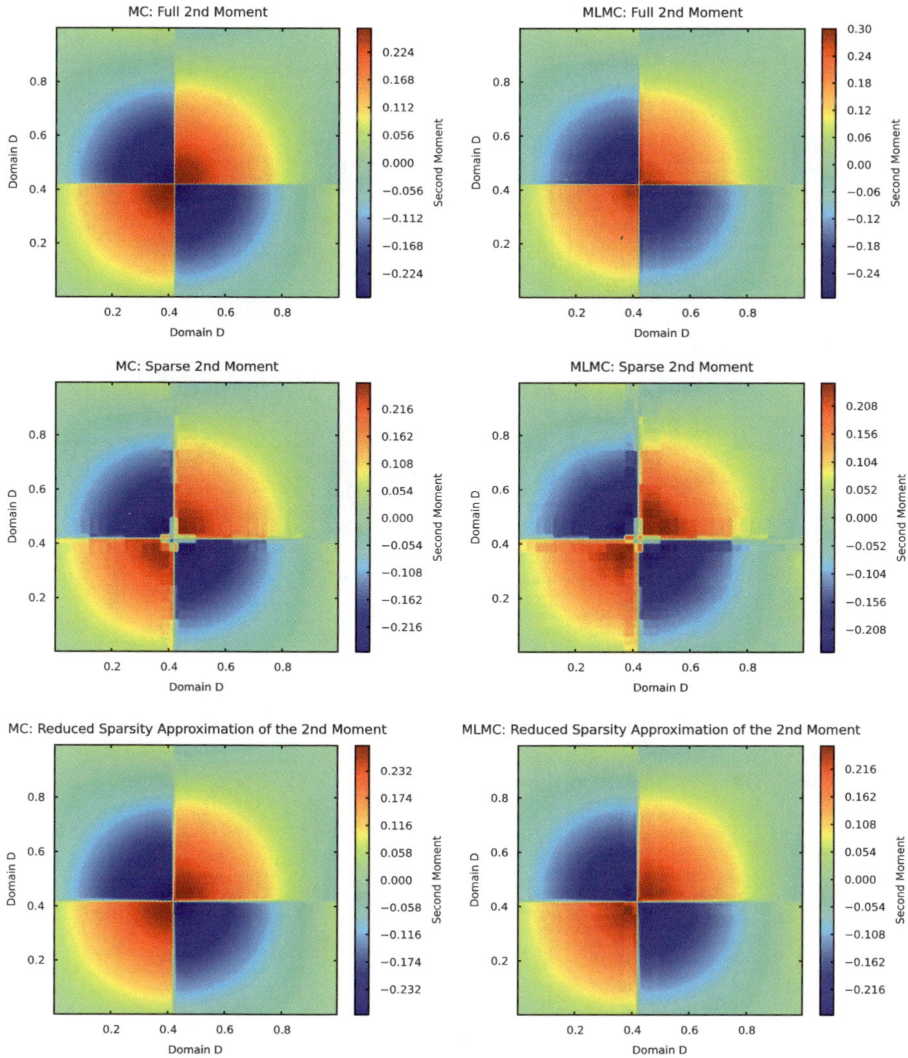


**Fig. 29** Estimated second moment of Burgers' equation (114) for $\alpha = 0.078$

acceptable the discontinuity in the second moment coincides with a cell boundary. This is not the case for $\alpha = 0.078$, where one can observe some artefacts (Fig. 31).

### 6.8.4 Burgers' Equation: Initial Data with Random Phase

Next we study Burgers' equation with sinusoidal initial data which has uncertain initial phase.

$$u_0(x, \omega) = \sin(2\pi(x - 0.1 \cdot X(\omega))), \quad X \sim \mathscr{U}(0, 1) \tag{115}$$

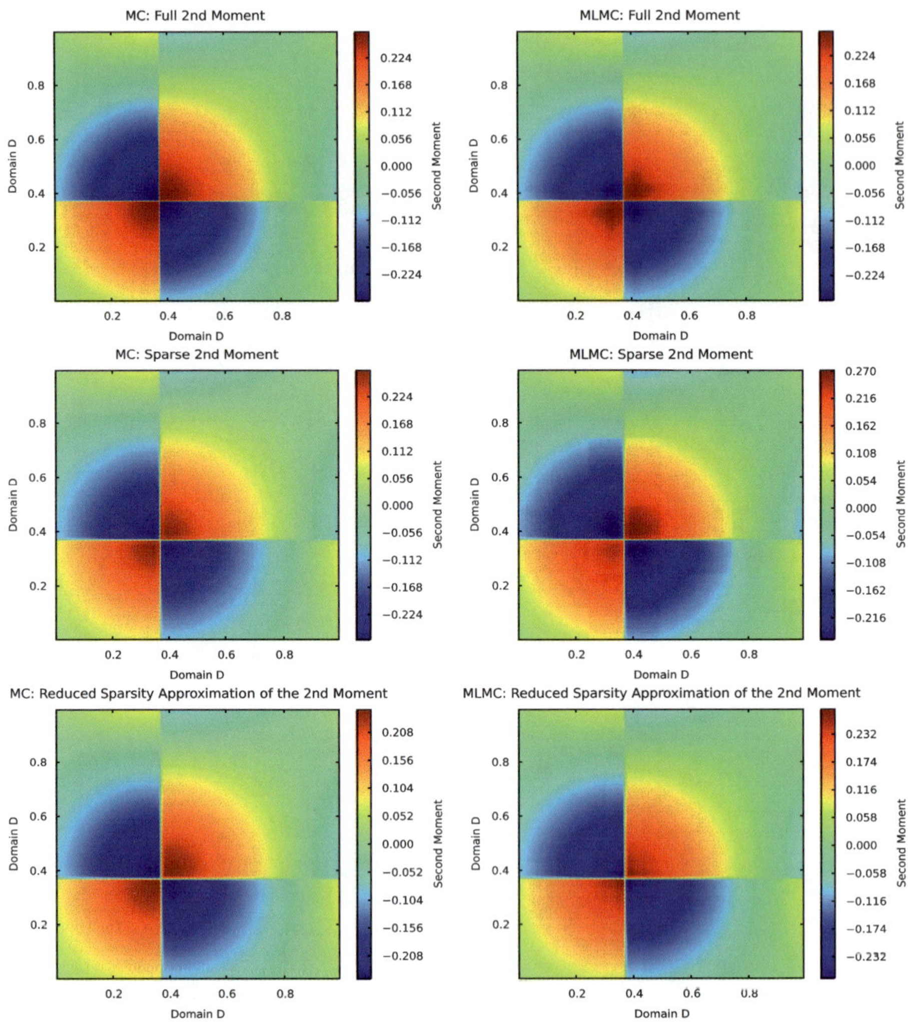


**Fig. 30** Estimated second moment of Burgers' equation (114) for $\alpha = 0.125$

The corresponding results for the two-point correlation function with $\alpha = 0.125$ are shown in Fig. 32.

### 6.8.5 Sparse Tensor Bubble Test

The next experiment will show that the sparse approximation of a deterministic bubble function

$$f(x) = \max(0, -500(x - \alpha)^2 + 1), \quad \alpha \in [0.3, 0.7]$$

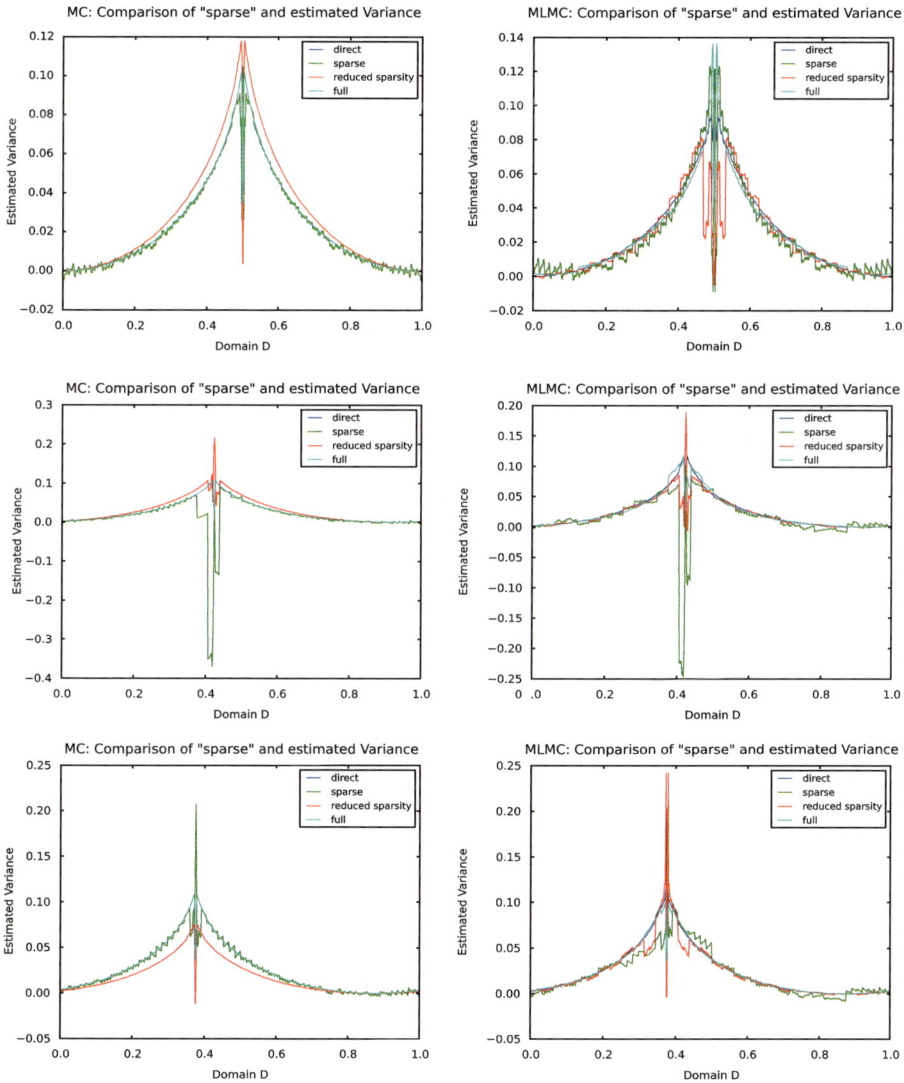


**Fig. 31** Comparison of different variance estimators for Burgers' equation (114): sparse, reduced sparsity, full and direct computation

has similar errors as the sparse variance estimate of Burgers' equation with random phase (Fig. 33). In particular, therefore, we conclude that *the errors in sparse tensor estimates of the two point correlation functions shown in Figs. 28–30 are not due to amplifications of Finite Volume discretization errors in the sparse tensor product construction, but occur generically*: they are, indeed, due to the compression of data in the sparse tensor product formation itself.

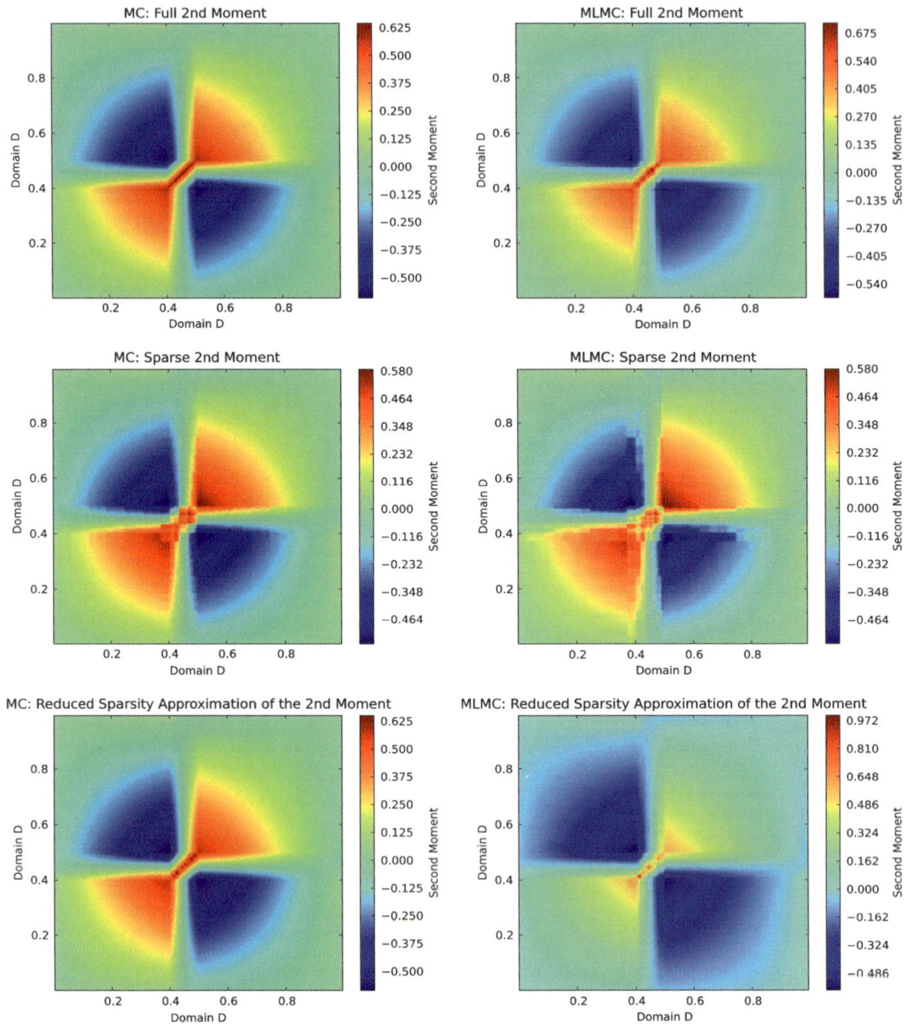


**Fig. 32** Estimated second moment of Burgers' equation (115) for $\alpha = 0.125$

Let $u$ be the piecewise constant map obtained from $f$ by taking cell averages on a uniform mesh with N cells, $\hat{u}^{(2)}$ and $\tilde{u}^{(2)}$ the approximations based on sparse tensor product and reduced sparsity tensor products, as defined above. The cell averages are computed using the midpoint rule on each cell. Figure 34 shows $u^2$, $\hat{u}^{(2)}(x,x)$ and $\tilde{u}^{(2)}(x,x)$ in a single plot. For all mesh sizes $N = 2^l, l = 5, \ldots, 10$ one can see negative spikes in both sparse second moment approximations, furthermore the positive peak in the sparse approximation can be way too weak. While the reduced sparsity tensor product approximation also has negative spikes, it is considerably more accurate than the regular sparse tensor product approximation, while still

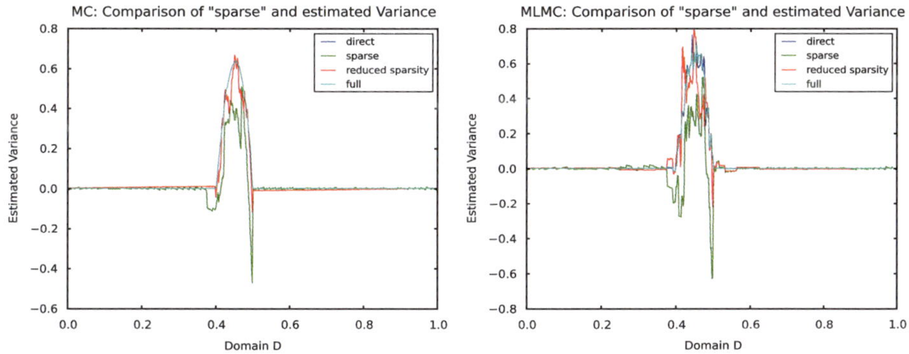


**Fig. 33** Comparison of different variance estimators: sparse, reduced sparsity, full and direct computation

preserving the asymptotic complexity scaling of the sparse tensor product estimate. We also observe in Fig. 34 that the use of sparse tensor approximations introduces significant errors which, in the specific example of variance computation, even lead to meaningless outputs such as negative values of the estimated variance.

# 7 MLMC Approximation of Probabilities

## *7.1 MLMC Estimation of Probabilities*

So far, in these notes, we addressed the analysis and implementation of MLMC-FVM solvers for the efficient solution of conservation laws with random inputs, and we focused on the efficient computation of first and higher order statistical moments of the random solution(s), such as mean field and variances.

Often, in applications, one is interested in probabilities of certain extremal events $\boldsymbol{E} \in \mathcal{F}$, rather than statistical moments, conditioned on the given random input data. Denoting by $\chi_{\boldsymbol{E}}(\omega)$ the indicator function of $\boldsymbol{E} \in \mathcal{F}$, the probability of interest is

$$\mathbb{P}(\boldsymbol{E}) = \int_{\omega \in \Omega} \chi_{\boldsymbol{E}}(\omega) d\mathbb{P}(\omega) . \tag{116}$$

One of the problems of interest (in order to assess the risk) would be the following: given a fixed sub-domain $\mathbf{C} \subset \mathbf{D} \subset \mathbb{R}^d$, for a fixed time $t \geq 0$, find the *probability* $p(\mathbf{U}) \in [0, 1]$ that a certain event $\boldsymbol{E} \in \mathcal{F}$ will occur. Here, we assume that the event $\boldsymbol{E}$ takes the generic form $\boldsymbol{E} = \{\omega \in \Omega : \chi(\mathbf{U}(\cdot, t; \omega)) = 1\}$, where $\chi(\cdot) : \mathbb{R}^m \to \{0, 1\}$ is a measurable function. Then, the probability of interest can be expressed as

$$p(\mathbf{U}) := \mathbb{P}(\{\chi_{\boldsymbol{E}} = 1\}) = \mathbb{E}[\chi(\mathbf{U}(\cdot, t, \omega))] . \tag{117}$$

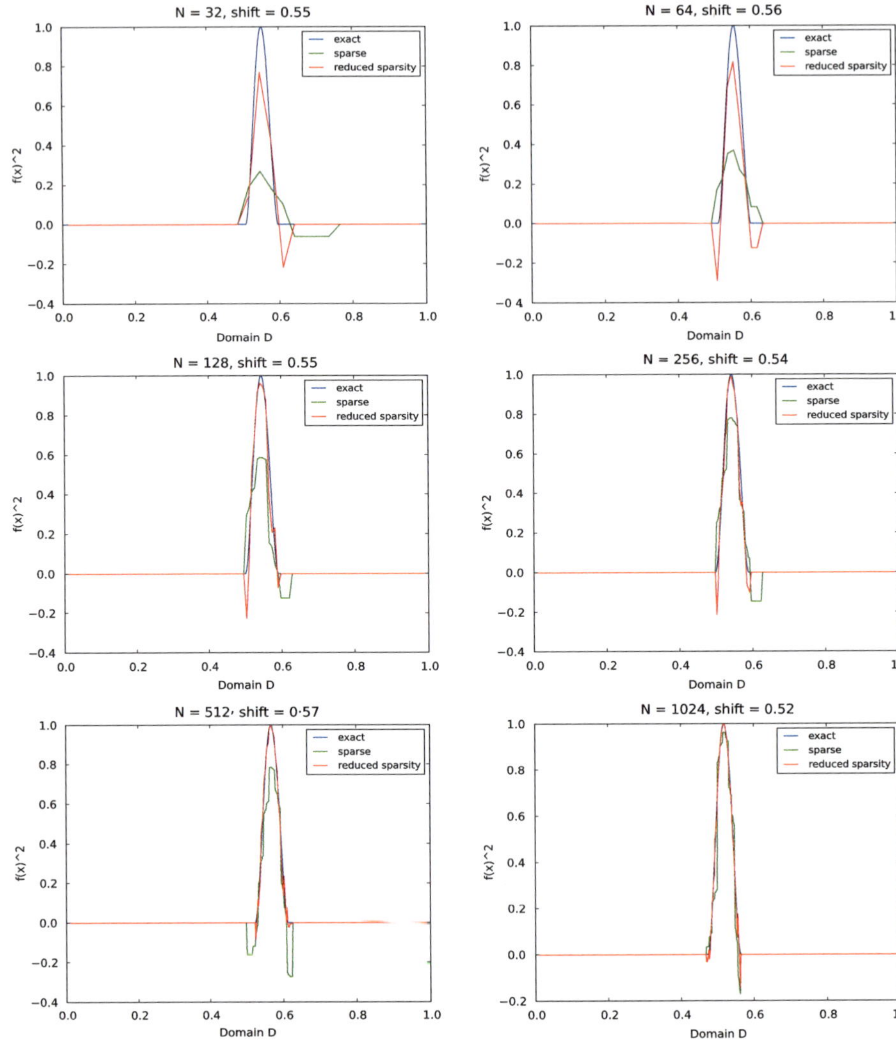


**Fig. 34** Diagonal of the sparse and reduced sparsity tensor product estimates along with the exact value, for different meshes and values for $\alpha$

Rather than developing here a general theory for the multi-level MC computation of such probabilities, we exemplify the main ideas for the Shallow Water Equations with uncertain bottom topography, i.e. (91). Here, one is often interested in the event "average water level $\frac{1}{|C|}\int_{x\in C} h(x,t,\omega)$ at time $t$ in subdomain $\mathbf{C} \subset \mathbf{D}$ (e.g. a neighborhood of the shoreline) *exceeds* some given *threshold* $h_{\max}$". Then

$$\chi_E(\mathbf{U}(\cdot,t,\omega)) := \chi_{(h_{\max},\infty)}\left(\frac{1}{|C|}\int_{x\in C} h(x,t,\omega)dx\right)$$

and $\mathbb{P}(\boldsymbol{E})$ is given by

$$\mathbb{P}(\boldsymbol{E}) = \mathbb{E}[\chi(\mathbf{U}(\cdot,t,\omega))] = \int_{\Omega} \chi_{(h_{\max},\infty)} \left( \frac{1}{|\boldsymbol{C}|} \int_{x \in \boldsymbol{C}} h(x,t,\omega)dx \right) d\mathbb{P}(\omega) \,. \tag{118}$$

The probability of interest $\mathbb{P}(\boldsymbol{E})$ in (118) is an integral w.r.t. the probability measure $\mathbb{P}$. Hence, the integral in (118) could be approximated numerically by a Monte-Carlo FVM estimator, i.e. by Monte-Carlo integration of an approximate Finite Volume solution at mesh level $\ell$, denoted by $\mathbf{U}_\ell$. The single-level Monte-Carlo Finite Volume estimator for $\mathbb{P}(\boldsymbol{E})$ with $M$ i.i.d. input data samples based on the FVM on mesh level $\ell$ is given by

$$p_M(\mathbf{U}_\ell) := \frac{1}{M} \sum_{i=1}^{M} \chi(\mathbf{U}_\ell^i(\cdot,t)) = \frac{1}{M} \sum_{i=1}^{M} \chi_{(h_{\max},\infty)} \left( \frac{1}{|\boldsymbol{C}|} \int_{x \in \boldsymbol{C}} h_\ell^i(x,t)dx \right), \tag{119}$$

where $h_\ell^i(\cdot,t)$, $i = 1,\dots,M$ are the MC samples approximated using the FVM scheme at mesh level $\ell$. The MLMC-FVM estimator combines MC-FVM estimators (119) on the nested family $\{\mathscr{T}_\ell\}_{\ell \geq 0}$ of FVM meshes, as before, and is given by:

$$p^L(\mathbf{U}_L) := p_{M_0}(\mathbf{U}_0) + \sum_{\ell=1}^{L} \left( p_{M_\ell}(\mathbf{U}_\ell) - p_{M_\ell}(\mathbf{U}_{\ell-1}) \right) . \tag{120}$$

## 7.2 *Shallow Water Equation in 2d: Perturbation of a Steady-State*

We consider the setup as in Sect. 6.3.2, i.e. we are interested in the two-dimensional shallow water equations where the uncertain initial perturbation (99) of the water surface is propagating over the uncertain bottom topography described in (94)–(95) and in (96). Our aim is to numerically approximate the probabilities as in (118) that the cell averaged water level $h + b$ will exceed the preset threshold $h_{\max} = 1.002$ in a subdomain $\boldsymbol{C}$ where $\boldsymbol{C} \in \mathscr{T}_L$ denotes a Finite Volume cell.

The results of the numerical simulation using the MLMC-FVM estimator (120) for the probability integral (118) are given in Fig. 35. There are 9 levels ($L = 8$) of FVM mesh resolution with the finest resolution (at the finest level $\ell = L$) being 2,048 cells. Rusanov flux with second order accurate well-balanced TECNO [15] reconstruction was used.

*Remark 7.* Firstly, we observe that $\chi_{\boldsymbol{E}}(\mathbf{U}_\ell^i(\cdot,t))$ can only attain values in $\{0,1\}$. Then, since the MC-FVM approximation $p_M(\mathbf{U}_\ell)$ of $p(\mathbf{U}_\ell)$ from (119) is a convex combination (with equal weights) of values from $\{0,1\}$, $p_M(\mathbf{U}_\ell)$ attains values in the interval [0, 1], which is consistent with the fact that $p_M(\mathbf{U}_\ell)$ is approximation of

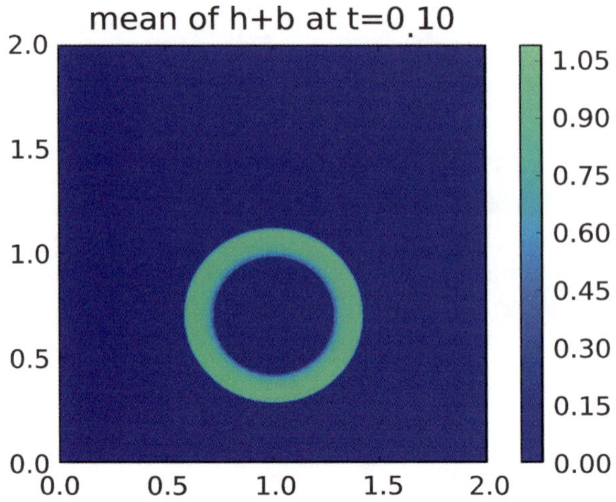


| $L$ | $M_L$ | Grid size | CFL | Cores | Runtime | Efficiency |
|---|---|---|---|---|---|---|
| 8 | 16 | 2,048 × 2,048 | 0.45 | 128 | 1:42:33 | 94.7 % |

**Fig. 35** MLMC-FVM approximation of the probabilities (118) that the random water level $h(\cdot, t, \omega) + b(\cdot, \omega)$ will exceed (at time $t = 0.1$) the preset maximal threshold $h_{\max} = 1.002$

the exact *probability*. However, such bounded range of values is no longer valid for the *Multi-Level* MC-FVM approximation $p^L(\mathbf{U}_L)$ in (120).

As a counter-example to prove this claim, consider only two levels, i.e. $L = 1$, one sample on the level $\ell = 0$, i.e. $M_0 = 1$, and two samples on the finer level $\ell = 1$, i.e. $M_1 = 2$. The coarsest (one-dimensional) mesh level is assumed to have only one cell, i.e. $\mathscr{T}_0 = \{C_1^0\}$, and the finer mesh level is assumed to have two equal cells, i.e. $\mathscr{T}_1 = \{C_1^1, C_2^1\}$. Then, assume we have random samples $\omega_1^0, \omega_1^1, \omega_2^1$ for random input data. Furthermore, we assume that the event values for these samples on different mesh resolutions are given by:

$$\begin{aligned}
&\chi_E(\mathbf{U}_0(C_1^0, \omega_1^0)) = 1, && \\
&\chi_E(\mathbf{U}_1(C_1^1, \omega_1^1)) = 1, && \chi_E(\mathbf{U}_1(C_1^1, \omega_2^1)) = 1, \\
&\chi_E(\mathbf{U}_1(C_2^1, \omega_1^1)) = 1, && \chi_E(\mathbf{U}_1(C_2^1, \omega_2^1)) = 0, \\
&\chi_E(\mathbf{U}_0(C_2^1, \omega_1^1)) = 1, && \chi_E(\mathbf{U}_0(C_2^1, \omega_2^1)) = 0.
\end{aligned}$$

Then, the corresponding MC-FVM estimates $p_{M_0}(\mathbf{U}_0)$, $p_{M_1}(\mathbf{U}_1)$ and $p_{M_1}(\mathbf{U}_0)$ are

$$p_{M_0}(\mathbf{U}_0) = 1, \quad p_{M_1}(\mathbf{U}_0(C_1^1)) = 1, \quad p_{M_1}(\mathbf{U}_0(C_2^1)) = \frac{1}{2}, \quad p_{M_1}(\mathbf{U}_0) = \frac{1}{2}1 = \frac{1}{2},$$

which lead to the following MLMC-FVM estimate (120) for each cell $C_1^1, C_2^1$

$$p^L(\mathbf{U}(C_1^1)) = 1 + 1 - \frac{1}{2} = 1\frac{1}{2}, \qquad p^L(\mathbf{U}(C_2^1)) = 1 + \frac{1}{2} - \frac{1}{2} = 1. \qquad (121)$$

Clearly $p^L(\mathbf{U}(C_1^1)) > 1$. The counter-example resulting in $p^L(\mathbf{U}(C_1^1)) < 0$ can be obtained analogously. Notice, that even if we *reuse* samples from the coarsest mesh level $\ell = 0$ in the finer mesh level $\ell = 1$, i.e. we set $\omega_1^1 = \omega_1^0$, the counter-example is still consistent, since $\omega_1^1 \neq \omega_2^1$.

The above considerations do not mean that the MLMC-FVM approximation of $p(\mathbf{U}_\ell)$ is wrong or inconsistent with the exact value; as $L \to \infty$ the MLMC-FVM approximation $p^L(\mathbf{U}_L)$ still *converges* to the exact value $p(\mathbf{U}_L)$. In order to avoid confusion, however, in Fig. 35 the approximated values of $p^L(\mathbf{U}_L)$ were clipped to the interval $[0, 1]$. Analogously, in variance plots throughout this manuscript, the variance was clipped to $[0, \infty)$. The consequences of such clipping are numerically *negligible*; however, the awareness of this *non-monotonicity* of the MLMC-FVM estimator is essential for the interpretation of the numerical results. In particular; it is important to realize that in MLMC-FVM such undershoots might occur, which are *not* related to numerical instability, bias of the estimators, inappropriate FVM schemes or other implementational errors.

## 8 Conclusion

The issue of uncertainty quantification for physical and engineering applications that are modeled by hyperbolic systems of balance laws, has gained increasing attention in recent years. The inputs to hyperbolic systems of balance laws such as initial data, boundary conditions, source terms as well as fluxes are in general, prone to measurement errors and are therefore in principle uncertain. This data uncertainty propagates into the solution, making the design of efficient and robust methods for the efficient numerical quantification of uncertainty, a task that is of utmost significance in CSE.

In these notes, we have presented recent results on the design, analysis and implementation of efficient statistical sampling methods of the Monte Carlo (MC) and Multi-Level Monte Carlo (MLMC) type for quantifying uncertainty in the solutions of systems of random balance laws with random input data. The main conclusions from the results in these notes are as follows:

- Uncertain inputs such as random initial data, sources and fluxes are modeled within the classical probabilistic framework of Kolmogorov. The corresponding notion of *random entropy solutions* was introduced and shown to be wellposed for scalar multi-dimensional conservation laws with random initial data and with random fluxes. Furthermore, statistical regularity of the random entropy solutions was mathematically described.

- The MC-FVM and MLMC-FVM algorithms for systems of balance laws were presented. Efficient high-resolution finite volume schemes were used for the spatio-temporal discretizations of the balance laws and were combined with the MC and MLMC algorithms. The convergence and complexity analysis of the resulting schemes (in scalar case) were presented. In particular, we showed that the MLMC Finite Volume Method had the same asymptotic complexity (up to a logarithmic term) as a single deterministic finite volume solve.
- Details of implementation of the MC and MLMC Finite Volume Methods for systems of balance laws were described briefly. A static load balancing algorithm was presented. This static load balancing algorithm was demonstrated to scale the MLMC Finite Volume Methods with near optimal parallel efficiency to several thousands of parallel processing units.
- A large number of numerical experiments were presented to illustrate the efficiency of the MLMC-FVM method. These experiments included Euler and MHD equations with uncertain initial data (initial data as well as fluxes), shallow water equations with uncertain initial data and uncertain bottom topography, Buckley-Leverett equations with uncertain relative permeabilities and the compressible Euler equations with uncertain equations of state.
- A sparse tensor discretization framework was introduced to efficiently compute $k$-point statistical correlation functions in several space dimensions for the random entropy solutions. Preliminary numerical results illustrating this method were presented, and the errors introduced by sparse tensorization (which is essential for computational efficiency) were discussed.
- We introduced a novel non-intrusive technique to use the MLMC algorithm for computing approximate probabilities of statistical events of engineering interest, such as probabilities of extremal events in engineering risk and failure analyses based on random entropy solutions.

The main purpose of these notes was to demonstrate that the MLMC statistical sampling method is a powerful tool in the context of computational uncertainty quantification for hyperbolic systems of balance laws with random input data. Its advantages include:

- The method is completely non-intrusive. It can be readily used in conjunction with any spatio-temporal hierarchical discretization of the underlying conservation (balance) laws. In the present notes, we combined the MLMC method with high-resolution Finite Volume Methods. We hasten to add, however, that DG (discontinuous Galerkin) discretizations can also be used as spatio-temporal solvers.
- The MLMC-FVM approach is very flexible and can be used for different types of uncertain inputs such as random initial data, source terms or flux functions.
- The method is robust with respect to very low regularity (presence of discontinuities) of the underlying random entropy solutions.
- The method can deal with a very large number of sources of uncertainty. For instance, the computation for shallow water equations with uncertain bottom topography involved approximately 1,000 sources of uncertainty. To the best

of our knowledge, no other method (particularly deterministic methods such as stochastic Galerkin or stochastic collocation) can handle efficiently this many sources of uncertainty (i.e., high "stochastic dimensions") with solutions of low regularity and possibly nonsmooth dependence on random input parameters.

- Single-level MC estimators are *sample averages* of realizations, i.e. convex combinations of FVM numerical solutions. Therefore, single-level MC-FVM estimates naturally inherit non-oscillation, discrete maximum and monotonicity properties of the underlying FVM discretization schemes. Since MLMC-FVM estimators are *not* based on convex combinations of numerical FVM solutions, these (desirable) properties of FVM discretization schemes may be lost in MLMC-FVM estimators. The efficiency gain afforded by the multi-level methodology, however, is so substantial that in our view it outweighs these shortcomings; nevertheless, the awareness of these effects is warranted in the interpretation of MLMC-FVM simulation results.
- All properties of the MLMC methods mentioned above are directly inherited from the standard Monte Carlo methods; however, our analysis for scalar balance laws indicates that MLMC methods offer a better asymptotic rate of convergence. In particular, in all our numerical experiments we found that already in the engineering range of accuracy MLMC-FVM methods are several orders of magnitude more efficient than plain MC methods.
- When coupled with efficient implementation on massively parallel hardware architectures, the MLMC-FVM methods can handle complex flow problems with random input data.
- MLMC-FVM is also amenable to *algorithmic fault-tolerant* parallelization, where several *lost MC samples* (due to node or network failures) could be completely *ignored*. In [34], assuming that the *expected* number of failures is not too large, the MLMC-FVM estimate is proven to converge (i.e., the *expected* MLMC-FVM error converges) even though only the remaining batch of MC "survived" samples are used to assemble the MLMC-FVM estimator.

Given these advantages, the MLMC Finite Volume Method appears as a powerful general purpose technique for quantifying uncertainty in solutions of random systems of balance laws in engineering practice.

**Acknowledgements** The authors wish to express their gratitude to Mr. Luc Grosheintz, a student in the ETH Zürich MSc Applied Mathematics curriculum for performing the numerics for the sparse two-point correlation computations reported in Sect. 6.8.

The authors thank the systems support at ETH Zürich parallel Compute Cluster BRUTUS [49] for their support in the production runs for the present paper, and the staff at the Swiss National Supercomputing center (CSCS) [50] at Lugano for their assistance in the large scale Euler and MHD simulations.

## References


1. R. Abgrall. *A simple, flexible and generic deterministic approach to uncertainty quantification in non-linear problems.* Rapport de Recherche, INRIA, 2007.
2. ALSVID. Available from http://folk.uio.no/mcmurry/amhd.
3. ALSVID-UQ. Available from http://www.sam.math.ethz.ch/alsvid-uq.
4. K. Aziz and A. Settari. *Fundamentals of petroleum reservoir simulation.* Applied Science Publishers, London, 1979.
5. A. Barth, Ch. Schwab and N. Zollinger. *Multilevel MC Method for Elliptic PDEs with Stochastic Coefficients.* Numerische Mathematik, Volume **119(1)**, pp. 123–161, 2011.
6. T. J. Barth. *Numerical methods for gas-dynamics systems on unstructured meshes.* An Introduction to Recent Developments in Theory and Numerics of Conservation Laws, pp. 195–285. Lecture Notes in Computational Science and Engineering volume 5, Springer, Berlin. Eds: D. Kroner, M. Ohlberger, and Rohde, C., 1999.
7. P. D. Bates, S. N. Lane and R. I. Ferguson. *Parametrization, Validation and Uncertainty analysis of CFD models of fluvial and flood hydraulics in natural environments.* Computational Fluid Dynamics: Applications in environmental hydraulics, John Wiley and sons, pp. 193–212, 2005.
8. Q. Y. Chen, D. Gottlieb and J. S. Hesthaven. *Uncertainty analysis for steady flow in a dual throat nozzle.* J. Comput. Phys, **204**, pp. 378–398, 2005.
9. B. Cockburn and C-W. Shu. *TVB Runge-Kutta local projection discontinuous Galerkin finite element method for conservation laws. II. General framework.* Math. Comput., **52**, pp. 411–435, 1989.
10. Constantine M. Dafermos. *Hyperbolic Conservation Laws in Continuum Physics (2nd Ed.).* Springer Verlag, 2005.
11. Josef Dick, Franes Y. Kuo and Ian H. Sloan. *High dimensional integration: the Quasi Monte-Carlo way.* Acta Numerica, to appear, 2013.
12. R. Eymard, T. Gallouët, and R. Herbin. *Finite volume methods*, in Handbook of numerical analysis, Vol. **VII**, pp. 713–1020, North-Holland, Amsterdam, 2000.
13. P. F. Fisher and N. J. Tate. *Causes and consequences of error in digital elevation models.* Prog. in Phy. Geography, **30(4)**, pp. 467–489, 2006.
14. G. Fishman. *Monte Carlo.* Springer, 1996.
15. U.S. Fjordholm, S. Mishra, and E. Tadmor. *Well-balanced, energy stable schemes for the shallow water equations with varying topology.* J. Comput. Phys., **230**, pp. 5587–5609, 2011.
16. F. Fuchs, A. D. McMurry, S. Mishra, N. H. Risebro and K. Waagan. *Approximate Riemann solver based high-order finite volume schemes for the Godunov-Powell form of ideal MHD equations in multi-dimensions.* Comm. Comput. Phys., **9**, pp. 324–362, 2011
17. M. Giles. *Improved multilevel Monte Carlo convergence using the Milstein scheme.* Preprint NA-06/22, Oxford computing lab, Oxford, U.K, 2006.
18. M. Giles. *Multilevel Monte Carlo path simulation.* Oper. Res., **56**, pp. 607–617, 2008.
19. E. Godlewski and P.A. Raviart. *Hyperbolic Systems of Conservation Laws.* Mathematiques et Applications, Ellipses Publ., Paris, 1991.
20. S. Gottlieb, C. W. Shu and E. Tadmor. *High order time discretizations with strong stability property.* SIAM. Review, **43**, pp. 89–112, 2001.
21. A. Harten, B. Engquist, S. Osher and S. R. Chakravarty. *Uniformly high order accurate essentially non-oscillatory schemes.* J. Comput. Phys., pp. 231–303, 1987s.
22. S. Heinrich. *Multilevel Monte Carlo methods.* Large-scale scientific computing, Third international conference LSSC 2001, Sozopol, Bulgaria, 2001, Lecture Notes in Computer Science, Vol **2170**, pp. 58–67, Springer Verlag, 2001.
23. P. L'Ecuyer and F. Panneton. *Fast Random Number Generators Based on Linear Recurrences Modulo 2: Overview and Comparison.* Proceedings of the 2005 Winter Simulation Conference, pp. 110–119, IEEE press, 2005.

24. P. L'Ecuyer and F. Panneton. *Fast Random Number Generators Based on Linear Recurrences Modulo 2.* ACM Trans. Math. Software, **32**, pp. 1–16, 2006.
25. R.A. LeVeque. *Numerical Solution of Hyperbolic Conservation Laws.* Cambridge Univ. Press 2005.
26. R. LeVeque, D. George and M. Berger. *Tsunami modeling with adaptively refined finite volume methods.* Acta Numerica, **20**, pp. 211–289, 2011.
27. G. Lin, C.H. Su and G. E. Karniadakis. *The stochastic piston problem.* PNAS **101**, pp. 15840–15845, 2004.
28. X. Ma and N. Zabaras. *An adaptive hierarchical sparse grid collocation algorithm for the solution of stochastic differential equations.* J. Comp. Phys, **228**, pp. 3084–3113, 2009.
29. M. Matsumoto and T. Nishimura. *Mersenne Twister: a 623-dimensionally equidistributed uniform pseudorandom number generator.* ACM Trans. Modeling and Computer Simulation, **8**, pp. 3–30, Jan. 1998.
30. S. Mishra and Ch. Schwab. *Sparse tensor multi-level Monte Carlo finite volume methods for hyperbolic conservation laws with random initial data.* Math. Comp. **280(81)**, pp. 1979–2018, 2012.
31. S. Mishra, N.H. Risebro, Ch. Schwab and S. Tokareva. *Numerical solution of scalar conservation laws with random flux functions.* SAM Technical Report No. 2012-35, in review, 2012. Also available from http://www.sam.math.ethz.ch/sam_reports/index.php?id=2012-35.
32. S. Mishra, Ch. Schwab and J. Šukys. *Multi-level Monte Carlo finite volume methods for nonlinear systems of conservation laws in multi-dimensions.* J. Comp. Phys., **231(8)**, pp. 3365–3388, 2012.
33. S. Mishra, Ch. Schwab, and J. Šukys. *Multi-level Monte Carlo Finite Volume methods for shallow water equations with uncertain topography in multi-dimensions.* SIAM J. Sci. Comput., **34(6)**, pp. B761–B784, 2012.
34. S. Pauli and P. Arbenz and Ch. Schwab. *Intrinsic fault tolerance of multi level Monte Carlo methods.* SAM Technical Report 2012-24, Seminar für Angewandte Mathematik ETH Zürich, 2012. Also available from http://www.sam.math.ethz.ch/sam_reports/index.php?id=2012-24.
35. G. Poette, B. Després and D. Lucor. *Uncertainty quantification for systems of conservation laws.* J. Comput. Phys. **228**, pp. 2443–2467, 2009.
36. G.D. Prato and J. Zabcyk, *Stochastic Equations in infinite dimensions*, Cambridge Univ. Press, 1991.
37. G. Schmidlin. *Fast solution algorithms for integral equations in $\mathbb{R}^3$.* PhD dissertation ETH Zürich No. 15016, 2003.
38. G. Schmidlin and Ch. Schwab. *Wavelet Galerkin BEM on unstructured meshes by aggregation.* LNCSE **20**, pp. 369–278, Springer Lecture Notes in CSE, Springer Verlag, Berlin Heidelberg New York, 2002.
39. Ch. Schwab and S. Tokareva. *High order approximation of probabilistic shock profiles in hyperbolic conservation laws with uncertain initial data.* ESAIM: Mathematical Modelling and Numerical Analysis, ESAIM: M2AN **47**, 807–835 (2013) DOI: 10.1051/m2an/2012060, www.esaim-m2an.org.
40. C. W. Shu and S. Osher. *Efficient implementation of essentially non-oscillatory schemes - II.* J. Comput. Phys., **83**, pp. 32–78, 1989.
41. C. W. Shu. *Essentially non-oscillatory and weighted essentially non-oscillatory schemes for hyperbolic conservation laws.* ICASE Technical report, NASA, 1997.
42. J. Šukys, S. Mishra, and Ch. Schwab. *Static load balancing for multi-level Monte Carlo finite volume solvers.* PPAM 2011, Part I, LNCS **7203**, pp. 245–254. Springer, Heidelberg, 2012.
43. J. Tryoen, O. Le Maitre, M. Ndjinga and A. Ern. *Intrusive projection methods with upwinding for uncertain non-linear hyperbolic systems.* Preprint, 2010.
44. T. von Petersdorff and Ch. Schwab. *Sparse Finite Element Methods for Operator Equations with Stochastic Data*, Applications of Mathematics **51(2)**, pp. 145–180, 2006.
45. X. Wan and G. E. Karniadakis. *Long-term behaviour of polynomial chaos in stochastic flow simulations.* Comput. Meth. Appl. Mech. Engg. **195**, pp. 5582–5596, 2006.

46. B. P. Welford. *Note on a Method for Calculating Corrected Sums of Squares and Products.* Technometrics, **4**, pp. 419–420, 1962.
47. J. A. S. Witteveen, A. Loeven, H. Bijl *An adaptive stochastic finite element approach based on Newton-Cotes quadrature in simplex elements.* Comput. Fluids, **38**, pp. 1270–1288, 2009.
48. D. Xiu and J. S. Hesthaven. *High-order collocation methods for differential equations with random inputs.* SIAM J. Sci. Comput., **27**, pp. 1118–1139, 2005.
49. *Brutus,* ETH Zürich, de.wikipedia.org/wiki/Brutus_(Cluster).
50. *Cray XE6,* Swiss National Supercomputing Center (CSCS), Lugano, www.cscs.ch.
51. *MPI: A Message-Passing Interface Standard.* Version 2.2, 2009, available from: http://www.mpi-forum.org/docs/mpi-2.2/mpi22-report.pdf.
52. *Open MPI: Open Source High Performance Computing.* Available from http://www.open-mpi.org/.

# Essentially Non-oscillatory Stencil Selection and Subcell Resolution in Uncertainty Quantification

**Jeroen A.S. Witteveen and Gianluca Iaccarino**

**Abstract** The essentially non-oscillatory stencil selection and subcell resolution robustness concepts from finite volume methods for computational fluid dynamics are extended to uncertainty quantification for the reliable approximation of discontinuities in stochastic computational problems. These two robustness principles are introduced into the simplex stochastic collocation uncertainty quantification method, which discretizes the probability space using a simplex tessellation of sampling points and piecewise higher-degree polynomial interpolation. The essentially non-oscillatory stencil selection obtains a sharper refinement of discontinuities by choosing the interpolation stencil with the highest polynomial degree from a set of candidate stencils for constructing the local response surface approximation. The subcell resolution approach achieves a genuinely discontinuous representation of random spatial discontinuities in the interior of the simplexes by resolving the discontinuity location in the probability space explicitly and by extending the stochastic response surface approximations up to the predicted discontinuity location. The advantages of the presented approaches are illustrated by the results for a step function, the linear advection equation, a shock tube Riemann problem, and the transonic flow over the RAE 2822 airfoil.

J.A.S. Witteveen (✉)
Center for Turbulence Research, Stanford University, Building 500, Stanford, CA 94305-3035, USA

Center for Mathematics and Computer Science (CWI), Science Park 123, 1098 XG Amsterdam, The Netherlands
e-mail: jeroen.witteveen@cwi.nl

G. Iaccarino
Mechanical Engineering, Stanford University, Building 500, Stanford, CA 94305-3035, USA
e-mail: jops@stanford.edu

H. Bijl et al. (eds.), *Uncertainty Quantification in Computational Fluid Dynamics*, Lecture Notes in Computational Science and Engineering 92,
DOI 10.1007/978-3-319-00885-1_7, 

# 1 Introduction

Non-intrusive uncertainty quantification (UQ) is applicable to many branches of the computational sciences and it has a strong foundation in mathematical approximation and interpolation theories. The unique contribution that computational fluid dynamics (CFD) can bring to the field of numerical methods for computationally intensive stochastic problems is the robust approximation of discontinuities in the probability space. There is extensive experience in the finite volume method (FVM) community with original robustness concepts for the reliable solution of discontinuities in the form of shock waves and contact surfaces in flow fields. Here, we extend two of these robustness principles to the probability space and demonstrate their effectiveness in test problems from stochastic CFD. The two considered concepts are the essentially non-oscillatory (ENO) stencil selection and the subcell resolution approach, which are introduced into the multi-element simplex stochastic collocation (SSC) UQ method.

The ENO spatial discretization [11] achieves an essentially non-oscillatory approximation of the solution of hyperbolic conservation laws. Non-oscillatory means, in this context, that the number of local extrema in the solution does not increase with time. The ENO scheme obtains this property using an adaptive-stencil approach with a uniform polynomial degree for reconstructing the spatial fluxes. Each spatial cell $X_j$ is assigned $r$ stencils $\{S_{j,i}\}_{i=1}^{r}$ of degree $p$, all of which include the cell $X_j$ itself. Out of this set of candidate stencils $\{S_{j,i}\}$, the stencil $S_j$ is selected for cell $X_j$ that results in the interpolation $w_j(\mathbf{x})$ which is smoothest in some sense based on an indicator of smoothness $\mathrm{IS}_{j,i}$. In this way, a cell next to a discontinuity is adaptively given a stencil consisting of the smooth part of the solution, which avoids Gibbs-like oscillations in physical space. Attention has been paid to the efficient implementation of ENO schemes by Shu and Osher [24, 25]. Figure 1a shows an example of the ENO stencil selection in a FVM discretization of a discontinuity in one spatial dimension using piecewise quadratic polynomials.

The notion of subcell resolution in FVM originated from Harten [12] to prevent the smearing of contact discontinuities in the solution of hyperbolic conservation laws in the physical space $X$. It is based on the observation that the location of a discontinuity $x_{\mathrm{disc}}$ within a spatial cell $X_j$ can be derived from the computed cell-averaged value $\bar{w}_j$ approximating a flow quantity $u(x)$. In an ENO scheme [11], the ENO reconstructions of $u(x)$ in the cells to the left and the right of the discontinuous cell $X_j$, $w_{j-1}(x)$ and $w_{j+1}(x)$, are then extended up to an approximation of the discontinuity location $x_{\mathrm{disc}}$ in $X_j$ such that their integral matches the cell average $\bar{w}_j$, see Fig. 1b. This allows for resolving discontinuities in the interior of the cells instead of restricting them to the cell face locations. The concept can be extended to multiple spatial dimensions using the dimensional splitting approach.

The SSC method [32, 33] is based on a simplex tessellation of the probability space with sampling points at the vertexes of the simplex elements. The polynomial approximation in the simplexes $\Xi_j$ is built using higher degree interpolation stencils $S_j$, with local polynomial degree $p_j$, consisting of samples in the vertexes of

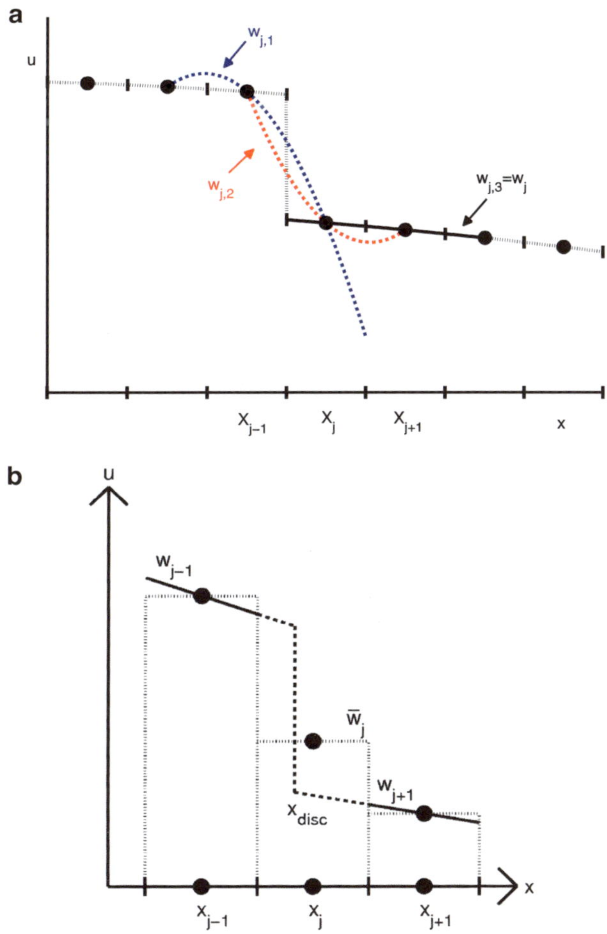


**Fig. 1** FVM discretizations of the physical space for the approximation $w_j(x)$ of the flow conditions in one spatial dimension. (**a**) ENO stencil selection. (**b**) Subcell resolution

surrounding simplexes. The degree $p_j$ is controlled by a local extremum conserving (LEC) limiter, which reduces $p_j$ and the stencil size to avoid overshoots in the interpolation of the samples where necessary. The limiter, therefore, leads to a non-uniform polynomial degree that reduces to a linear interpolation in simplexes which contain a discontinuity and that increases away from singularities. SSC employs adaptive refinement measures based on the hierarchical surplus and the geometrical properties of the simplexes to identify the location of discontinuities. However, the limiter can result in an excessive reduction of the polynomial degree also at significant distances away from a discontinuity. Since the polynomial degree affects the refinement criteria, this can also deteriorate the effectiveness of the refinement to sharply resolve singularities. The SSC approach can also handle correlated random input parameters and non-hypercube probability spaces with constraints effectively.

In order to obtain a more accurate solution of nonlinear response surfaces, ENO-type stencil selection is introduced into the SSC–ENO method [34]. For each simplex $\Xi_j$, $r_j$ stencils $\{S_{j,i}\}_{i=1}^{r_j}$ are constructed that contain $\Xi_j$, and the stencil $S_j$ is selected that results in the smoothest interpolation $w_j(\boldsymbol{\xi})$. The polynomial degree $p_{j,i}$ of the candidate stencils $S_{j,i}$ that are accepted by the LEC limiter, is used here as the indicator of smoothness $\mathrm{IS}_{j,i}$. A simplex $\Xi_j$ near a discontinuity, therefore, achieves a higher order approximation by assigning $\Xi_j$ a higher degree interpolation stencil $S_j$ that does not contain the discontinuity. The higher polynomial degree $p_j$ leads also to lower values of the refinement measures in $\Xi_j$, which restricts the refinement more to the simplexes that contain the discontinuity. The stencil selection does not affect the linear approximation in the latter simplexes, since the LEC limiter rejects the higher degree stencils that contain these elements.

However, in problems where the location of a discontinuity in the physical space is random, adaptive refinement in the probability space proves ineffective. For each point $x$ in the physical space, the spatial discontinuity namely results in a jump at a different location $\xi$ in the probability space. Random discontinuity locations also result in staircase approximations for the mean and standard deviation fields that converge with first-order accuracy only. This coincides with an underprediction of the maximum standard deviation as well. In a non-intrusive stochastic approach, the staircase behavior is caused by the discontinuity locally crossing a sampling point in the probability space and the lack of resolution of the discontinuity location in between the samples.

Therefore, we introduce the concept of subcell resolution into the SSC–SR method [35] by extracting the discontinuity location $x_{\mathrm{disc}}$ in the physical space from each of the deterministic simulations for the sampled random parameter values $\xi_k$. These physical discontinuity locations $x_{\mathrm{disc}}$ are interpolated in the stochastic dimensions to derive a relation for the location of the discontinuity $\xi_{\mathrm{disc}}$ in the probability space as a function of the spatial coordinate $x$. In the discontinuous cells, the interpolations $w_j(\xi)$ of the neighboring cells $\Xi_j$ are then extended from both sides up to the predicted discontinuity location $\xi_{\mathrm{disc}}$. This leads to a genuinely discontinuous representation of the jump in the interior of the cells in the probability space, which avoids the underprediction of the standard deviation without the need for linear interpolation and adaptive sampling near the discontinuity. It also avoids the staircase approximation of the statistical moments because of the continuous dependence of the discontinuity location $\xi_{\mathrm{disc}}$ in the probability space on the spatial coordinate $x$. For multiple random parameters, it leads to a multi-dimensional representation of the discontinuous front without the need for dimensional splitting.

The SSC method with the stencil selection and subcell resolution extensions are of the multi-element type of UQ methods. Multi-element UQ methods discretize the stochastic dimensions using multiple subdomains comparable to spatial discretizations in physical space. These local methods [3, 18, 28] can be based on Stochastic Galerkin (SG) projections of Polynomial Chaos (PC) expansions [14, 36] in each of the subdomains. Other methods [2, 13, 17] use a Stochastic Collocation (SC) approach [4,37] to construct the local polynomial approximations based on sampling at quadrature points in the elements. These methods commonly use sparse grids of Gauss quadrature rules in hypercube subdomains combined with solution-based refinement measures [19] for resolving nonlinearities. Because of the hypercube elements, these methods are most effective in capturing discontinuities that are aligned with one of the stochastic coordinates.

Abgrall [1] and Barth [5] have extended FVM directly to discretize the combined physical and probability spaces using the ENO scheme. We consider here the spatial and stochastic dimensions separately to reduce the dimensionality of the problems. Subcell resolution in stochastic methods has also been proposed by Ghosh and Ghanem [15] in the form of basis enrichment in the polynomial chaos expansion. Their approach is, however, based on incorporating a priori knowledge about the discontinuity location by selecting appropriate enrichment functions. A solution for the staircase approximation of the statistics in case of random spatial discontinuities has also been proposed by Barth [6] using image enhancement postprocessing techniques in the combined discretization of the physical and probability spaces.

The improved performance of the ENO stencil selection and subcell resolution in the SSC method is demonstrated in application to a step function, the linear advection equation, a shock tube problem, and the transonic flow over the RAE 2822 airfoil. A shock tube problem is considered to study the solution of a system of hyperbolic conservation laws with uncertain discontinuity locations in space. Uncertainty analysis of hyperbolic systems has received relatively little attention [16]. The example involves Sod's Riemann problem [26] for the Euler equations of inviscid compressible gas dynamics. This Riemann problem was used by Poëtte et al. [22] to illustrate their PC method based on an entropic variable in an example with steep fronts and shocks. The problem was also studied by Tryoen et al. [27] using upwinding in a multi-resolution approach based on local SG projections of the PC expansion. Abgrall [1] used shock tube like test cases to demonstrate the application of his method to the Euler equations.

In transonic airfoil flows, a stair-like solution profile and an underprediction of the standard deviation have been observed by Simon et al. [23] in the region of the shock movement. They conclude that these problems can be alleviated by selecting a sufficiently high number of sampling points with respect to the spatial resolution at the expense of significant computational costs in higher dimensional problems. A non-intrusive stochastic projection method has also been applied to transonic flow over an airfoil by Chassaing and Lucor [7] with randomness in the free-stream conditions. Other transonic airfoil flows have previously been considered in [29,30].

At this location, there is also pointed to [10] to guide the reader in comparing and contrasting its technical content with the current chapter. Both chapters consider approaches to propagate uncertainty in application to CFD models with adaptivity in the stochastic space to allow for UQ in the presence of strong shocks in the response in application to different transonic airfoils. Both the Adaptive Stochastic Finite Elements (ASFE) method included in [10] and SSC approximate the model response by a piecewise polynomial interpolation on a simplex grid in the stochastic space. However, ASFE uses deterministic sampling locations limited to quadratic Newton-Cotes quadrature rules. SSC is based on randomized sampling and higher-degree interpolation stencils for higher efficiency in multiple stochastic dimensions. Its efficiency is further enhanced using adaptive refinement measures based on local error estimates with one new sample per refinement instead of the weighted curvature of the response surface and multiple new samples at a simplex refinement. This higher flexibility of the SSC approach also allows the extensions to the ENO stencil selection and the subcell resolution presented in this chapter. SSC is implemented for arbitrary dimensionality and applied to the transonic airfoil case with two uncertainties, while the application of ASFE is restricted to randomness in the Mach number only. Barth [6] uses FVM concepts to approximate discontinuities in the probability space comparable to the ENO stencil selection and subcell resolution principles extended to probability space here. However, his approach treats both the physical and probability spaces with the same methodology. SSC only discretizes the probability space to reduce the dimensionality of the stochastic problem and to allow for more flexibility in combining it with different spatial discretizations.

This chapter is outlined as follows. The SSC method is introduced in Sect. 2. The ENO stencil selection and the subcell resolution extensions are presented in Sects. 3 and 4. In Sects. 5–8, the applications to the test function, the linear advection problem, the Riemann problem in the shock tube, and the transonic airfoil flow are considered. The main conclusions are drawn in Sect. 9.

## 2 Simplex Stochastic Collocation

Consider the following computationally intensive problem subject to $n_\xi$ second-order random parameters $\boldsymbol{\xi} = \{\xi_1, \ldots, \xi_{n_\xi}\} \subset \Xi$ with a known input probability density $f_\xi(\boldsymbol{\xi})$ in the parameter space $\Xi \subset \mathbb{R}^{n_\xi}$

$$\mathscr{L}(\mathbf{x}, t, \boldsymbol{\xi}; u(\mathbf{x}, t, \boldsymbol{\xi})) = \mathscr{S}(\mathbf{x}, t, \boldsymbol{\xi}). \tag{1}$$

where the operator $\mathscr{L}$ and the source term $\mathscr{S}$ are defined on the domain $X \times T \times \Xi$ with output quantity of interest $u(\mathbf{x}, t, \boldsymbol{\xi})$, space $\mathbf{x} \in X \subset \mathbb{R}^{n_x}$, $n_x = \{1, 2, 3\}$, and time $t \in T \subset \mathbb{R}$. The latter two arguments are dropped to simplify the notation. Second-order random parameters are random parameters with finite variance, which includes most practical cases [14]. The solution of (1) is a random event with the set of outcomes $\Omega$ of the probability space $(\Omega, \mathscr{F}, P)$ with $\mathscr{F} \subset 2^\Omega$ the $\sigma$-algebra of events and $P$ a probability measure.

The simplex stochastic collocation (SSC) method [32, 33] computes the statistics and the probability distribution of $u(\boldsymbol{\xi})$ in a non-intrusive way by discretizing the probability space using a simplex tessellation of $n_s$ samples points $\boldsymbol{\xi}_k$, with $k = 1, \ldots, n_s$. A series of $n_s$ deterministic problems (1) is then solved to compute the samples $\mathbf{v} = \{v_1, \ldots, v_{n_s}\}$, with $v_k = u(\boldsymbol{\xi}_k)$, for the parameter values $\boldsymbol{\xi}_k$ that correspond to the vertexes of the $n_e$ simplexes $\Xi_j$ in probability space with $j = 1, \ldots, n_e$. The response surface $u(\boldsymbol{\xi})$ is approximated by a piecewise polynomial interpolation $w(\boldsymbol{\xi})$ of the samples $\mathbf{v}$ using a polynomial chaos [14, 36] expansion $w_j(\boldsymbol{\xi})$ in each of the simplexes $\Xi_j$

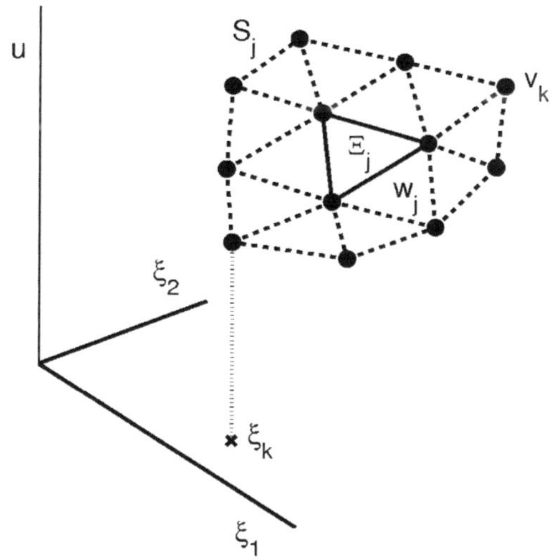


**Fig. 2** Approximation of the response surface $u(\boldsymbol{\xi})$ by the interpolation $w_j(\boldsymbol{\xi})$ of the samples $v_k$ at a stencil $S_j$ of sampling points $\boldsymbol{\xi}_k$ for the simplex $\Xi_j$ in a two-dimensional probability space with $\boldsymbol{\xi} = \{\xi_1, \xi_2\}$

$$w_j(\boldsymbol{\xi}) = \sum_{i=0}^{P_j} c_{j,i} \boldsymbol{\Psi}_{j,i}(\boldsymbol{\xi}), \tag{2}$$

for $\boldsymbol{\xi} \in \Xi_j$, where $\boldsymbol{\Psi}_{j,i}$ are the basis polynomials, $c_{j,i}$ are the coefficients, and $P_j + 1 = (n_\xi + p_j)!/(n_\xi! p_j!)$ is the number of expansion terms, with $p_j$ the local polynomial degree of $w_j(\boldsymbol{\xi})$. The coefficients $c_{j,i}$ are computed by interpolating a stencil $S_j$ out of the $n_s$ samples $\mathbf{v} = \{v_1, \ldots, v_{n_s}\}$. For a piecewise linear interpolation $w_j(\boldsymbol{\xi})$ with $p_j = 1$, the stencil $S_j = \{\boldsymbol{\xi}_{k_{j,0}}, \ldots, \boldsymbol{\xi}_{k_{j,N_j}}\}$ consists of the $N_j + 1 = n_\xi + 1$ vertexes of the simplex $\Xi_j$, with $k_{j,l} \in \{1, \ldots, n_s\}$ for $j = 1, \ldots, n_e$ and $l = 0, \ldots, N_j$. Higher degree stencils $S_j$ with $N_j \geq P_j$ are constructed by adding vertexes $\boldsymbol{\xi}_k$ of surrounding simplexes to the stencil according to a nearest neighbor search based on the Euclidean distance to the center of the simplex $\Xi_j$ in parameter space $\Xi$. The center of $\Xi_j$ is defined as the average of the vertex locations of $\Xi_j$. The relation $N_j = P_j$ is used here or least squares approximation can be used to construct the interpolation for $N_j > P_j$. The interpolation procedure (2) in the probability space is denoted by the interpolation operator $\mathscr{I}$, for which holds $w(\mathbf{x}, \boldsymbol{\xi}) = \mathscr{I}(\mathbf{v}(\mathbf{x}))$. The notation is visualized in Fig. 2 for an example of a response surface approximation in a two-dimensional probability space with $n_\xi = 2$.

The polynomial degree $p_j$ is chosen as high as possible with respect to the total number of available samples $n_s$ with $N_j + 1 \leq n_s$. The robustness of the interpolation $w_j(\boldsymbol{\xi})$ of the samples $\mathbf{v}_j = \{v_{k_{j,0}}, \ldots, v_{k_{j,n_\xi}}\}$ in the simplex $\Xi_j$ is guaranteed by the local extremum conserving (LEC) limiter that reduces the stencil size $N_j + 1$, and $p_j$, in case of overshoots until $w_j(\boldsymbol{\xi})$ satisfies

$$\min_{\boldsymbol{\xi} \in \Xi_j} w_j(\boldsymbol{\xi}) = \min \mathbf{v}_j \;\wedge\; \max_{\boldsymbol{\xi} \in \Xi_j} w_j(\boldsymbol{\xi}) = \max \mathbf{v}_j. \tag{3}$$

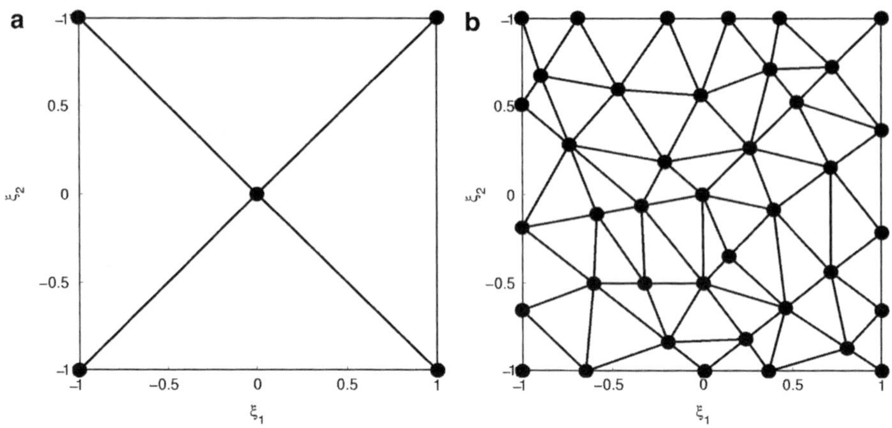


**Fig. 3** Simplex tessellation of the two-dimensional parameter space $\Xi$, where the sampling points $\boldsymbol{\xi}_k$ and the simplexes $\Xi_j$ are denoted by the *closed circles* and the *lines*, respectively. (**a**) Initial discretization. (**b**) Discretization with $n_s = 40$, measure $\bar{\Xi}$

The LEC limiter (3) is applied to all simplexes in a stencil $S_j$ and always holds for $p_j = 1$. The $i$th moment $\mu_{u_i}$ of $u(\boldsymbol{\xi})$ is then computed as a summation of integrals over the $n_e$ simplexes $\Xi_j$ using a Monte Carlo evaluation with $n_{mc} \gg n_s$ integration points $\boldsymbol{\xi}_{mc_k}$

$$\mu_{u_i}(\mathbf{x}) \approx \sum_{j=1}^{n_e} \int_{\Xi_j} w_j(\mathbf{x}, \boldsymbol{\xi})^i f_{\xi}(\boldsymbol{\xi}) d\boldsymbol{\xi} \approx \sum_{k=1}^{n_{mc}} w(\mathbf{x}, \boldsymbol{\xi}_{mc_k})^i, \tag{4}$$

with $w(\mathbf{x}, \boldsymbol{\xi}_{mc_k}) = w_j(\mathbf{x}, \boldsymbol{\xi}_{mc_k})$ for $\boldsymbol{\xi}_{mc_k} \in \Xi_j$. The motivation for using Monte Carlo instead of quadrature for integrating the polynomials over the simplexes is that the integrals in (4) are weighted by the probability density $f_{\xi}(\boldsymbol{\xi})$ of the random input parameters $\boldsymbol{\xi}$. This weighting can be different for each simplex in case of non-uniform probability distributions. Different quadrature weights then need to be computed first for all simplexes using, for example, again a Monte Carlo simulation for each weight.

The initial discretization of the probability space consists of a simplex tessellation of sampling points $\boldsymbol{\xi}_k$ at the corners of a hypercube probability space and one sample in the interior, see Fig. 3a for a two-dimensional example on the domain $[-1, 1]^{n_\xi}$. The sampling strategy is then based on splitting the longest edge of the simplex $\Xi_j$ with the highest value of a refinement measure in each refinement step in two by adding a sampling point $\boldsymbol{\xi}_k$. The random sampling point is located at least one third of the edge length away from the endpoints to ensure a sufficient spread of the samples. In a one-dimensional probability space, the new sampling point $\boldsymbol{\xi}_k$ is used to split the cell $\Xi_j$ into two cells of equal size. The tesselation is updated by computing the sample $v_{k_{new}} = u(\boldsymbol{\xi}_{k_{new}})$ and by making a Delaunay triangulation of the new set of sampling points $\boldsymbol{\xi}_k$ or by splitting the simplexes $\Xi_j$ that contain the refined edge in two. The following refinement measure $e_j$ is used based on the geometrical properties of the simplex $\Xi_j$ and the local polynomial degree $p_j$

$$e_j = \bar{\Omega}_j \bar{\Xi}_j^{2O_j}, \tag{5}$$

where the probability $\bar{\Omega}_j$, the normalized volume $\bar{\Xi}_j$, and the estimated order of convergence $O_j$ of $\Xi_j$ are defined as

$$\bar{\Omega}_j = \int_{\Xi_j} f_{\xi}(\boldsymbol{\xi}) \mathrm{d}\boldsymbol{\xi}, \qquad \bar{\Xi}_j = \frac{1}{\bar{\Xi}} \int_{\Xi_j} \mathrm{d}\boldsymbol{\xi}, \qquad O_j = \frac{p_j + 1}{n_{\xi}}, \tag{6}$$

with $\bar{\Xi} = \sum_{j=1}^{n_e} \bar{\Xi}_j$. Measure (5) accounts for both the interpolation accuracy and the probabilistic weighting in the moment integrals (4). It also leads to solution-based refinement through the reduction of $p_j$ at discontinuities by the LEC limiter (3). The size $\bar{\Xi}_j$ of the simplexes can be used as the refinement measure in order to obtain uniform or volumetric refinement, see Fig. 3b for an example with $n_s = 40$.

## 3 Essentially Non-oscillatory Stencil Selection

The essentially non-oscillatory (ENO) type stencil selection is introduced into the SSC–ENO method [34] below as alternative for the nearest neighbor construction of the stencils. Attention is also paid to the efficient implementation of the algorithm.

### 3.1 *Interpolation Stencil Selection*

The nearest neighbor construction of the interpolation stencils $S_j$ combined with the LEC limiter (3) results in one stencil $S_j$ for each simplex $\Xi_j$. If the stencils $S_j$ are not restricted to the nearest neighbor sampling points $\boldsymbol{\xi}_k$, then multiple stencils $S_{j,i}$ may be possible for simplex $\Xi_j$ that satisfy the LEC limiter. The stencil $S_{j,i}$ that leads to the smoothest interpolation $w_{j,i}(\boldsymbol{\xi})$ is then selected for a more accurate approximation of $u(\boldsymbol{\xi})$.

The first $n_{\xi} + 1$ sampling points $\boldsymbol{\xi}_k$ of each stencil $S_{j,i} = \{\boldsymbol{\xi}_{k_{j,0}}, \dots, \boldsymbol{\xi}_{k_{j,n_{\xi}}}\}$ consist of the vertexes of the simplex $\Xi_j$. This stencil corresponds to the piecewise linear interpolation. The higher degree stencils of $N_{j,i} + 1$ sampling points

$$S_{j,i} = \{\boldsymbol{\xi}_{k_{j,0}}, \dots, \boldsymbol{\xi}_{k_{j,n_{\xi}}}, \dots, \boldsymbol{\xi}_{k_{j,N_{j,i}}}\}, \tag{7}$$

can be constructed by adding, in principle, any combination of $N_{j,i} - n_{\xi}$ samples for any $p_{j,i}$ out of the remaining sampling points $\boldsymbol{\xi}_k$, with $k \in \{1, \dots, n_s\} \backslash \{k_{j,0}, \dots, k_{j,n_{\xi}}\}$ and each sampling point appearing only once in the stencil $S_{j,i}$. Out of these stencils, only a set of $r_j$ candidate stencils $\{S_{j,i}\}_{i=1}^{r_j}$ is

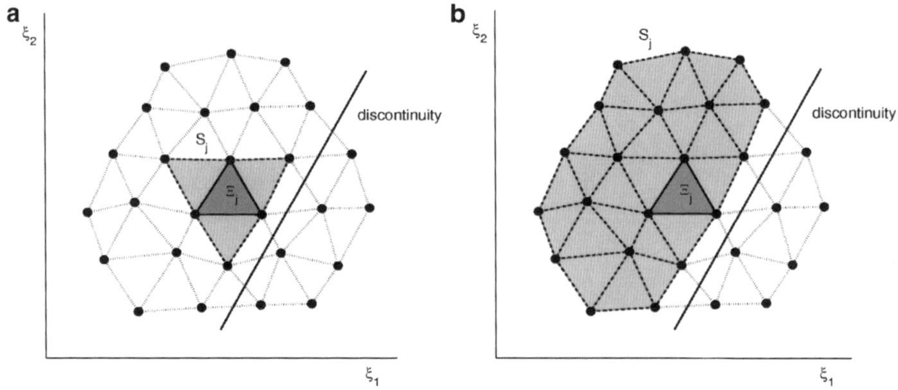


**Fig. 4** Selection of the interpolation stencil $S_j$ for the simplex $\Xi_j$ near a discontinuity in a two-dimensional probability space. (**a**) Nearest neighbor stencil. (**b**) Selected stencil

accepted of which the interpolation $w_{j,i}(\boldsymbol{\xi})$ satisfies the LEC limiter. The stencil $S_j$ for $\Xi_j$ is selected from this set $\{S_{j,i}\}$ based on an indicator of smoothness $\mathrm{IS}_{j,i}$ for each of the candidates. Since the stencils $S_{j,i}$ have a non-uniform polynomial degree $p_{j,i}$, the degree of the stencils accepted by the LEC limiter is here used as the indicator of smoothness $\mathrm{IS}_{j,i} = p_{j,i}$. The stencil with the highest polynomial degree is then assigned to $\Xi_j$ in order to obtain the highest order approximation

$$S_j = S_{j,i}, \qquad \text{with } i = \underset{i^* \in \{1,\ldots,r_j\}}{\arg\max}\ p_{j,i^*}. \tag{8}$$

If multiple stencils have the same smoothness $p_{j,i}$, then out of these stencils the one with the minimum average Euclidean distance of the sampling points $\boldsymbol{\xi}_k$ to the center of $\Xi_j$ is chosen.

A two-dimensional example is given in Fig. 4 of the stencil selection for the simplex $\Xi_j$ close to a discontinuity, of which the location is denoted by the diagonal line. The nearest neighbor stencil $S_j$ for $\Xi_j$ only leads to a quadratic interpolation $w_j(\boldsymbol{\xi})$ with $N_j + 1 = 6$, since higher degree stencils cross the discontinuity and are rejected by the LEC limiter. On the other hand, stencil selection can result in a stencil $S_j$, with a higher polynomial degree $p_j$, that contains all sampling points $\boldsymbol{\xi}_k$ in the smooth region at one side of the discontinuity.

## 3.2 *Efficient Implementation*

Constructing all possible stencils $S_{j,i}$ for all simplexes $\Xi_j$ can become impractical as its complexity increases binomially with the number of samples $n_s$. Therefore, we restrict the stencil selection to a subset of these stencils by employing the multi-element character of the approach. We allow only nearest neighbor stencils of other

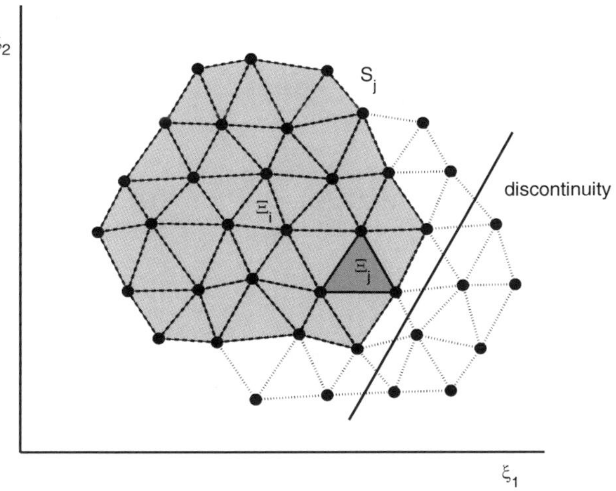


**Fig. 5** Efficiently selected stencil $S_j$ for the simplex $\Xi_j$ by adopting the nearest neighbor stencil of the simplex $\Xi_i$ in a two-dimensional probability space

simplexes that contain $\Xi_j$ to be assigned to the simplex $\Xi_j$, if that leads to a higher polynomial degree than its own stencil.

To that end, the nearest neighbor stencils $\tilde{S}_j$, with interpolation $\tilde{w}_j(\boldsymbol{\xi})$ and degree $\tilde{p}_j$, are first constructed for each simplex $\Xi_j$ as described in Sect. 2. This results in a set of $n_e$ stencils $\{\tilde{S}_j\}_{j=1}^{n_e}$ for all simplexes $\Xi_j$. Next, it needs to be determined for each stencil $\tilde{S}_j$ which simplexes $\Xi_i$ are part of the stencil. A stencil $\tilde{S}_j$ is considered to contain another simplex $\Xi_i$, if $\tilde{S}_j$ contains all vertexes $\{\boldsymbol{\xi}_{k_{i,0}}, \dots, \boldsymbol{\xi}_{k_{i,n_\xi}}\}$ of $\Xi_i$, which is always true for $i = j$. A set of $\tilde{r}_j$ candidate stencils $\{\tilde{S}_{j,i}\}_{i=1}^{\tilde{r}_j}$ for simplex $\Xi_j$ is then collected from the nearest neighbor stencils that contain $\Xi_j$. The stencil $S_j = \tilde{S}_{j,i}$, and the interpolation $w_j(\boldsymbol{\xi}) = \tilde{w}_{j,i}(\boldsymbol{\xi})$, with the highest degree $\tilde{p}_{j,i}$ is selected from $\{\tilde{S}_{j,i}\}_{i=1}^{\tilde{r}_j}$ as in (8). If none of the stencils $\{\tilde{S}_{j,i}\}$ has a higher degree than the nearest neighbor stencil $\tilde{S}_j$, i.e. $\tilde{p}_{j,i} \leq \tilde{p}_j$ for all $i = 1, \dots, \tilde{r}_j$, then the original stencil $\tilde{S}_j$ is automatically maintained, since the sampling points of the nearest neighbor stencil have, by definition, the smallest average Euclidean distance to the center of $\Xi_j$.

This efficient SSC–ENO stencil selection algorithm results in virtually no additional computational costs compared to SSC with nearest neighbor stencils, since no additional stencils or interpolations are constructed. Existing nearest neighbor stencils are assigned only to other simplexes, if that increases the local polynomial degree. The algorithm can, therefore, only improve the polynomial degree, $p_j \geq \tilde{p}_j$, because $\{\tilde{S}_{j,i}\}$ always contains the stencil $\tilde{S}_j$. The impact of the stencil selection on the increase of the polynomial degree $p_j$ in the smooth regions of the solution, decreases the refinement measure $e_j$ in the simplexes in which the solution is smooth, since $\bar{\Xi}_j < 1$. This results in more focused refinement of the simplexes that contain nonlinearities.

Figure 5 shows the adoption of the nearest neighbor stencil of another simplex $\Xi_i$ by the simplex $\Xi_j$ in the two-dimensional example. Because the resulting stencil

$S_j$ is asymmetrical with respect to $\Xi_j$, it leads to a higher polynomial degree $p_j$ than its nearest neighbor stencil of Fig. 4a. The efficiently selected stencil does not necessarily contain all the sampling points on one side of the discontinuity.

# 4 Subcell Resolution

The subcell resolution in the SSC–SR method [35] is presented for spatial discontinuities and discontinuous derivatives with a random location in combination with a sampling strategy for the resulting SSC–SR algorithm.

## *4.1 Discontinuous Representation*

The subcell resolution is first introduced for a one-dimensional physical space $x \in X$ with $n_x = 1$ and later described for multiple spatial dimensions. Assume that $u(x, \boldsymbol{\xi})$ contains a discontinuity, of which the location $x_{\text{disc}}(\boldsymbol{\xi})$ in the physical space $X$ is a function of the stochastic dimensions $\boldsymbol{\xi}$. The samples $\mathbf{v}(x)$ for the flow quantity $u(x, \boldsymbol{\xi})$ are then used to extract $n_s$ realizations $\mathbf{v}_{\text{disc}} = \{v_{\text{disc}_1}, \dots, v_{\text{disc}_{n_s}}\}$ for the physical discontinuity location $x_{\text{disc}}(\boldsymbol{\xi})$ at the sampling points $\boldsymbol{\xi}_k$. This is referred to as the extraction operation $\mathscr{E}$ that returns $\mathbf{v}_{\text{disc}} = \mathscr{E}(\mathbf{v}(x))$ with $v_{\text{disc}_k} = x_{\text{disc}}(\boldsymbol{\xi}_k)$. In Fig. 6a, an example of the extraction of $v_{\text{disc}_k}$ from the sample $v_k(x)$ for the sampling point $\xi_k$ is given for one stochastic dimension $\xi$. The set of realizations $v_{\text{disc}_k}$ for $n_s = 5$ sampling points $\xi_k$ is shown in Fig. 6b by the dots in the plot of the discontinuity location $x_{\text{disc}}(\xi)$ in the physical space as function of the random parameter $\xi$.

The specific method for the extraction $\mathscr{E}$ of the physical discontinuity locations $v_{\text{disc}_k}$ from the deterministic flow fields for the local flow quantity $v_k(x)$ can depend on the type of representation of the discontinuity that is used by the spatial discretization method. The discontinuity locations are explicitly resolved, for example, in subcell resolution FVM discretizations in the physical space and front tracking methods or level set approaches. Shock sensors in hybrid shock capturing methods or adaptive mesh refinement strategies can also be used to identify the discontinuity locations. Otherwise, approaches can be used based on local maxima in the gradient magnitude of the solutions $v_k(x)$, such as the shock detector proposed by Harten [12]. These different extraction operators $\mathscr{E}$ are illustrated in the numerical examples in Sects. 7 and 8 for FVM and front tracking discretizations in physical space.

The realizations $\mathbf{v}_{\text{disc}}$ for the physical discontinuity location are interpolated over the probability space to the function $w_{\text{disc}}(\boldsymbol{\xi})$ to obtain an approximation of the discontinuity location $x_{\text{disc}}(\boldsymbol{\xi})$ in the physical space as function of the stochastic coordinates $\boldsymbol{\xi}$, see Fig. 6b. The piecewise higher-degree polynomial interpolation $w_{\text{disc}}(\boldsymbol{\xi}) = \mathscr{I}(\mathbf{v}_{\text{disc}})$ is obtained using the interpolation operator $\mathscr{I}$ (2)

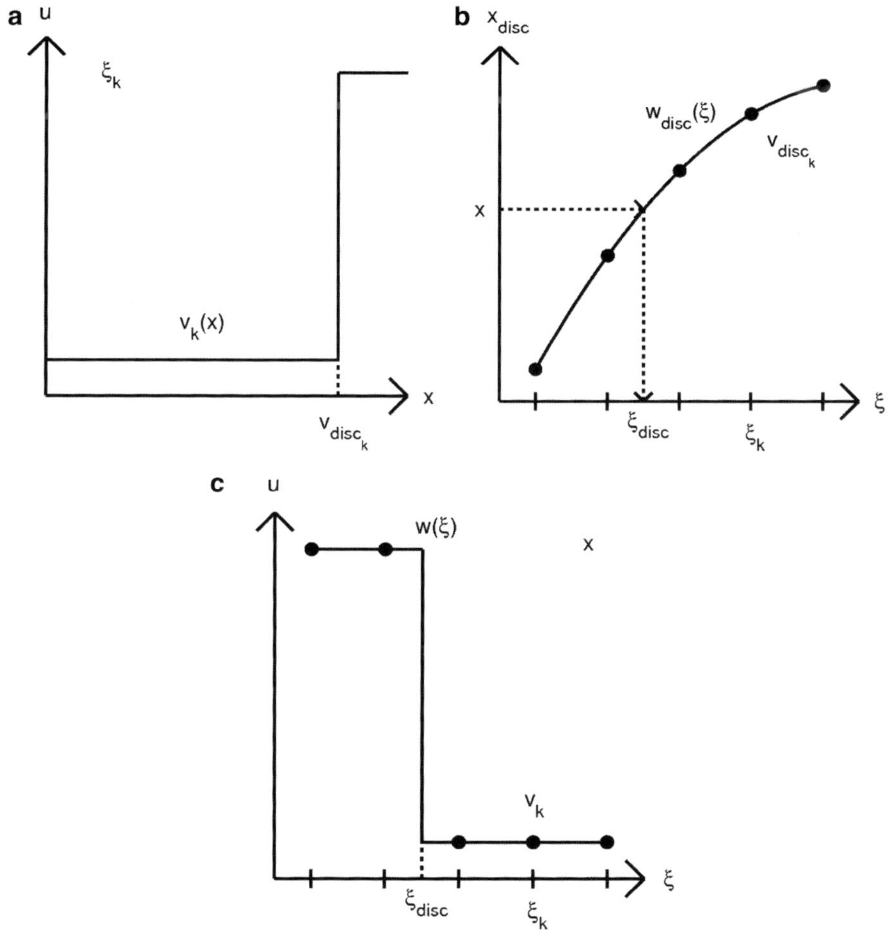


**Fig. 6** Example of the subcell resolution approach for a discontinuity in a one-dimensional probability space. (**a**) Extraction $\mathscr{E}$ of $v_{\text{disc}_k}$ from the sample $v_k(x) = u(x, \xi_k)$ for the sampling point $\xi_k$. (**b**) Intersection of $x_{\text{disc}}(\xi) = x$ with the interpolation $w_{\text{disc}}(\xi)$ of $\mathbf{v}_{\text{disc}}$ at $\xi_{\text{disc}}$. (**c**) Reconstruction of $w(\xi)$ with a discontinuity at $\xi_{\text{disc}}$ for the physical location $x$

$$w_{\text{disc}_j}(\boldsymbol{\xi}) = \sum_{i=0}^{P_{\text{disc}_j}} c_{\text{disc}_{j,i}} \Psi_{j,i}(\boldsymbol{\xi}), \tag{9}$$

with $w_{\text{disc}}(\boldsymbol{\xi}) = w_{\text{disc}_j}(\boldsymbol{\xi})$ for $\boldsymbol{\xi} \in \Xi_j$ and the $P_{\text{disc}_j} + 1$ coefficients $c_{\text{disc}_{j,i}}$ determined by the interpolation of a stencil $S_{\text{disc}_j}$ of the realizations $v_{\text{disc}_k}$. The location of the discontinuity in the probability space for a certain point $x$ in the physical space can then be described by the hypersurface $\Xi_{\text{disc}}(x) \subset \Xi$. This discontinuous hypersurface $\Xi_{\text{disc}}(x)$ in the probability space is given by the intersection of $w_{\text{disc}}(\boldsymbol{\xi})$ with the hyperplane $x_{\text{disc}}(\boldsymbol{\xi}) = x$, such that for all points $\boldsymbol{\xi}_{\text{disc}} \in \Xi_{\text{disc}}(x)$ holds $w_{\text{disc}}(\boldsymbol{\xi}_{\text{disc}}) = x$. Therefore, $\Xi_{\text{disc}}$ contains all combinations of random parameter values $\boldsymbol{\xi}_{\text{disc}}$ for which the discontinuity in the physical space is predicted to be located at $x$. In the example of Fig. 6b with a one-dimensional probability space, the set $\Xi_{\text{disc}}(x)$ consists of a single point $\xi_{\text{disc}}$ for which $w_{\text{disc}}(\xi_{\text{disc}}) = x$. For multiple random parameters $\boldsymbol{\xi}$, the intersection $\Xi_{\text{disc}}(x)$ of $w_{\text{disc}}(\boldsymbol{\xi})$ with $x_{\text{disc}}(\boldsymbol{\xi}) = x$ is a piecewise higher-degree function that is able to capture nonlinear curvatures of discontinuous hypersurfaces in the probability space.

The parameter space $\Xi$ is next divided into two subdomains $\Xi^-(x)$ and $\Xi^+(x)$ separated by $\Xi_{\text{disc}}(x)$, for which holds $w_{\text{disc}}(\boldsymbol{\xi}^-) < x$ with $\boldsymbol{\xi}^- \in \Xi^-(x)$, $w_{\text{disc}}(\boldsymbol{\xi}^+) > x$ with $\boldsymbol{\xi}^+ \in \Xi^+(x)$, and $\Xi = \Xi^-(x) \cup \Xi^+(x)$. The discontinuous simplexes $\Xi_j$ that contain $\Xi_{\text{disc}}(x)$ are identified by

$$w_{\text{disc}_j}(\boldsymbol{\xi}^-) < x \text{ for some } \boldsymbol{\xi}^- \in \Xi_j \;\wedge\; w_{\text{disc}_j}(\boldsymbol{\xi}^+) > x \text{ for some } \boldsymbol{\xi}^+ \in \Xi_j. \quad (10)$$

The interpolation $w_j(x, \boldsymbol{\xi})$ (2), in the simplexes $\Xi_j$ that satisfy (10), is replaced by a discontinuous representation of the response surface $u(x, \boldsymbol{\xi})$. To that end, the simplex $\Xi_j$ is divided into two regions $\Xi_j^-(x) \subset \Xi^-(x)$ and $\Xi_j^+(x) \subset \Xi^+(x)$ with $\Xi_j = \Xi_j^-(x) \cup \Xi_j^+(x)$. The interpolation $w_j(x, \boldsymbol{\xi})$ in $\Xi_j^-(x)$ is replaced by the approximation $w_j^-(x, \boldsymbol{\xi}) = w_{i^-}(x, \boldsymbol{\xi})$ of the simplex $\Xi_{i^-}$ closest to $\Xi_j$ for which holds $\Xi_{i^-} \subset \Xi^-$. The nearest cell $\Xi_{i^-}$ is defined as the simplex that has the most vertexes $\boldsymbol{\xi}_k$ in common with $\Xi_j$ and that has the highest polynomial degree $p_{i^-}$ out of these cells. The region $\Xi_j^+(x)$ is assigned the different interpolation $w_j^+(x, \boldsymbol{\xi}) = w_{i^+}(x, \boldsymbol{\xi})$ of the nearest simplex $\Xi_{i^+} \subset \Xi^+$ with $i^-, i^+ \in \{1, \ldots, n_{\text{e}}\}/j$. The notation is illustrated in Fig. 7 for the case of a two-dimensional probability space. This leads for SSC–SR to the response surface approximation $w(x, \boldsymbol{\xi})$ given by

$$w(x, \boldsymbol{\xi}) = \begin{cases} w_j(x, \boldsymbol{\xi}), \; \boldsymbol{\xi} \in \Xi_j, & \Xi_j \subset \Xi^- \vee \Xi_j \subset \Xi^+, \\ w_j^-(x, \boldsymbol{\xi}), \; \boldsymbol{\xi} \in \Xi_j^-, & \Xi_j \not\subset \Xi^- \wedge \Xi_j \not\subset \Xi^+, \\ w_j^+(x, \boldsymbol{\xi}), \; \boldsymbol{\xi} \in \Xi_j^+, & \Xi_j \not\subset \Xi^- \wedge \Xi_j \not\subset \Xi^+, \end{cases} \quad (11)$$

which is discontinuous at the predicted discontinuity location $\Xi_{\text{disc}}(x)$, see Fig. 6c for the result in the one-dimensional probability space. Integrating $w(x, \boldsymbol{\xi})$ over the parameter space $\Xi$ yields an approximation of the statistical moments of $u$ at the spatial point $x$. In order to obtain the spatial fields for the mean $\mu_{\text{w}}(x)$ and the standard deviation $\sigma_{\text{w}}(x)$, the approximations $\Xi_{\text{disc}}(x)$ and $w(x, \boldsymbol{\xi})$ are constructed for each point $x$ in the spatial discretization of $X$.

In multiple spatial dimensions, $n_{\text{x}} \in \{2, 3\}$, the location of the discontinuity is described by the discontinuous surface $X_{\text{disc}}(\boldsymbol{\xi}) \subset X$ in the physical space. Therefore, instead of using the discontinuity location $x_{\text{disc}}(\boldsymbol{\xi})$, the signed Euclidean distance $d_{\text{disc}}(\mathbf{x}, \boldsymbol{\xi})$ from the discontinuity $X_{\text{disc}}(\boldsymbol{\xi})$ to a point $\mathbf{x}$ in the physical space is parameterized for $n_{\text{x}} > 1$. The sign is obtained using the cross product, at the point on the discontinuity $X_{\text{disc}}(\boldsymbol{\xi})$ closest to $\mathbf{x}$, between the tangent of

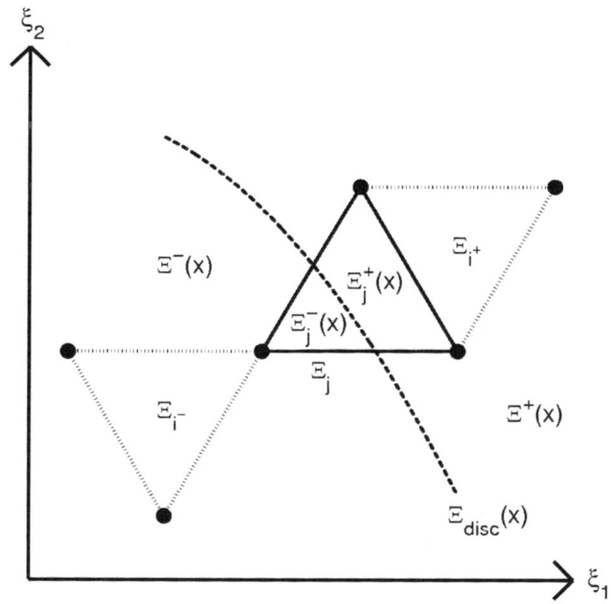


**Fig. 7** Division of the simplex $\Xi_j$ in the two-dimensional parameter space $\Xi$ by the discontinuous front $\Xi_{\text{disc}}(x)$ into $\Xi_j^-(x)$ and $\Xi_j^+(x)$ with the nearest simplexes $\Xi_{i^-}$ and $\Xi_{i^+}$ in $\Xi^-(x)$ and $\Xi^+(x)$, respectively

$X_{\text{disc}}(\boldsymbol{\xi})$ and the vector to $\mathbf{x}$, which is illustrated in Sect. 8 for the RAE 2822 airfoil. The subcell resolution method is then equal to the approach presented in this section for $n_x = 1$ by substituting $d_{\text{disc}}(\mathbf{x}, \boldsymbol{\xi})$ for $x_{\text{disc}}(\boldsymbol{\xi})$ and the location of the discontinuous hypersurface $\Xi_{\text{disc}}(\mathbf{x})$ is given by $d_{\text{disc}}(\mathbf{x}, \boldsymbol{\xi}) = 0$. Since $d_{\text{disc}}(\mathbf{x}, \boldsymbol{\xi})$ depends on the reference point $\mathbf{x}$ in the physical space, the realizations $v_{\text{disc}_k}(\mathbf{x})$ of $d_{\text{disc}}(\mathbf{x}, \boldsymbol{\xi})$ and the interpolation $w_{\text{disc}}(\mathbf{x}, \boldsymbol{\xi})$ become a function of $\mathbf{x}$, such that the interpolation step $w_{\text{disc}}(\mathbf{x}, \boldsymbol{\xi}) = \mathscr{I}(\mathbf{v}_{\text{disc}}(\mathbf{x}))$ is also repeated for each $\mathbf{x}$ in the spatial discretization. For a one-dimensional physical space, $r = 1$, the distance reduces to $d_{\text{disc}}(x, \boldsymbol{\xi}) = x_{\text{disc}}(\boldsymbol{\xi}) - x$ such that parameterizing the discontinuity location $x_{\text{disc}}(\boldsymbol{\xi}) = X_{\text{disc}}(\boldsymbol{\xi})$ is sufficient, which is independent of $x$. If multiple discontinuities are present in the spatial field of $u(x, \boldsymbol{\xi})$ then the subcell resolution algorithm is applied to each physical discontinuity. A non-monotonic function for $w_{\text{disc}}(x, \boldsymbol{\xi})$ also results in multiple discontinuities in $w(x, \boldsymbol{\xi})$ at certain values of $x$.

At physical points $x$ where the hypersurface $\Xi_{\text{disc}}(x)$ is located close to the boundary of the parameter space $\Xi$, no simplexes $\Xi_i$ may lie entirely in the region on one side of $\Xi_{\text{disc}}(x)$. For instance, $\Xi^-$ might not contain any simplexes $\Xi_{i^-}$ for updating the interpolation $w_j^-(x, \boldsymbol{\xi})$ to $w_{i^-}(x, \boldsymbol{\xi})$ in $\Xi_j^-(x)$ of the simplex $\Xi_j$ that contains $\Xi_{\text{disc}}(x)$. Such an example is given in Fig. 8 for the initial number of $n_s = 5$ samples in a two-dimensional probability space. In that case, a constant function is used for $w_j^-(x, \boldsymbol{\xi})$. The constant value for $w_j^-(x, \boldsymbol{\xi})$ is the arithmetic average of the

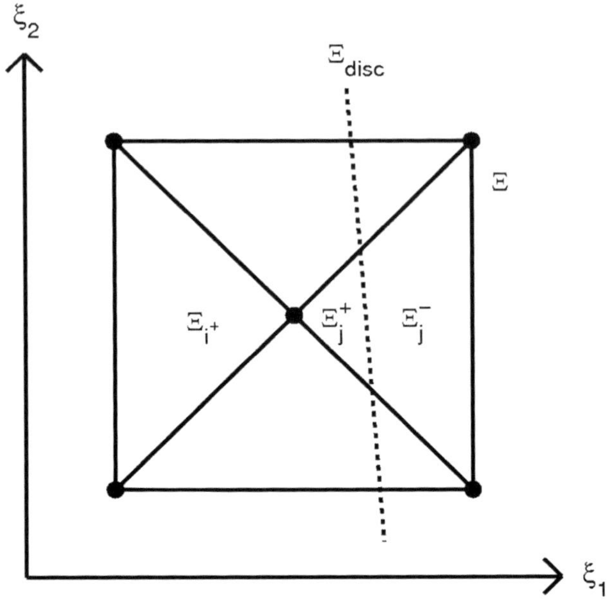


**Fig. 8** Example of the initial discretization of a two-dimensional parameter space $\Xi$ with $n_s = 5$, where no simplex $\Xi_i-$ is available to update the interpolation $w_j^-(x, \boldsymbol{\xi})$ in $\Xi_j^-(x)$

samples $v_k(x)$ at the vertexes of $\Xi_j$ in $\Xi_j^-$. In any other cases, the interpolation $w_j(x, \boldsymbol{\xi})$ (2) is simply retained.

## 4.2 Discontinuous Derivatives

Discontinuities in the first derivatives of a continuous response surface $u(x, \boldsymbol{\xi})$ can also be treated by the subcell resolution algorithm to avoid a local reduction in the polynomial degree $p_j$ of the approximation $w(x, \boldsymbol{\xi})$. The extraction step $\mathbf{v}_{\text{disc}} = \mathscr{E}(\mathbf{v}(x))$ determines, in that case, the realizations $\mathbf{v}_{\text{disc}}$ of the location of the discontinuous derivatives in the physical space. A method $\mathscr{E}$ for detecting kinks in the samples $\mathbf{v}(x)$ is illustrated for the example in Sect. 7. The interpolation $w_{\text{disc}}(\boldsymbol{\xi})$ of the realizations $\mathbf{v}_{\text{disc}}$ is then used to determine the hypersurface $\Xi_{\text{disc}}(x)$ describing the location of the discontinuous derivatives in the probability space in order to construct the different approximations $w_j^-(x, \boldsymbol{\xi})$ and $w_j^+(x, \boldsymbol{\xi})$ in the simplexes $\Xi_j$ that contain $\Xi_{\text{disc}}(x)$ as in (11). This leads to a response surface approximation $w(x, \boldsymbol{\xi})$ with discontinuous derivatives at $\Xi_{\text{disc}}(x)$, which is however not necessarily continuous at $\Xi_{\text{disc}}(x)$. The size of this, in general small, discontinuity at $\Xi_{\text{disc}}(x)$ decreases with an increasingly accurate approximation of the smooth responses on both sides of $\Xi_{\text{disc}}(x)$ as the number of samples $n_s$ increases. An example with a discontinuous derivative in a one-dimensional probability space is given in Fig. 9, which is equivalent to the discontinuous example in Fig. 6a, c. Figure 6b would be the same for both cases.

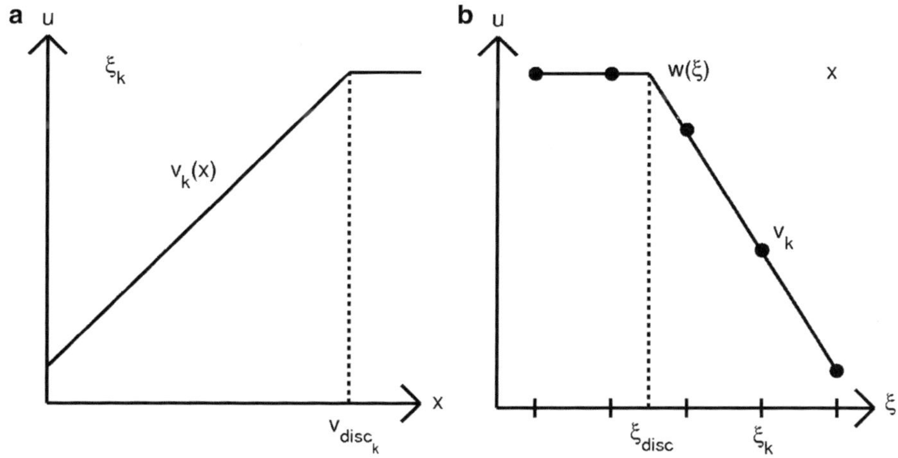


**Fig. 9** Example of the subcell resolution approach for a discontinuous derivative in a one-dimensional probability space. (**a**) Extraction $\mathscr{E}$ of $v_{\text{disc}_k}$ from the sample $v_k(x)$ for the sampling point $\xi_k$. (**b**) Reconstruction of $w(\xi)$ with a discontinuous derivative at $\xi_{\text{disc}}$ for the physical location $x$

In the case that $\Xi_{\text{disc}}(x)$ represents a discontinuous derivative and no $\Xi_i-$ is available for updating $w_j^-(x,\boldsymbol{\xi})$ in $\Xi_j^-(x)$, because $\Xi_{\text{disc}}(x)$ is located close to the boundary of $\Xi$, then a linear function is used for $w_j^-(x,\boldsymbol{\xi})$, instead of a constant value in case $\Xi_{\text{disc}}(x)$ is a discontinuity. The updated interpolation $w_j^+(x,\boldsymbol{\xi})$ in $\Xi_j^+(x)$ on the other side of $\Xi_{\text{disc}}(x)$ can be used in combination with the continuity of $u(x,\boldsymbol{\xi})$ at $\Xi_{\text{disc}}(x)$ to update $w_j^-(x,\boldsymbol{\xi})$ using a linear approximation. The first step is then to determine the values of $w_j^+(x,\boldsymbol{\xi})$ at the locations where $\Xi_{\text{disc}}(x)$ intersects the edges of $\Xi_j$. These values are used in combination with the samples $v_k(x)$ at the vertexes $\boldsymbol{\xi}_k$ of $\Xi_j$ in $\Xi_j^-$ to construct virtual values in the other vertexes of $\Xi_j$ by linear extrapolation over the edges. This is visualized in Fig. 10 by the open circles in the approximation of the response $u(x,\boldsymbol{\xi})$ in a two-dimensional probability space. The virtual values are used together with $v_k(x)$ at the vertexes $\boldsymbol{\xi}_k$ in $\Xi_j^-$ to construct a linear update for $w_j^-(x,\boldsymbol{\xi})$ using (2) with $p_j = 1$. If multiple virtual values are estimated for one vertex, because the vertex in $\Xi_j^+$ is the endpoint of multiple edges of $\Xi_j$ that cross $\Xi_{\text{disc}}(x)$, then their arithmetic average is used.

## 4.3 Sampling Strategy

The potentially constant approximation of $w_j^-(x,\boldsymbol{\xi})$ near the boundary of $\Xi$, or the linear function in case of a discontinuous derivative, is the lowest local polynomial degree of the otherwise higher-degree response surface approximation $w(x,\boldsymbol{\xi})$. Therefore, a sampling strategy is used that reduces the size of these regions in the

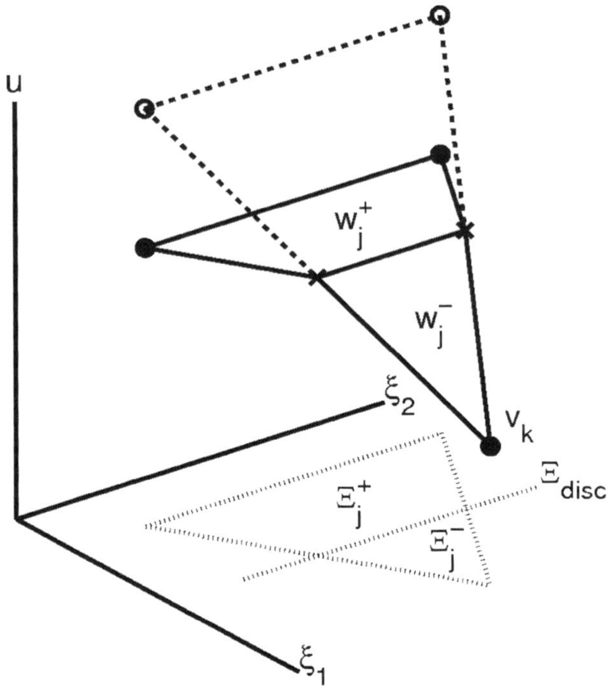


**Fig. 10** Subcell resolution for a discontinuous derivative $\Xi_{\mathrm{disc}}(x)$ located close to the boundary of a two-dimensional parameter space $\Xi$ with a linear update of $w_j^-(x,\boldsymbol{\xi})$ using $w_j^+(x,\boldsymbol{\xi})$ and the continuity at $\Xi_{\mathrm{disc}}(x)$

parameter space $\Xi$. It is based on the maximum $e_{\max_j}$ of the refinement measure $e_j(x)$ (5) for the simplexes $\Xi_j$ in the refinement procedure

$$e_{\max_j} = \max_{x \in X} e_j(x), \qquad e_j(x) = \bar{\Omega}_j \bar{\Xi}_j^{2O_j(x)}, \tag{12}$$

where $O_j(x)$ is a function of the spatial coordinate $x$ through $p_j(x)$. The polynomial degree $p_j(x)$ of the discontinuous simplexes $\Xi_j$ is defined as the probabilistically weighted average of the polynomial degrees over its two subdomains $\Xi_j^-$ and $\Xi_j^+$

$$p_j(x) = \frac{1}{\bar{\Omega}_j}\left[\int_{\Xi_j^-} p_j^-(x) f_{\xi}(\boldsymbol{\xi})\mathrm{d}\boldsymbol{\xi} + \int_{\Xi_j^+} p_j^+(x) f_{\xi}(\boldsymbol{\xi})\mathrm{d}\boldsymbol{\xi}\right], \tag{13}$$

with $p_j^-(x)$ and $p_j^+(x)$ the polynomial degrees of $w_j^-(x,\boldsymbol{\xi})$ and $w_j^+(x,\boldsymbol{\xi})$ in $\Xi_j^-$ and $\Xi_j^+$, respectively. The subcell resolution refinement focuses automatically near the boundary of $\Xi$ at simplexes with a low $p_j(x)$, instead of at the discontinuity

$\Xi_{\text{disc}}(x)$, as is demonstrated in the numerical examples. The response surface approximation $w_{\text{disc}}(x, \boldsymbol{\xi})$ for the discontinuity location $x_{\text{disc}}(\boldsymbol{\xi})$ can be included in the refinement measure (12) through $p_{\text{disc}_j}$ to resolve nonlinearities in $w_{\text{disc}}(x, \boldsymbol{\xi})$ as well.

Details of the implementation are given below to conclude the description of the methods before presenting the results. The simulations are initialized by generating the MC sampling points and the samples in the initial discretization of the probability space. A Delaunay triangulation is made and the first interpolation of the samples is constructed in the simplexes based on the nearest neighbor interpolation stencils. This is equivalent to assigning the value of the interpolants to the MC sampling points. The stencil selection is then performed in the SSC–ENO approach and the interpolation is potentially updated. An approximation of the moments is obtained from the ensemble statistics of the MC data. The refinement measures of the simplexes can be computed if the accuracy is unsatisfactory and a new sample can be computed, after which the triangulation and interpolation process is repeated.

The interpolation of the discontinuity locations over the probability space in case of SSC–SR is performed in the refinement cycle prior to the loop of the points in the spatial discretization. It is determined on which side of the discontinuity the sampling points and the MC points are located by comparing the interpolant of the discontinuity location at that sampling point with the considered spatial point. The discontinuous simplexes are identified by those elements that contain points at both sides of the discontinuity and the continuous interpolant is updated in those simplexes with the discontinuous one. The implementation of the methods is general for arbitrary dimensionality, multiple output quantities of interest, and different input distributions. The Delaunay triangulation can be replaced for increased computational efficiency in higher dimensions by splitting the simplexes that contain the new sampling point into two smaller simplexes.

## 5 Discontinuous Test Function

Results of the SSC–ENO method are compared to those of SSC without stencil selection to isolate the effect of the introduced stencil selection. The analysis focuses on the impact of the stencil selection on the local polynomial degree and the adaptive refinement for a step function in one stochastic dimension. The discontinuous test case is a combination of the Heaviside step function and a smooth background function. This example requires both a robust interpolation of the sharp discontinuity and a higher order approximation of the smooth parts of the solution. The step response is given by

$$u(\boldsymbol{\xi}) = u_{\text{step}}(\boldsymbol{\xi}) + u_{\text{back}}(\boldsymbol{\xi}), \tag{14}$$

with $u_{\text{step}}(\boldsymbol{\xi})$ the following Heaviside function

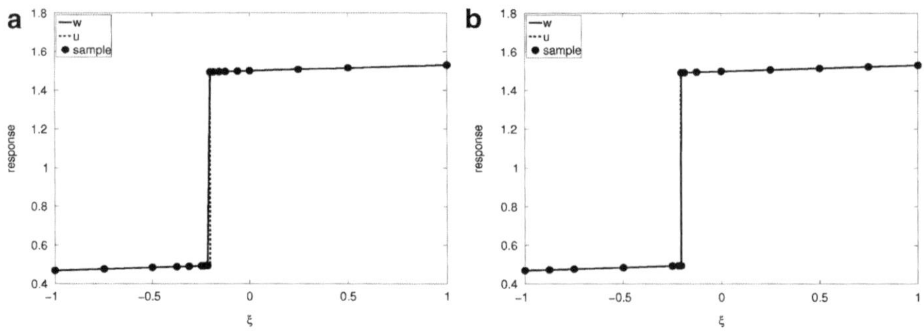


**Fig. 11** Response surface approximation with the adaptive refinement measure $e_j$ and $n_s = 17$ for the one-dimensional step function. (**a**) SSC. (**b**) SSC–ENO

$$u_{\text{step}}(\boldsymbol{\xi}) = H\left(\frac{\boldsymbol{\xi} \cdot \boldsymbol{\xi}^*}{\sqrt{\boldsymbol{\xi}^* \cdot \boldsymbol{\xi}^*}} + \xi_1^*\right) = \begin{cases} 0, & \frac{\boldsymbol{\xi} \cdot \boldsymbol{\xi}^*}{\sqrt{\boldsymbol{\xi}^* \cdot \boldsymbol{\xi}^*}} + \xi_1^* < 0, \\ 1, & \text{otherwise,} \end{cases} \tag{15}$$

and $u_{\text{back}}(\boldsymbol{\xi})$ the background function

$$u_{\text{back}}(\boldsymbol{\xi}) = \frac{1}{2} + \frac{1}{\pi} \arctan(c\xi_1). \tag{16}$$

This definition of the Heaviside function $u_{\text{step}}(\boldsymbol{\xi})$ (15) is used to give the discontinuity an arbitrary location and orientation in multi-dimensional cases that does not align with one of the stochastic coordinate axes, with $\boldsymbol{\xi}^* = \{\xi_1^*, \ldots, \xi_{n_\xi}^*\} \in [-0.5; 0.5]^{n_\xi}$ a vector of arbitrary numbers. The background function $u_{\text{back}}(\boldsymbol{\xi})$ is normalized to a range between 0 and 1, with parameter value $c = 0.1$. The random parameters $\boldsymbol{\xi}$ are uniformly distributed on the domain $U(-1, 1)$.

The approximation $w(\xi)$ of the step response $u(\xi)$ (14) with $n_\xi = 1$ random parameter $\xi$ and $n_s = 17$ samples is shown in Fig. 11. The SSC–ENO stencil selection in combination with the adaptive refinement measure $e_j$ impacts the sampling. SSC–ENO leads to a significantly narrower clustering of the sampling points around the discontinuity than SSC without stencil selection. This results in a notably sharper resolution of the discontinuity by SSC–ENO.

The corresponding polynomial degree in the elements is given in Fig. 12. The location of the step is denoted by the vertical dashed line. For SSC, the degree is reduced to $p_j = 1$ at the step and increases slowly with distance away from the discontinuity. SSC–ENO leads to a uniform high polynomial degree right up to the discontinuity in both smooth regions. The high polynomial degree in all elements that do not contain the step reduces the refinement measure $e_j$ in those elements relatively to $e_j$ in the discontinuous element. The refinement is, therefore, more focused in the latter element, which leads to a substantially finer resolution of the discontinuity.

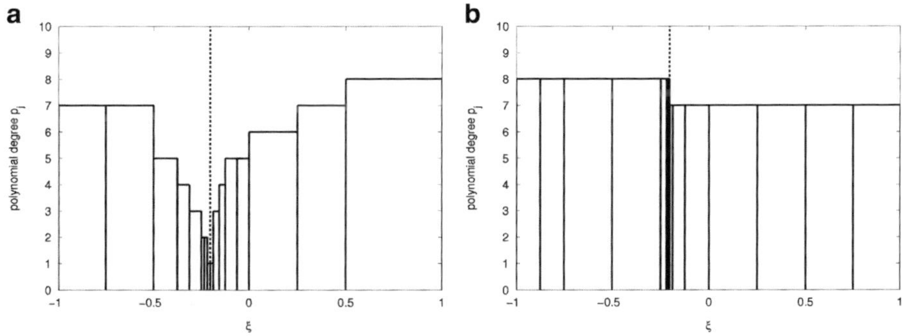


**Fig. 12** Local polynomial degree $p_j$ with the adaptive refinement measure $e_j$ and $n_s = 17$ for the one-dimensional step function, where the *vertical dashed line* denotes the discontinuity location. (**a**) SSC. (**b**) SSC–ENO

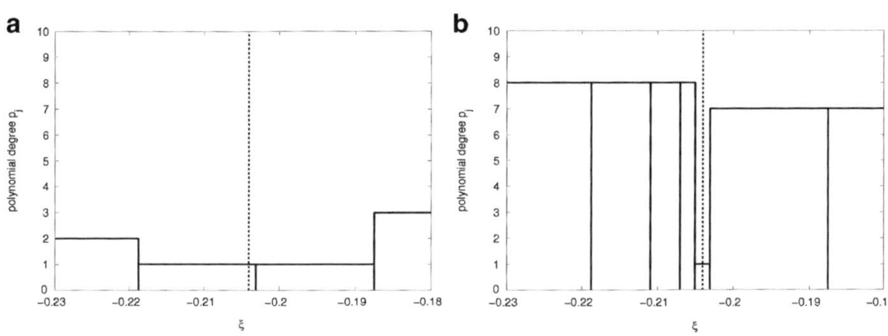


**Fig. 13** Zoom of the local polynomial degree $p_j$ near the discontinuity with the adaptive refinement measure $e_j$ and $n_s = 17$ for the one-dimensional step function, where the *vertical dashed line* denotes the discontinuity location. (**a**) SSC. (**b**) SSC–ENO

The zoom of the singularity in Fig. 13 shows that the size of the element that contains the discontinuity is a factor eight smaller for SSC–ENO than for SSC. In SSC–ENO, the polynomial degree is reduced from the maximum degree to $p_j = 1$ in the discontinuous element only, which ensures the robust approximation of the discontinuity without overshoots.

The quality of the response surface approximation is assessed by considering the root mean square (RMS) error $\varepsilon_{\text{rms}}$ between $w(\boldsymbol{\xi})$ and $u(\boldsymbol{\xi})$. The resulting convergence of $\varepsilon_{\text{rms}}$ is shown in Fig. 14 for SSC and SSC–ENO with refinement measures $\bar{\Xi}_j$, $e_j$, and $\varepsilon_j$. The adaptive refinement measures $e_j$ and $\varepsilon_j$ result for SSC in only a slightly lower error than uniform refinement with measure $\bar{\Xi}_j$. The effectiveness of the adaptive measures is greatly enhanced by SSC–ENO, for which the error is up to two orders of magnitude lower compared to $\bar{\Xi}_j$ at $n_s = 17$. This is a result of the concentration of the refinement in the discontinuous element.

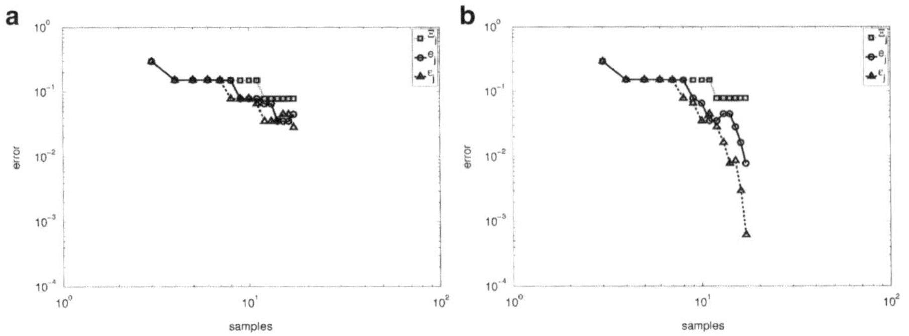

**Fig. 14** RMS error convergence for the one-dimensional step function. (**a**) SSC. (**b**) SSC–ENO

# 6 Linear Advection Equation

The results for the SSC–SR method to a linear advection problem with discontinuous initial conditions are compared to those of the SSC–ENO approach and Monte Carlo (MC) simulation. The linear advection equation in two spatial dimensions for the convected quantity $u(x, y, t)$ is

$$\frac{\partial u}{\partial t} + a\frac{\partial u}{\partial x} + b\frac{\partial u}{\partial y} = 0, \tag{17}$$

with the advection velocities $a$ and $b$ in the $x$ and $y$–directions, respectively. The initial conditions are given by

$$u(x, y, 0) = H\left((x - x_0) + (y - y_0)\right) = \begin{cases} 0, & x + y < x_0 + y_0, \\ 1, & x + y \geq x_0 + y_0, \end{cases} \tag{18}$$

where $H$ is the Heaviside step function, and $x_0$ and $y_0$ describe the initial discontinuity location. The analytical solution of (17) and (18) is $u(x, y, t) = H\left((x - x_0 - at) + (y - y_0 - bt)\right)$.

Randomness is considered in the advection velocity $a$, given by a uniform distribution on the interval $\mathcal{U}(-0.5; 0.5)$, with one spatial coordinate $x$ and a deterministic value for the initial discontinuity location of $x_0 = 0$. The initial condition and the deterministic solution at $t = 1$ are shown in Fig. 15 for $a = 0.5$. The randomness in $a$ leads to a random discontinuity location $x_{\mathrm{disc}}$ in the physical space with $x_{\mathrm{disc}} = a$ at $t = 1$.

The resulting profiles for the mean $\mu_{\mathrm{u}}(x)$ and standard deviation $\sigma_{\mathrm{u}}(x)$ of $u$ are given in Figs. 16 and 17 for SSC–ENO and SSC–SR. Uniform, or volumetric, sampling is used in this example for both approaches based on the refinement measure $\bar{\Xi}_j$. The sampling strategy of Sect. 4.3 is explored for the other test cases. In Fig. 16a, the SSC–ENO method clearly gives a staircase approximation of the

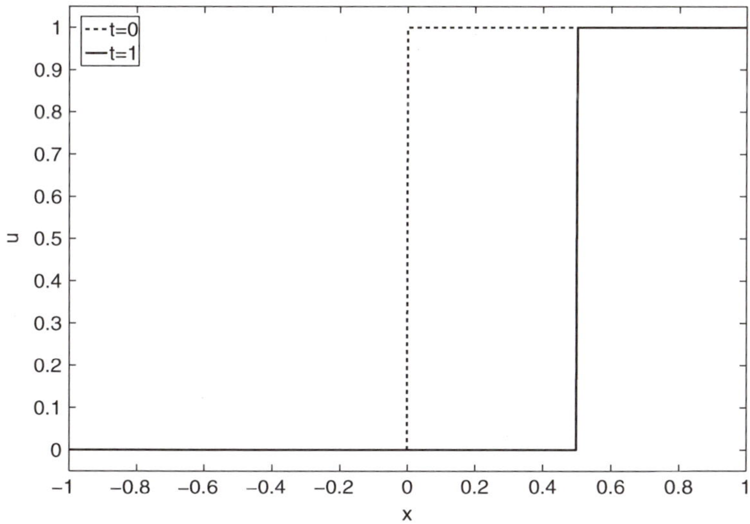


**Fig. 15** Initial conditions and deterministic solutions at $t = 1$ for the linear advection equation in one spatial dimension with advection velocity $a = 0.5$

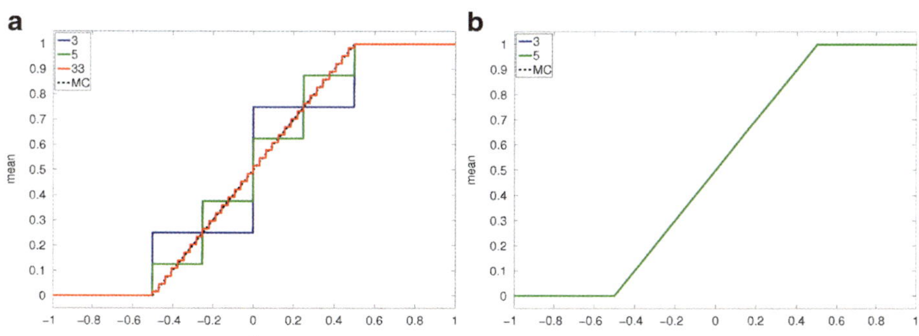


**Fig. 16** Mean $\mu_u(x)$ of $u$ for the linear advection equation in one spatial dimension with the random advection velocity $a$. (**a**) SSC–ENO with $n_s = \{3, 5, 33\}$. (**b**) SSC–SR with $n_s = \{3, 5\}$

mean $\mu_u(x)$ with the number of steps equal to the number of sampling points $n_s = \{3, 5, 33\}$. The results with the increasing number of smaller discontinuities approach the converged linear MC solution obtained with $n_{mc} = 5 \cdot 10^4$ samples. The equidistant steps occur when the spatial discontinuity crosses one of the uniformly spaced sampling points in the one-dimensional probability space. The staircase behavior is, therefore, typical for non-intrusive methods and not specific to SSC–ENO only. It would even appear in the MC solution at an inadequate number of samples of $n_{mc} < n_x$, with $n_x$ the spatial resolution of $n_x = 1 \cdot 10^3$ points used here. In contrast, SSC–SR achieves a continuous solution for $\mu_u(x)$ in Fig. 16b, which is already converged to the MC result for $n_s = 3$ and $n_s = 5$ samples.

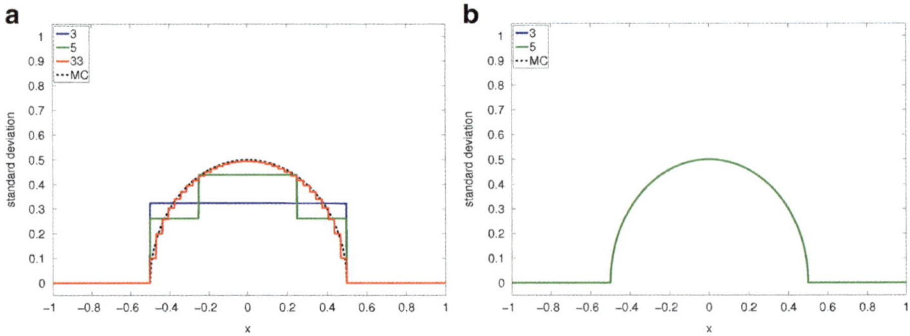


**Fig. 17** Standard deviation $\sigma_u(x)$ of $u$ for the linear advection equation in one spatial dimension with the random advection velocity $a$. (**a**) SSC–ENO with $n_s = \{3, 5, 33\}$. (**b**) SSC–SR with $n_s = \{3, 5\}$

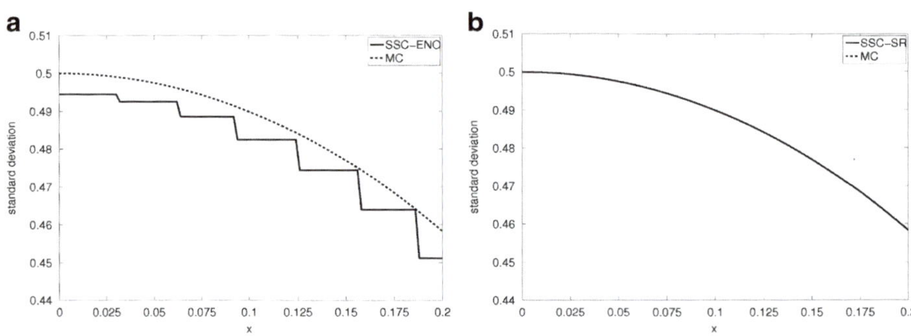


**Fig. 18** Zoom of the standard deviation $\sigma_u(x)$ of $u$ on $x \in [0; 0.2]$ for the linear advection equation in one spatial dimension with the random advection velocity $a$. (**a**) SSC–ENO with $n_s = 33$. (**b**) SSC–SR with $n_s = 3$

The staircase approximation of SSC–ENO and the converged continuous solution of SSC–SR can also be observed in Fig. 17 for $\sigma_u(x)$. The standard deviation of SSC–ENO converges from below to the MC maximum of $\sigma_{u_{\max}} = 0.500$ at $x = 0$. A zoom of $\sigma_u(x)$ on the interval $x \in [0; 0.2]$, in Fig. 18, reveals that for $n_s = 33$ samples SSC–ENO still underestimates the maximum standard deviation with $\sigma_{u_{\max}} = 0.495$. It also shows the truly continuous solution of SSC–SR for $\sigma_u(x)$ with $n_s = 3$.

The behavior of the SSC–ENO and SSC–SR solutions can be understood by inspecting the response surface of $u$ in the probability space as function of the random parameter $a$ for an arbitrary $x$-location, $x = 0.1$, in Fig. 19. The randomness in the step location $x_{\text{disc}}$ in the physical space appears in the MC response surface in the probability space for $x = 0.1$ as a discontinuity at $a_{\text{disc}} = 0.1$. SSC–ENO results in a linear interpolation of the samples at the jump, which converges only slowly with the increasing number of samples $n_s = 3$ and $n_s = 5$. The actual location of the discontinuity in between two samples is, therefore, not reflected in

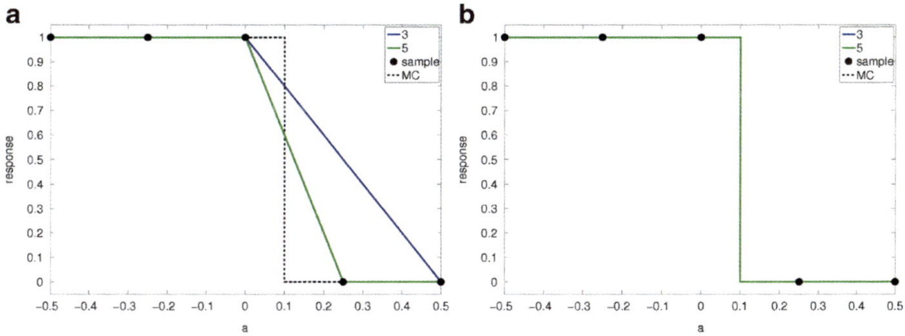


**Fig. 19** Response surface approximations for $u$ at $x = 0.1$ with $n_s = \{3, 5\}$ samples as function of the random advection velocity $a$ for the linear advection equation in one spatial dimension. (**a**) SSC–ENO. (**b**) SSC–SR

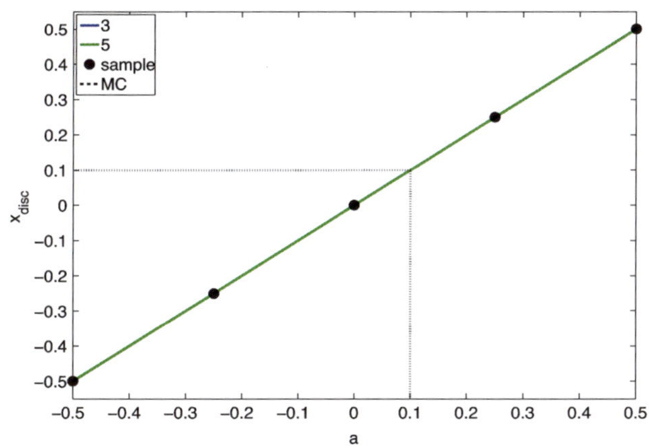


**Fig. 20** Response surface approximation $w_{disc}$ for the physical discontinuity location $x_{disc}$ as function of the random advection velocity $a$ by SSC–SR with $n_s = \{3, 5\}$ samples for the linear advection equation in one spatial dimension

the SSC–ENO approximation, which leads to the plateaus in the solutions for $\mu_u(x)$ and $\sigma_u(x)$.

On the contrary, SSC–SR gives a sharp representation of the jump in the discontinuous cell. It extrapolates the approximations in the adjacent cells for $n_s = 5$ up to the estimate of the discontinuity location $a_{disc}$. For $n_s = 3$, the approximation to the right of the discontinuity is a constant function through the rightmost sample at $a = 0.5$, because in that case the discontinuous cell has no neighboring cell to the right. This corresponds to the exact piecewise constant solution of the linear advection equation, which will not be the case for the nonlinear problems in the next sections.

The location of the discontinuity $a_{disc}$ in the probability space is estimated by SSC–SR as shown in Fig. 20 for $n_s = \{3, 5\}$. The closed circles denote the realizations $v_{\mathrm{disc}_k} = x_0 + a_k t$ of the discontinuity location $x_{\mathrm{disc}}$ in the physical space for the deterministic solutions at the sampling points $a_k$. The interpolation $w_{\mathrm{disc}}(a)$ of the linear relation between $x_{\mathrm{disc}}$ and $a$ is exact for $n_s = 3$. For the example of Fig. 19 at $x = 0.1$, the discontinuity location $a_{\mathrm{disc}} = 0.1$ in the probability space results from the intersection of $w_{\mathrm{disc}}(a)$ with the horizontal line $x_{\mathrm{disc}} = 0.1$.

## 7 Shock Tube Problem

The shock tube problem involves Sod's Riemann problem for the Euler equations of one-dimensional unsteady inviscid flow without heat conduction. The governing system of hyperbolic equations is given in conservation formulation by

$$\frac{\partial U}{\partial t} + \frac{\partial F(U)}{\partial x} = 0, \tag{19}$$

with the state vector $U(x,t)$ and flux vector $F(x,t)$

$$U = \begin{pmatrix} \rho \\ \rho u \\ \rho E \end{pmatrix}, \qquad F = \begin{pmatrix} \rho u \\ \rho u^2 + p \\ \rho u H \end{pmatrix}, \tag{20}$$

and initial conditions $U(x,0) = U_0(x)$. For a perfect gas, the density $\rho(x,t)$, velocity $u(x,t)$, static pressure $p(x,t)$, total energy $E(x,t)$, and enthalpy $H(x,t)$ are related as $E = (1/(\gamma-1))p/\rho + u^2/2$ and $H = E + p/\rho$, with ratio of specific heats $\gamma = c_p/c_v$ [8]. Sod's Riemann problem [26] is characterized by the initial conditions $U_0(x)$ consisting of two uniform states at the left and the right of $x_0 = 0$

$$\begin{cases} u_{\mathrm{left}} = 0, \\ p_{\mathrm{left}} = 1, \\ \rho_{\mathrm{left}} = 1, \end{cases} \qquad \begin{cases} u_{\mathrm{right}} = 0, \\ p_{\mathrm{right}} = 0.1, \\ \rho_{\mathrm{right}} = 0.125. \end{cases} \tag{21}$$

The pressure $p_{\mathrm{left}}$ of the initial left state and the location $x_0$ of the initial discontinuity are assumed to be uncertain. The uncertainty is given by two uniform distributions on the domains $p_{\mathrm{left}} \in [0.9; 1.1]$ and $x_0 \in [-0.025; 0.025]$. The output quantities of interest are the density $\rho$ at $x = 0.82$ and on the entire spatial domain. The problem is here confined to a closed shock tube on a finite spatial domain $x \in [-0.2; 2]$ with reflective walls at the boundaries, as considered deterministically in [31]. The Euler equations (19) are solved up to $t = 1$ using a second order front tracking method [31], which tracks the location of waves in the flow solution and solves local Riemann problems to simulate their interactions. It resolves shock waves and contact surfaces as true discontinuities unaffected by numerical diffusion, which results in sharp jumps in the physical and probability spaces. Rarefaction

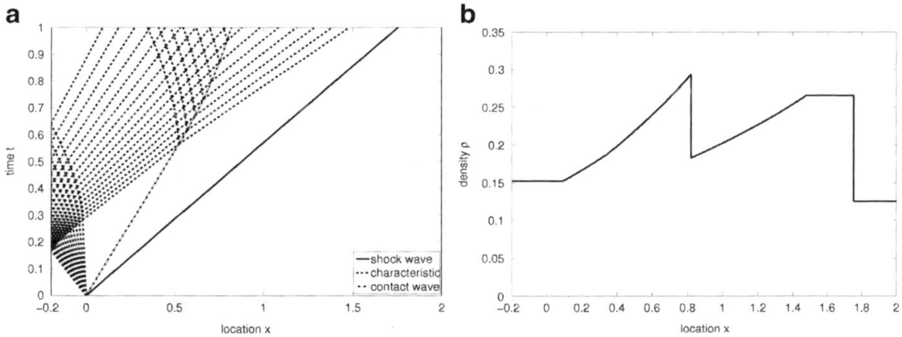


**Fig. 21** Sod's Riemann problem in a closed shock tube with deterministic initial conditions. (**a**) Space-time, $n_{\mathrm{f}} = 16$. (**b**) Density $\rho$ at $t = 1$, $n_{\mathrm{f}} = 64$

waves are discretized by a series of characteristics and second order convergence is obtained using a piecewise linear reconstruction of the rarefaction wave solution. Based on a convergence study, the rarefaction wave is here discretized using $n_{\mathrm{f}} = 64$ characteristic fronts.

The space–time solution of the deterministic problem is shown in Fig. 21a in terms of the wave paths for $n_{\mathrm{f}} = 16$. A left running rarefaction wave, a contact discontinuity, and a right running shock wave emanate from the discontinuity in the initial conditions at $x_0 = 0$. The rarefaction wave reflects from the left boundary and interacts with the contact discontinuity in the interior of the domain. The corresponding profile of the density $\rho$ at $t = 1$ is given in Fig. 21b for $n_{\mathrm{f}} = 64$. Next to the shock wave, the contact surface also results in a discontinuity and three points with a discontinuous derivative in the density field.

The uncertainty in $p_{\mathrm{left}}$ and $x_0$ leads to a jump in the response surface for the density $\rho$ at an $x$–location near the contact discontinuity, $x = 0.82$. The SSC discretization of the two-dimensional probability space is shown in Fig. 22a in terms of the tessellation of $n_{\mathrm{s}} = 100$ sampling points. The adaptive refinement algorithm clusters the sampling points near the discontinuity that runs diagonally through the probability space. The SSC–ENO method obtains a significantly higher density of the sampling points near the jump in Fig. 22b for the same number of samples. This results in a sharper resolution of the discontinuity and a larger ratio in size between the cells near the singularity and those that discretize the continuous regions. The improved effectiveness of the adaptive refinement is caused by the increase of the local polynomial degree $p_j$ in the smooth cells $\Xi_j$ by the stencil selection and the resulting concentration of the sampling in the cells that contain the discontinuity. SSC–ENO predicts a mean pressure of $\mu_\rho = 0.231$ with a standard deviation of $\sigma_\rho = 0.0543$. The coarser discretization of the discontinuity by SSC leads to an underprediction of the standard deviation with $\sigma_\rho = 0.0534$.

The SSC–ENO response surface approximation for $\rho$ as a function of $p_{\mathrm{left}}$ and $x_0$ with the simplex tessellation is shown in Fig. 23a. The response shows two continuous regions separated by a discontinuity that varies in strength and that

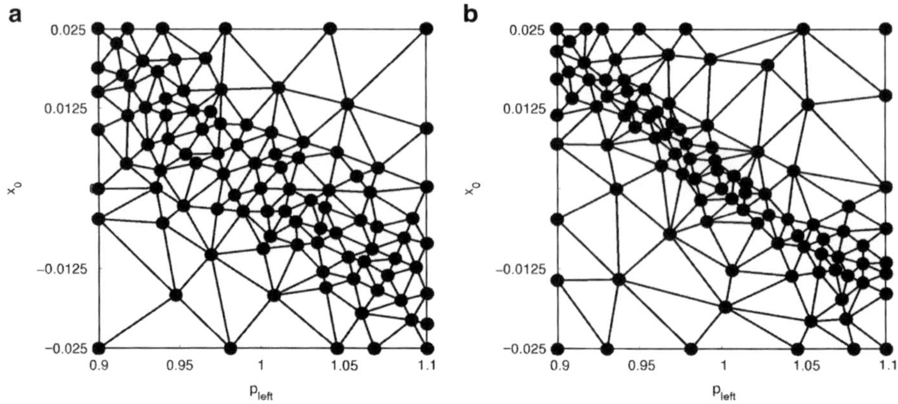


**Fig. 22** Discretization of the parameter space $\Xi$ for the density $\rho$ at $x = 0.82$ and $t = 1$ with $n_s = 100$ for Sod's Riemann problem in a closed shock tube with uncertain $p_{\text{left}}$ and $x_0$. (**a**) SSC. (**b**) SSC–ENO

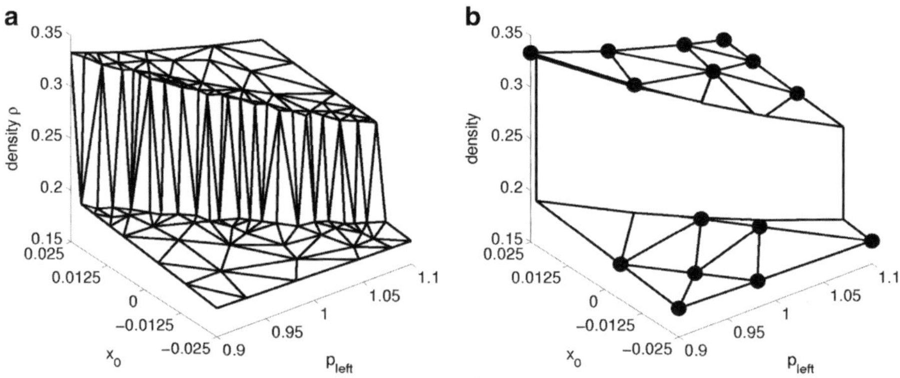


**Fig. 23** Response surface approximations for the density $\rho$ at $x = 0.82$ and $t = 1$ for Sod's Riemann problem in a closed shock tube with uncertain $p_{\text{left}}$ and $x_0$. (**a**) SSC–ENO with $n_s = 100$. (**b**) SSC–SR with $n_s = 15$.

is slightly curved. SSC–ENO gives a robust approximation of the discontinuity without overshoots because of the linear interpolation in the small simplexes that contain the singularity. Nevertheless, the subcell resolution of the SSC–SR method already achieves a more accurate response surface approximation using uniform sampling with only $n_s = 15$ samples in Fig. 23b. The jump is captured as a true discontinuity by extrapolating the interpolations $w_j(\boldsymbol{\xi})$ from both sides into the discontinuous cells up to the predicted singularity location. This leads to a piecewise higher-degree approximation that resolves the two smooth regions and the curved discontinuity of varying strength in between.

The discontinuity location is approximated using SSC–SR by interpolating the contact discontinuity locations $x_{\text{contact}}$ extracted from the deterministic simulations,

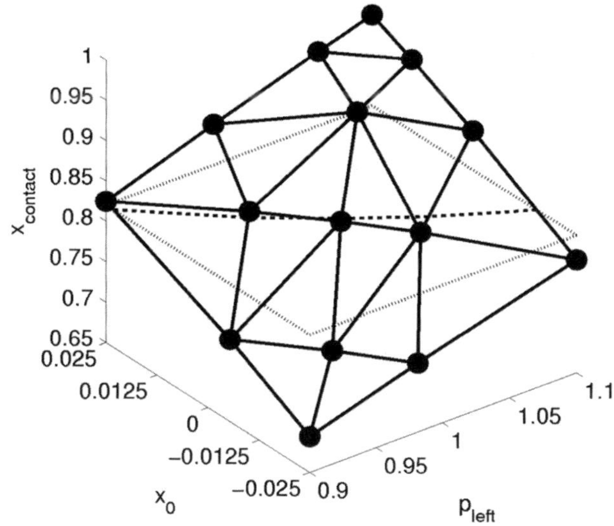


**Fig. 24** SSC–SR response surface approximation for the contact discontinuity location $x_{\text{contact}}$ at $t = 1$ with $n_s = 15$ for Sod's Riemann problem in a closed shock tube with uncertain $p_{\text{left}}$ and $x_0$.

as shown in Fig. 24. The interpolation of $x_{\text{contact}}$ as a function of $p_{\text{left}}$ and $x_0$ is performed using the SSC–ENO algorithm. The resulting jump line for $x = 0.82$ in the $(p_{\text{left}}, x_0)$–plane is then given by the intersection of the surface for $x_{\text{contact}}$ with the horizontal plane at $x_{\text{contact}} = 0.82$. The jump line approximation consists of a piecewise higher-order polynomial that is able to capture its curvature. The statistical moments predicted by SSC–SR are $\mu_\rho = 0.231$ and $\sigma_\rho = 0.0557$. The mean value matches that of SSC–ENO, but the standard deviation is underpredicted by SSC–ENO because of the linear approximation of the discontinuity. It corresponds to an output coefficient of variation of $\text{CoV}_\rho = 24.1\,\%$.

The output uncertainty in the entire density profile on $x \in [-0.2; 2]$ is depicted in Figs. 25 and 26 in terms of the convergence for the mean $\mu_\rho(x)$ and the standard deviation $\sigma_\rho(x)$ of SSC–ENO with $n_s = \{10, 20, 100\}$ and SSC–SR with $n_s = \{10, 15, 20\}$. The mean density $\mu_\rho(x)$ shows the smearing of the shock and contact waves compared to the deterministic solution, which is caused by the random location of the discontinuities. This also produces the local maxima of the standard deviation $\sigma_\rho(x)$ in the discontinuous regions. The results for SSC–SR are indistinguishable and converged to a maximum standard deviation of $\sigma_{\rho,\max} = 0.0730$ at $x = 1.754$ for $n_s = 15$.

SSC–ENO results in a staircase approximation in the discontinuous regions. With an increasing number of samples, the solution converges to a smooth representation with a larger number of smaller jumps. However, due to the absence of viscosity in the physical problem, the approximation maintains a staircase character which leads to first-order accuracy. It also results in the convergence to the maximum standard deviation from below which causes an underprediction of the maximum standard

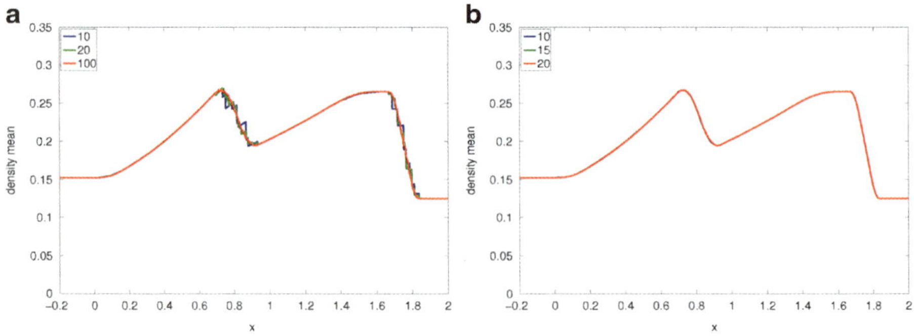


**Fig. 25** Mean $\mu_\rho(x)$ of the density $\rho$ for Sod's Riemann problem in a closed shock tube with uncertain initial pressure $p_{\text{left}}$ and diaphragm location $x_0$. (**a**) SSC-ENO with $n_s = \{10, 20, 100\}$. (**b**) SSC-SR with $n_s = \{10, 15, 20\}$

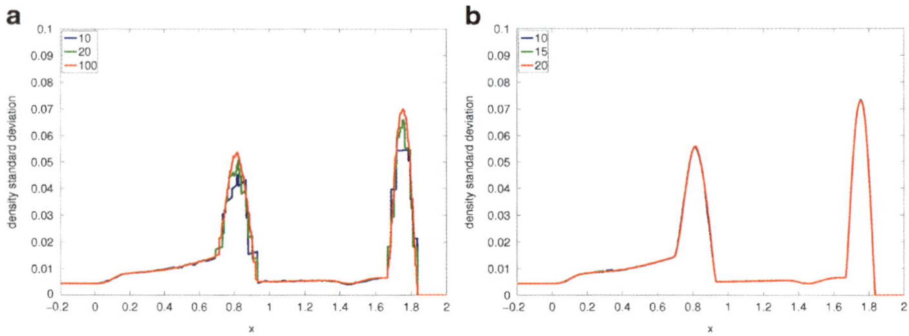


**Fig. 26** Standard deviation $\sigma_\rho(x)$ of the density $\rho$ for Sod's Riemann problem in a closed shock tube with uncertain initial pressure $p_{\text{left}}$ and diaphragm location $x_0$. (**a**) SSC-ENO with $n_s = \{10, 20, 100\}$. (**b**) SSC-SR with $n_s = \{10, 15, 20\}$

deviation at underresolved sample sizes. It leads to an underprediction of the maximum output uncertainty with $\sigma_{\rho,\max} = 0.0700$ by 4.16 % even for $n_s = 100$. Refinement measure $\bar{\Xi}_j$ is used here for SSC–ENO, since the discontinuities have different locations in probability space for each spatial point $x$.

The 90 and 100 % probability intervals are compared to the mean density profile in Fig. 27a for SSC–SR with $n_s = 15$. The variation in the discontinuity locations is captured by the 100 % interval, which is broadest in these regions and asymmetrical around the mean caused by the highly nonlinear propagation of the input uncertainty. The varying interval size near the contact discontinuity is caused by the physical variation of the density jump strength in the interaction region. There is no uncertainty at the right boundary, because the region at the right of the shock wave lies outside the domain of influence of $p_{\text{left}}$ and $x_0$. Figure 27b shows in a zoom of the shock region that the mean crosses the 90 % line. This is another sign of the high nonlinearity of the problem. At that spatial location, the response

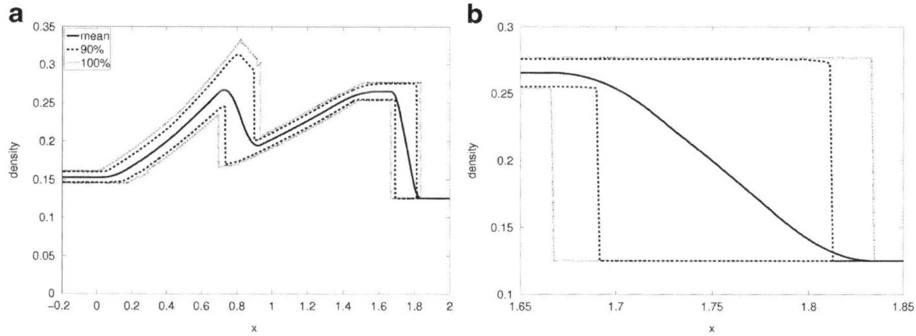


**Fig. 27** Mean, and 90 and 100 % uncertain intervals for the density $\rho$ at $t = 1$ by SSC–SR with the measure $\bar{\Xi}_j$ and $n_s = 100$ for Sod's Riemann problem in a closed shock tube with uncertain $p_{\text{left}}$ and $x_0$. (**a**) Whole spatial domain. (**b**) Zoom of the shock region

surface resembles a step function, of which less than 5 % falls at one side of the discontinuity. Therefore, the mean is located outside the 90 % confidence interval, but never outside the 100 % interval.

# 8 Transonic Flow Over the RAE 2822 Airfoil

It is shown that SSC–ENO converges to the same solution as SSC–SR for the discontinuous test function, the linear advection equation, and the shock tube problem in Sects. 5–7. The transonic airfoil problem is intended to compare the performance of the two methods for a computationally more challenging case at a number of samples that is realistic in practical situations. It is here assumed here that a computational budget of no more than $n_s = 50$ samples is available, for example, in a robust design optimization cycle of a complex aerospace application.

Non-uniform probability distributions are considered in a FVM discretization of multiple spatial dimensions for the transonic flow over the RAE 2822 airfoil [9]. The randomness in this NODESIM–CFD test case [20] is given by independent normal distributions for the free-stream Mach number $M_\infty$ and the angle of attack $\alpha$ with the mean values 0.734 and 2.79°, and standard deviations 0.005 and 0.1°, respectively. The flow problem is solved using an upwind discretization of the inviscid Euler equations in *FLUENT* to obtain a sharp discontinuity in the flow field up to the pressure distribution on the airfoil surface. The deterministic results for the two-dimensional spatial discretization with $5 \cdot 10^4$ cells are shown in Fig. 28 in terms of the static pressure field around the airfoil and the distribution of the pressure coefficient $C_p$ over the surface

$$C_p = \frac{p - p_\infty}{\frac{1}{2}\rho_\infty u_\infty^2}, \tag{22}$$

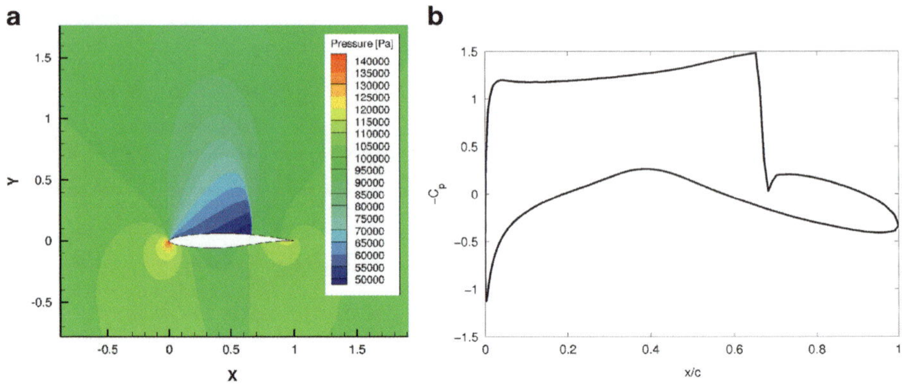


**Fig. 28** Deterministic results for the transonic flow over the RAE 2822 airfoil. (**a**) Static pressure field $p$. (**b**) Surface pressure coefficient $-C_\mathrm{p}$

where the subscript $\infty$ denotes the free-stream conditions of the pressure, the density, and the velocity. A transonic shock wave forms above the airfoil, which results in a discontinuity in the surface pressure distribution. The undershoot downstream of the shock wave is caused by the expansion present after an inviscid shock in a transonic flow. The mesh size is chosen based on a convergence study for this pressure coefficient profile of the deterministic problem for the nominal boundary conditions. For structures, Pettit and Beran [21] showed that in computing higher statistical moments, greater requirements are levied on the grid to be converged. It is important that the grid is adequate for each one of the cases as is sampled at different Mach numbers and at different angles of attack to make sure that the grid convergence does not influences the interpretation of the results.

The shock location $x_\mathrm{shock}$ along the airfoil is parameterized by SSC–SR for resolving the stochastic surface pressure distribution. The shock sensor of Harten [12] is used to extract $x_\mathrm{shock}$ from each of the samples, based on the maximum of the gradient magnitude of the pressure coefficient $|\mathrm{d}C_\mathrm{p}/\mathrm{d}x|$ in the shock region. A discrete resolution of the shock location, limited to the spatial cell faces, is avoided by defining $x_\mathrm{shock}$ as the extremum of a parabolic fit through the maximum of $|\mathrm{d}C_\mathrm{p}/\mathrm{d}x|$ and the values at its two neighboring spatial points. This approach is illustrated for the nominal flow conditions in Fig. 29a, for which a shock location of $x_\mathrm{shock} = 0.667$ is found.

The extraction step $\mathscr{E}$ is repeated on each layer of cells above the airfoil to obtain the two-dimensional shape of the shock wave $X_\mathrm{shock}$ for resolving the stochastic pressure field. In that case, the signed distance $d_\mathrm{shock}$ between the points in the spatial mesh $(x, y)$ and the closest point on the shock wave is parameterized instead of the shock location $x_\mathrm{shock}$. The sign of $d_\mathrm{shock}$ is obtained from the cross product $\mathbf{d}_\mathrm{shock} \times \mathbf{t}_\mathrm{shock}$ between the vector $\mathbf{d}_\mathrm{shock}$ from the point on the shock to the spatial point $(x, y)$ and the tangent vector $\mathbf{t}_\mathrm{shock}$ on the shock wave, with $d_\mathrm{shock} = \|\mathbf{d}_\mathrm{shock}\|$. To that end, the third component of the vector $\mathbf{d}_\mathrm{shock} \times \mathbf{t}_\mathrm{shock}$ is considered, since it changes sign when the reference point $(x, y)$ crosses the shock wave $X_\mathrm{shock}$.

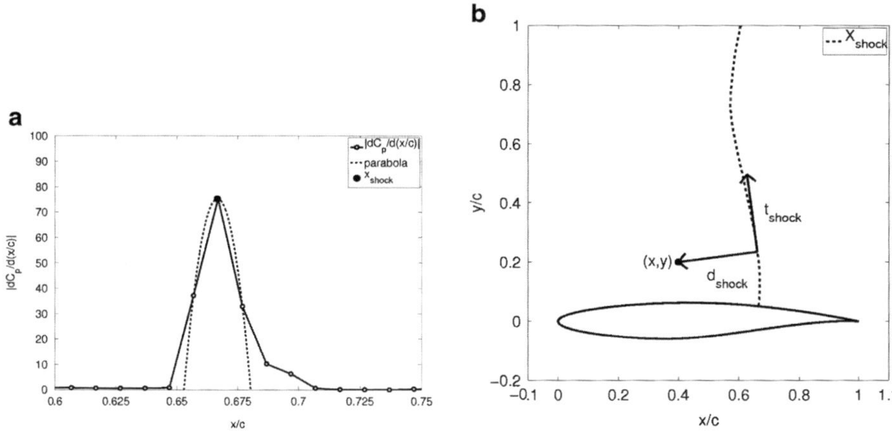


**Fig. 29** The SSC–SR algorithm for the transonic flow over the RAE 2822 airfoil at the nominal flow conditions. (**a**) Extraction of $x_{\text{shock}}$ from $|\mathrm{d}C_{\text{p}}/\mathrm{d}(x/c)|$. (**b**) Orientation of $\mathbf{d}_{\text{shock}}$, $\mathbf{t}_{\text{shock}}$, and $X_{\text{shock}}$

The orientation of the vectors $\mathbf{d}_{\text{shock}}$ and $\mathbf{t}_{\text{shock}}$ with respect to the shock $X_{\text{shock}}$ is shown in Fig. 29b for the nominal flow conditions. An alternative for three spatial dimensions is to define the sign of $d_{\text{shock}}$ by whether the point $(x, y)$ is located upstream or downstream of the shock $X_{\text{shock}}$. The unbounded range of the normal distribution is treated by truncating the probability space beyond the last MC integration point used for calculating the statistical moments as in [33]. The probabilistic weighting by the normal distribution is accounted for by the sampling strategy (12).

The mean of the pressure coefficient $\mu_{C_{\text{p}}}(x)$ along the airfoil is given in Fig. 30 as function of the $x$ coordinate normalized by the chord length $c$ for SSC–ENO and SSC–SR with $n_{\text{s}} = \{5, 17, 50\}$. The steps in the staircase approximation of SSC–ENO have different strengths in this case due to the non-uniform distribution of the probability over the parameter space. The SSC–ENO solution smoothens for an increasing number of samples, partly because of the presence of numerical diffusion in FVM for solving the Euler equations. This works in the following way. The staircase approximation of the mean will remain visible for all sample sizes in an inviscid problem with a purely inviscid solution on a sufficiently fine spatial mesh. Only the number of steps increases and their individual strength decreases as has been shown in the previous examples. In the current case, the staircase approximation starts to disappear on the given spatial mesh at a certain number of samples even with the relatively minor numerical diffusion of the grid converged solutions. The increasingly small steps of the converging staircase approximation are in that case no longer resolved on the spatial mesh and easily smeared by minor numerical diffusion. This has also been discussed in [23]. However, SSC–ENO initially overestimates the length of the region over which the shock wave is smeared in the mean pressure distribution compared to the SSC–SR results, which

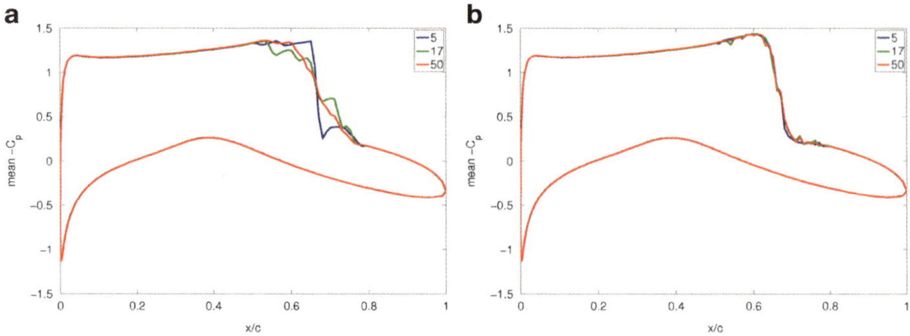


**Fig. 30** Mean pressure coefficient $\mu_{C_p}(x)$ along the surface with $n_s = \{5, 17, 50\}$ for the transonic flow over the RAE 2822 airfoil with random free-stream Mach number $M_\infty$ and angle of attack $\alpha$. (**a**) SSC–ENO. (**b**) SSC–SR

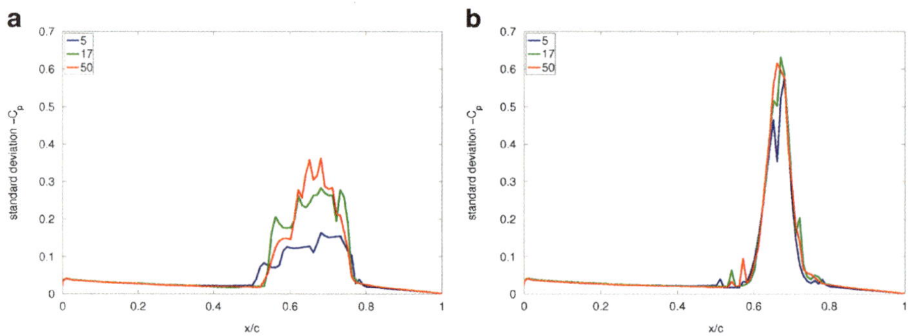


**Fig. 31** Standard deviation of the pressure coefficient $\sigma_{C_p}(x)$ along the surface with $n_s = \{5, 17, 50\}$ for the transonic flow over the RAE 2822 airfoil with random free-stream Mach number $M_\infty$ and angle of attack $\alpha$. (**a**) SSC–ENO. (**b**) SSC–SR

show good convergence already for $n_s = 5$. This overestimation of the smearing of the shock is not primarily caused by the numerical diffusion in the FVM model. It is caused by the relatively large effect of samples at low probabilities in the tails of the normal distributions on the interpolation of the response surface at small sample sizes. This broadens the spatial region in which the shock wave has a significant effect on the mean. The effect reduces with increasing samples size as can be seen by comparing the SSC–ENO results for $n_s = 17$ and $n_s = 50$. The SSC–ENO results are therefore trending to the same result as SSC–SR.

The standard deviation of the pressure coefficient $\sigma_{C_p}(x)$ along the upper surface in Fig. 31 is significantly underpredicted in the shock region by SSC–ENO with a maximum of $\sigma_{C_p,\max} = 0.362$ for $n_s = 50$. It converges only slowly to the SSC–SR solution of $\sigma_{C_p,\max} = 0.616$ at $n_s = 50$, which corresponds to an underprediction by 41.2 %. Convergence is therefore not established with this method for $ns = 50$. In contrast, the SSC–SR method already gives an accurate prediction for only $n_s = 5$ samples that largely coincides with the approximation of $n_s = 50$. On the other

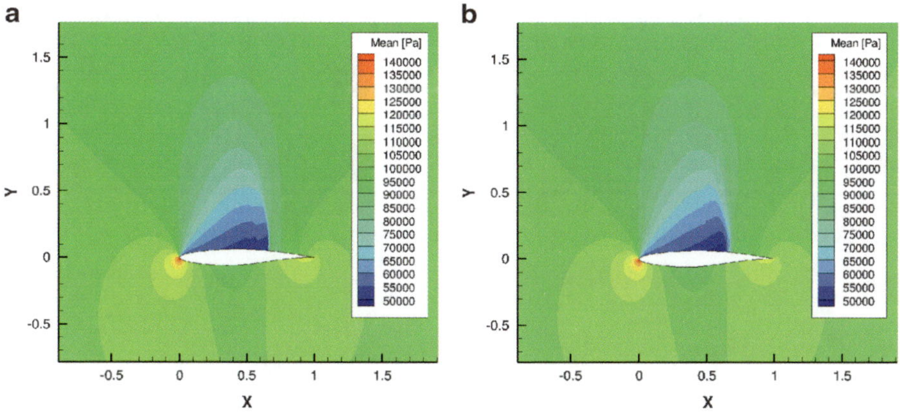


**Fig. 32** Mean pressure field $\mu_p(x, y)$ with $n_s = 5$ for the transonic flow over the RAE 2822 airfoil with random free-stream Mach number $M_\infty$ and angle of attack $\alpha$. (**a**) SSC–ENO. (**b**) SSC–SR

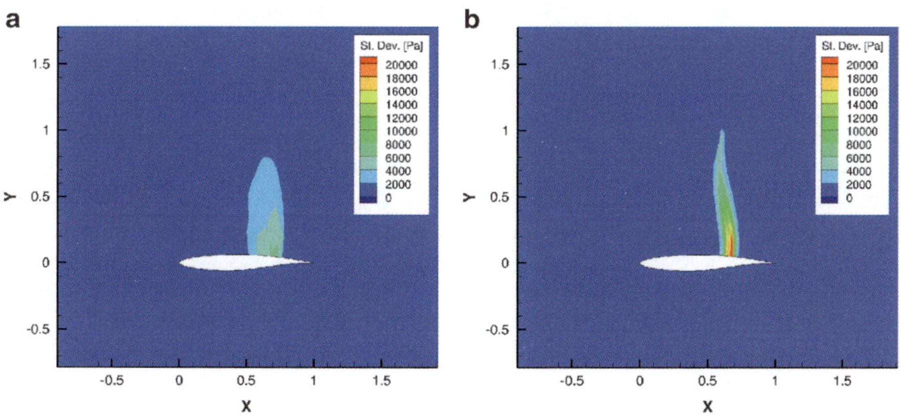


**Fig. 33** Standard deviation of the pressure field $\sigma_p(x, y)$ with $n_s = 5$ for the transonic flow over the RAE 2822 airfoil with random free-stream Mach number $M_\infty$ and angle of attack $\alpha$. (**a**) SSC–ENO. (**b**) SSC–SR

hand, the region around the shock wave in which $\sigma_{C_p}(x)$ is elevated is overpredicted by SSC–ENO. Both effects are caused by the underresolution of the discontinuity in the probability space by the piecewise linear approximation of the discontinuity by SSC–ENO. The linear function leads to a lower standard deviation through the underprediction of the gradients in the response surface and to a longer shock region through the smearing of the discontinuity in the probability space. The normal input distributions increase these two effects due to the concentration of the probability in a small region of the probability space, which makes the sharp resolution of the discontinuity in that region even more important.

The mean $\mu_{\mathrm{p}}(x, y)$ and standard deviation $\sigma_{\mathrm{p}}(x, y)$ of the pressure field around the RAE 2822 airfoil are given in Figs. 32 and 33 for SSC–ENO and SSC–SR with $n_{\mathrm{s}} = 5$ samples, which correspond on the upper surface with the results of Figs. 30 and 31. SSC–ENO does not resolve the smearing of the shock wave in the mean for $n_{\mathrm{s}} = 5$ compared to the deterministic solution of Fig. 28a as SSC–SR does. For this minimal number of samples, SSC–SR also captures already the detailed spatial structure of the local standard deviation field $\sigma_{\mathrm{p}}(x, y)$, while SSC–ENO gives a qualitative indication of the region with increased values of $\sigma_{\mathrm{p}}(x, y)$ only.

## 9 Conclusions

The Simplex Stochastic Collocation (SSC) method obtains robust and non-intrusive solutions of uncertainty quantification problems in computational fluid dynamics. It is based on a simplex tessellation discretization of the probability space and piecewise polynomial interpolation of higher-degree stencils of samples at the vertexes of the simplexes.

Essentially Non-Oscillatory (ENO) type stencil selection is introduced into the SSC method to achieve an accurate approximation of discontinuities in probability space. The stencil selection for simplex $\Xi_j$ chooses the stencil $S_j$ with the highest polynomial degree $p_j$ that is accepted by the Local Extremum Conserving (LEC) limiter. This results in an increase of the local polynomial degree in the smooth regions and a concentration of the refinement in the simplexes that contain the discontinuity. The efficient implementation of the algorithm assigns only nearest neighbor stencils to other simplexes without constructing new stencils or interpolations.

A subcell resolution approach is also introduced into the SSC method for solving stochastic problems with randomness in the location of spatial discontinuities. The presented SSC–SR method is based on extracting the discontinuity location $X_{\mathrm{disc}}(\boldsymbol{\xi})$ in the physical space from each of the deterministic solutions. The realizations of the physical distance $d_{\mathrm{disc}}(\mathbf{x}, \boldsymbol{\xi})$ to the discontinuity $X_{\mathrm{disc}}(\boldsymbol{\xi})$ are interpolated over the stochastic dimensions to predict the location of the discontinuity $\Xi_{\mathrm{disc}}(\mathbf{x})$ in the probability space. The stochastic response surface approximations are then extended from both sides up to the discontinuous hypersurface $\Xi_{\mathrm{disc}}(x)$. This results in a truly discontinuous representation of random spatial discontinuities in the interior of the cells discretizing the stochastic dimensions.

The increased refinement effectiveness of the resulting SSC–ENO method for a step function shows that the local polynomial degree is reduced to a linear interpolation in only a thin layer of simplexes that contain the discontinuity. The more concentrated sampling around the jump can lead to a reduction of the size of the discontinuous simplexes by a factor eight and a decrease of the error by two orders of magnitude.

The application to a linear advection problem shows that SSC–SR avoids the staircase approximation of the mean and the standard deviation by the SSC–ENO method without subcell resolution, because of the continuous dependence of the discontinuity location $\Xi_{\text{disc}}(\mathbf{x})$ in the probability space on the spatial coordinates $\mathbf{x}$. It also prevents the underestimation of the maximum standard deviation by matching the exact solution already for the initial number of samples.

The uncertainty in the shock tube problem results in an output coefficient of variation for the density of 23.5 % in the interaction region of the contact and rarefaction wave. The large and asymmetrical uncertainty intervals, near the smeared discontinuities in the mean sense, indicate a robust approximation of the highly nonlinear propagation of the uncertainty in those regions. SSC–SR results in a converged solution for $n_s = 15$ samples compared to an underprediction of the maximum standard deviation $\sigma_{\rho,\max}$ by 4.16 % for SSC–ENO with $n_s = 100$.

The impact of the random free-stream conditions on the transonic flow around the RAE 2822 airfoil are accurately resolved in the surface pressure distribution, and the mean and standard deviation pressure fields for a minimal number of $n_s = 5$ samples. The non-uniform input probability distributions lead to an even more significant underprediction of $\sigma_{C_p,\max}$ by 41.2 % for SSC–ENO with $n_s = 50$.

A larger number of stochastic dimensions could be introduced in the form of unsteady speed and angle of attack fluctuations, body shape, etc. The scaling of the presented methods up to five stochastic dimensions is reported in [34,35]. The effectiveness of the current implementation of the efficient ENO stencil selection tends to decrease with dimensionality. The subcell resolution remains as effective in higher dimensions for planar discontinuities. These trends are expected to extend to higher dimensional probability spaces with ten dimensions. The approximation of discontinuities in general using adaptive methods might not be feasible in these higher dimensional probability spaces, because of the increasing computational costs.

**Acknowledgements** This work was supported by the Netherlands Organization for Scientific Research (NWO) and the European Union Marie Curie Cofund Action under Rubicon grant 680-50-1002.

## References

1. Abgrall R (2010) A simple, Flexible and generic deterministic approach to uncertainty quantifications in nonlinear problems: application to fluid flow problems. In: Proceedings of the 5th European conference on computational fluid dynamics, ECCOMAS CFD, Lisbon, Portugal
2. Agarwal N, Aluru NR (2009) A domain adaptive stochastic collocation approach for analysis of MEMS under uncertainty. Journal of Computational Physics 228: 7662–7688
3. Babuška I, Tempone R, Zouraris GE (2004) Galerkin finite elements approximation of stochastic finite elements. SIAM Journal on Numerical Analysis 42: 800–825
4. Babuška I, Nobile F, Tempone R (2007) A stochastic collocation method for elliptic partial differential equations with random input data. SIAM Journal on Numerical Analysis 45: 1005–1034
5. Barth T (2012) On the propagation of statistical model parameter uncertainty in CFD calculations. Theoretical and Computational Fluid Dynamics 26: 435–457

6. Barth T (2011) UQ methods for nonlinear conservation laws containing discontinuities. In: AVT-193 Lecture Series on Uncertainty Quantification, RTO-AVT-VKI Short Course on Uncertainty Quantification, Stanford, California
7. Chassaing J-C, Lucor D (2010) Stochastic investigation of flows about airfoils at transonic speeds. AIAA Journal 48: 938–950
8. Chorin AJ, Marsden JE (1979) A mathematical introduction to fluid mechanics. Springer-Verlag, New York
9. Cook PH, McDonald MA (1979) Firmin MCP, Aerofoil RAE 2822 – pressure distributions, and boundary layer and wake measurements. Experimental data base for computer program assessment, AGARD report AR 138
10. Dwight RP, Witteveen JAS, Bijl H (this issue) Adaptive uncertainty quantification for computational fluid dynamics. In: Uncertainty quantification, Lecture notes in computational science and engineering, Springer
11. Harten A, Osher S (1987) Uniformly high-order accurate nonoscillatory schemes I. SIAM Journal on Numerical Analysis 24: 279–309
12. Harten A (1989) ENO schemes with subcell resolution. Journal of Computational Physics 83: 148–184
13. Foo J, Wan X, Karniadakis GE (2008) The multi-element probabilistic collocation method (ME-PCM): error analysis and applications. Journal of Computational Physics 227: 9572–9595
14. Ghanem RG, Spanos PD (1991) Stochastic finite elements: a spectral approach. Springer-Verlag, New York
15. Ghosh D, Ghanem R (2008) Stochastic convergence acceleration through basis enrichment of polynomial chaos expansions. International Journal on Numerical Methods in Engineering. 73: 162–184
16. Gottlieb D, Xiu D (2008) Galerkin method for wave equations with uncertain coefficients. Communications in Computational Physics 3: 505–518
17. Ma X, Zabaras N (2009) An adaptive hierarchical sparse grid collocation algorithm for the solution of stochastic differential equations. Journal of Computational Physics 228: 3084–3113
18. Le Maître OP, Najm HN, Ghanem RG, Knio OM (2004) Multi–resolution analysis of Wiener–type uncertainty propagation schemes. Journal of Computational Physics 197: 502–531
19. Mathelin L, Le Maître OP (2007) Dual-based a posteriori error estimate for stochastic finite element methods. Communications in Applied Mathematics and Computational Science 2: 83–115
20. Onorato G, Loeven GJA, Ghorbaniasl G, Bijl H, Lacor C (2010) Comparison of intrusive and non-intrusive polynomial chaos methods for CFD applications in aeronautics. In: Proceedings of the 5th European conference on computational fluid dynamics, ECCOMAS CFD, Lisbon, Portugal
21. Pettit CL, Beran PS (2006) Convergence studies of Wiener expansions for computational non-linear mechanics. In: Proceedings of the 8th AIAA non-deterministic approaches conference, Newport, Rhode Island, AIAA-2006-1993
22. Poëtte G, Després B, Lucor D (2009) Uncertainty quantification for systems of conservation laws, Journal of Computational Physics 228: 2443–2467
23. Simon F, Guillen P, Sagaut P, Lucor D (2010) A gPC-based approach to uncertain transonic aerodynamics. Computer Methods in Applied Mechanics and Engineering 199: 1091–1099
24. Shu C-W, Osher S (1988) Efficient implementation of essentially non-oscillatory shock-capturing schemes. Journal of Computational Physics 77: 439–471
25. Shu C-W, Osher S (1989) Efficient implementation of essentially non-oscillatory shock-capturing schemes II, Journal of Computational Physics 83: 32–78
26. Sod GA (1978) A survey of several finite difference methods for systems of nonlinear hyperbolic conservation laws. Journal of Computational Physics 27: 1–31
27. Tryoen J, Le Maître O, Ndjinga M, Ern A (2010) Intrusive Galerkin methods with upwinding for uncertain nonlinear hyperbolic systems. Journal of Computational Physics 229: 6485–6511
28. Wan X, Karniadakis GE (2005) An adaptive multi-element generalized polynomial chaos method for stochastic differential equations. Journal of Computational Physics 209: 617–642

29. Witteveen JAS, Bijl H (2009) A TVD uncertainty quantification method with bounded error applied to transonic airfoil flutter. Communications in Computational Physics 6: 406–432
30. Witteveen JAS, Loeven GJA, Bijl H (2009) An adaptive stochastic finite elements approach based on Newton-Cotes quadrature in simplex elements. Computers and Fluids 38: 1270–1288
31. Witteveen JAS (2010) Second order front tracking for the Euler equations. Journal of Computational Physics 229: 2719–2739
32. Witteveen JAS, Iaccarino G (2012) Simplex stochastic collocation with random sampling and extrapolation for nonhypercube probability spaces. SIAM Journal on Scientific Computing 34: A814–A838
33. Witteveen JAS, Iaccarino G (2012) Refinement criteria for simplex stochastic collocation with local extremum diminishing robustness. SIAM Journal on Scientific Computing 34: A1522–A1543
34. Witteveen JAS, Iaccarino G (2013) Simplex stochastic collocation with ENO-type stencil selection for robust uncertainty quantification. Journal of Computational Physics 239: 1–21
35. Witteveen JAS, Iaccarino G (submitted) Subcell resolution in simplex stochastic collocation for spatial discontinuities
36. Xiu D, Karniadakis GE (2002) The Wiener-Askey polynomial chaos for stochastic differential equations. SIAM Journal on Scientific Computing 24: 619–644
37. Xiu D, Hesthaven JS (2005) High-order collocation methods for differential equations with random inputs. SIAM Journal on Scientific Computing 27: 1118–1139

# *Editorial Policy*

1. Volumes in the following three categories will be published in LNCSE:

i) Research monographs
ii) Tutorials
iii) Conference proceedings

Those considering a book which might be suitable for the series are strongly advised to contact the publisher or the series editors at an early stage.

2. Categories i) and ii). Tutorials are lecture notes typically arising via summer schools or similar events, which are used to teach graduate students. These categories will be emphasized by Lecture Notes in Computational Science and Engineering. **Submissions by interdisciplinary teams of authors are encouraged.** The goal is to report new developments – quickly, informally, and in a way that will make them accessible to non-specialists. In the evaluation of submissions timeliness of the work is an important criterion. Texts should be well-rounded, well-written and reasonably self-contained. In most cases the work will contain results of others as well as those of the author(s). In each case the author(s) should provide sufficient motivation, examples, and applications. In this respect, Ph.D. theses will usually be deemed unsuitable for the Lecture Notes series. Proposals for volumes in these categories should be submitted either to one of the series editors or to Springer-Verlag, Heidelberg, and will be refereed. A provisional judgement on the acceptability of a project can be based on partial information about the work: a detailed outline describing the contents of each chapter, the estimated length, a bibliography, and one or two sample chapters – or a first draft. A final decision whether to accept will rest on an evaluation of the completed work which should include

- at least 100 pages of text;
- a table of contents;
- an informative introduction perhaps with some historical remarks which should be accessible to readers unfamiliar with the topic treated;
- a subject index.

3. Category iii). Conference proceedings will be considered for publication provided that they are both of exceptional interest and devoted to a single topic. One (or more) expert participants will act as the scientific editor(s) of the volume. They select the papers which are suitable for inclusion and have them individually refereed as for a journal. Papers not closely related to the central topic are to be excluded. Organizers should contact the Editor for CSE at Springer at the planning stage, see *Addresses* below.

In exceptional cases some other multi-author-volumes may be considered in this category.

4. Only works in English will be considered. For evaluation purposes, manuscripts may be submitted in print or electronic form, in the latter case, preferably as pdf- or zipped ps-files. Authors are requested to use the LaTeX style files available from Springer at http://www.springer.com/authors/book+authors/helpdesk?SGWID=0-1723113-12-971304-0 (Click on Templates → LaTeX → monographs or contributed books).

For categories ii) and iii) we strongly recommend that all contributions in a volume be written in the same LaTeX version, preferably LaTeX2e. Electronic material can be included if appropriate. Please contact the publisher.

Careful preparation of the manuscripts will help keep production time short besides ensuring satisfactory appearance of the finished book in print and online.

5. The following terms and conditions hold. Categories i), ii) and iii):

Authors receive 50 free copies of their book. No royalty is paid.
Volume editors receive a total of 50 free copies of their volume to be shared with authors, but no royalties.

Authors and volume editors are entitled to a discount of 33.3 % on the price of Springer books purchased for their personal use, if ordering directly from Springer.

6. Springer secures the copyright for each volume.

Addresses:

Timothy J. Barth
NASA Ames Research Center
NAS Division
Moffett Field, CA 94035, USA
barth@nas.nasa.gov

Michael Griebel
Institut für Numerische Simulation
der Universität Bonn
Wegelerstr. 6
53115 Bonn, Germany
griebel@ins.uni-bonn.de

David E. Keyes
Mathematical and Computer Sciences
and Engineering
King Abdullah University of Science
and Technology
P.O. Box 55455
Jeddah 21534, Saudi Arabia
david.keyes@kaust.edu.sa

and

Department of Applied Physics
and Applied Mathematics
Columbia University
500 W. 120 th Street
New York, NY 10027, USA
kd2112@columbia.edu

Risto M. Nieminen
Department of Applied Physics
Aalto University School of Science
and Technology
00076 Aalto, Finland
risto.nieminen@aalto.fi

Dirk Roose
Department of Computer Science
Katholieke Universiteit Leuven
Celestijnenlaan 200A
3001 Leuven-Heverlee, Belgium
dirk.roose@cs.kuleuven.be

Tamar Schlick
Department of Chemistry
and Courant Institute
of Mathematical Sciences
New York University
251 Mercer Street
New York, NY 10012, USA
schlick@nyu.edu

Editor for Computational Science
and Engineering at Springer:
Martin Peters
Springer-Verlag
Mathematics Editorial IV
Tiergartenstrasse 17
69121 Heidelberg, Germany
martin.peters@springer.com

# Lecture Notes in Computational Science and Engineering

1. D. Funaro, *Spectral Elements for Transport-Dominated Equations.*
2. H.P. Langtangen, *Computational Partial Differential Equations.* Numerical Methods and Diffpack Programming.
3. W. Hackbusch, G. Wittum (eds.), *Multigrid Methods V.*
4. P. Deuflhard, J. Hermans, B. Leimkuhler, A.E. Mark, S. Reich, R.D. Skeel (eds.), *Computational Molecular Dynamics: Challenges, Methods, Ideas.*
5. D. Kröner, M. Ohlberger, C. Rohde (eds.), *An Introduction to Recent Developments in Theory and Numerics for Conservation Laws.*
6. S. Turek, *Efficient Solvers for Incompressible Flow Problems.* An Algorithmic and Computational Approach.
7. R. von Schwerin, ***Multi Body System SIMulation.*** Numerical Methods, Algorithms, and Software.
8. H.-J. Bungartz, F. Durst, C. Zenger (eds.), *High Performance Scientific and Engineering Computing.*
9. T.J. Barth, H. Deconinck (eds.), *High-Order Methods for Computational Physics.*
10. H.P. Langtangen, A.M. Bruaset, E. Quak (eds.), *Advances in Software Tools for Scientific Computing.*
11. B. Cockburn, G.E. Karniadakis, C.-W. Shu (eds.), *Discontinuous Galerkin Methods.* Theory, Computation and Applications.
12. U. van Rienen, *Numerical Methods in Computational Electrodynamics.* Linear Systems in Practical Applications.
13. B. Engquist, L. Johnsson, M. Hammill, F. Short (eds.), *Simulation and Visualization on the Grid.*
14. E. Dick, K. Riemslagh, J. Vierendeels (eds.), *Multigrid Methods VI.*
15. A. Frommer, T. Lippert, B. Medeke, K. Schilling (eds.), *Numerical Challenges in Lattice Quantum Chromodynamics.*
16. J. Lang, *Adaptive Multilevel Solution of Nonlinear Parabolic PDE Systems.* Theory, Algorithm, and Applications.
17. B.I. Wohlmuth, *Discretization Methods and Iterative Solvers Based on Domain Decomposition.*
18. U. van Rienen, M. Günther, D. Hecht (eds.), *Scientific Computing in Electrical Engineering.*
19. I. Babuška, P.G. Ciarlet, T. Miyoshi (eds.), *Mathematical Modeling and Numerical Simulation in Continuum Mechanics.*
20. T.J. Barth, T. Chan, R. Haimes (eds.), *Multiscale and Multiresolution Methods.* Theory and Applications.
21. M. Breuer, F. Durst, C. Zenger (eds.), *High Performance Scientific and Engineering Computing.*
22. K. Urban, *Wavelets in Numerical Simulation.* Problem Adapted Construction and Applications.

23. L.F. Pavarino, A. Toselli (eds.), *Recent Developments in Domain Decomposition Methods.*

24. T. Schlick, H.H. Gan (eds.), *Computational Methods for Macromolecules: Challenges and Applications.*

25. T.J. Barth, H. Deconinck (eds.), *Error Estimation and Adaptive Discretization Methods in Computational Fluid Dynamics.*

26. M. Griebel, M.A. Schweitzer (eds.), *Meshfree Methods for Partial Differential Equations.*

27. S. Müller, *Adaptive Multiscale Schemes for Conservation Laws.*

28. C. Carstensen, S. Funken, W. Hackbusch, R.H.W. Hoppe, P. Monk (eds.), *Computational Electromagnetics.*

29. M.A. Schweitzer, *A Parallel Multilevel Partition of Unity Method for Elliptic Partial Differential Equations.*

30. T. Biegler, O. Ghattas, M. Heinkenschloss, B. van Bloemen Waanders (eds.), *Large-Scale PDE-Constrained Optimization.*

31. M. Ainsworth, P. Davies, D. Duncan, P. Martin, B. Rynne (eds.), *Topics in Computational Wave Propagation.* Direct and Inverse Problems.

32. H. Emmerich, B. Nestler, M. Schreckenberg (eds.), *Interface and Transport Dynamics.* Computational Modelling.

33. H.P. Langtangen, A. Tveito (eds.), *Advanced Topics in Computational Partial Differential Equations.* Numerical Methods and Diffpack Programming.

34. V. John, *Large Eddy Simulation of Turbulent Incompressible Flows.* Analytical and Numerical Results for a Class of LES Models.

35. E. Bänsch (ed.), *Challenges in Scientific Computing - CISC 2002.*

36. B.N. Khoromskij, G. Wittum, *Numerical Solution of Elliptic Differential Equations by Reduction to the Interface.*

37. A. Iske, *Multiresolution Methods in Scattered Data Modelling.*

38. S.-I. Niculescu, K. Gu (eds.), *Advances in Time-Delay Systems.*

39. S. Attinger, P. Koumoutsakos (eds.), *Multiscale Modelling and Simulation.*

40. R. Kornhuber, R. Hoppe, J. Périaux, O. Pironneau, O. Wildlund, J. Xu (eds.), *Domain Decomposition Methods in Science and Engineering.*

41. T. Plewa, T. Linde, V.G. Weirs (eds.), *Adaptive Mesh Refinement – Theory and Applications.*

42. A. Schmidt, K.G. Siebert, *Design of Adaptive Finite Element Software.* The Finite Element Toolbox ALBERTA.

43. M. Griebel, M.A. Schweitzer (eds.), *Meshfree Methods for Partial Differential Equations II.*

44. B. Engquist, P. Lötstedt, O. Runborg (eds.), *Multiscale Methods in Science and Engineering.*

45. P. Benner, V. Mehrmann, D.C. Sorensen (eds.), *Dimension Reduction of Large-Scale Systems.*

46. D. Kressner, *Numerical Methods for General and Structured Eigenvalue Problems.*

47. A. Boriçi, A. Frommer, B. Joó, A. Kennedy, B. Pendleton (eds.), *QCD and Numerical Analysis III.*

48. F. Graziani (ed.), *Computational Methods in Transport.*

49. B. Leimkuhler, C. Chipot, R. Elber, A. Laaksonen, A. Mark, T. Schlick, C. Schütte, R. Skeel (eds.), *New Algorithms for Macromolecular Simulation.*

50. M. Bücker, G. Corliss, P. Hovland, U. Naumann, B. Norris (eds.), *Automatic Differentiation: Applications, Theory, and Implementations.*

51. A.M. Bruaset, A. Tveito (eds.), *Numerical Solution of Partial Differential Equations on Parallel Computers.*

52. K.H. Hoffmann, A. Meyer (eds.), *Parallel Algorithms and Cluster Computing.*

53. H.-J. Bungartz, M. Schäfer (eds.), *Fluid-Structure Interaction.*

54. J. Behrens, *Adaptive Atmospheric Modeling.*

55. O. Widlund, D. Keyes (eds.), *Domain Decomposition Methods in Science and Engineering XVI.*

56. S. Kassinos, C. Langer, G. Iaccarino, P. Moin (eds.), *Complex Effects in Large Eddy Simulations.*

57. M. Griebel, M.A Schweitzer (eds.), *Meshfree Methods for Partial Differential Equations III.*

58. A.N. Gorban, B. Kégl, D.C. Wunsch, A. Zinovyev (eds.), *Principal Manifolds for Data Visualization and Dimension Reduction.*

59. H. Ammari (ed.), *Modeling and Computations in Electromagnetics: A Volume Dedicated to Jean-Claude Nédélec.*

60. U. Langer, M. Discacciati, D. Keyes, O. Widlund, W. Zulehner (eds.), *Domain Decomposition Methods in Science and Engineering XVII.*

61. T. Mathew, *Domain Decomposition Methods for the Numerical Solution of Partial Differential Equations.*

62. F. Graziani (ed.), *Computational Methods in Transport: Verification and Validation.*

63. M. Bebendorf, *Hierarchical Matrices.* A Means to Efficiently Solve Elliptic Boundary Value Problems.

64. C.H. Bischof, H.M. Bücker, P. Hovland, U. Naumann, J. Utke (eds.), *Advances in Automatic Differentiation.*

65. M. Griebel, M.A. Schweitzer (eds.), *Meshfree Methods for Partial Differential Equations IV.*

66. B. Engquist, P. Lötstedt, O. Runborg (eds.), *Multiscale Modeling and Simulation in Science.*

67. I.H. Tuncer, Ü. Gülcat, D.R. Emerson, K. Matsuno (eds.), *Parallel Computational Fluid Dynamics 2007.*

68. S. Yip, T. Diaz de la Rubia (eds.), *Scientific Modeling and Simulations.*

69. A. Hegarty, N. Kopteva, E. O'Riordan, M. Stynes (eds.), *BAIL* 2008 – *Boundary and Interior Layers.*

70. M. Bercovier, M.J. Gander, R. Kornhuber, O. Widlund (eds.), *Domain Decomposition Methods in Science and Engineering XVIII.*

71. B. Koren, C. Vuik (eds.), *Advanced Computational Methods in Science and Engineering.*

72. M. Peters (ed.), *Computational Fluid Dynamics for Sport Simulation.*

73. H.-J. Bungartz, M. Mehl, M. Schäfer (eds.), *Fluid Structure Interaction II - Modelling, Simulation, Optimization.*

74. D. Tromeur-Dervout, G. Brenner, D.R. Emerson, J. Erhel (eds.), *Parallel Computational Fluid Dynamics 2008.*

75. A.N. Gorban, D. Roose (eds.), *Coping with Complexity: Model Reduction and Data Analysis.*

76. J.S. Hesthaven, E.M. Rønquist (eds.), *Spectral and High Order Methods for Partial Differential Equations.*

77. M. Holtz, *Sparse Grid Quadrature in High Dimensions with Applications in Finance and Insurance.*

78. Y. Huang, R. Kornhuber, O.Widlund, J. Xu (eds.), *Domain Decomposition Methods in Science and Engineering XIX.*

79. M. Griebel, M.A. Schweitzer (eds.), *Meshfree Methods for Partial Differential Equations V.*

80. P.H. Lauritzen, C. Jablonowski, M.A. Taylor, R.D. Nair (eds.), *Numerical Techniques for Global Atmospheric Models.*

81. C. Clavero, J.L. Gracia, F.J. Lisbona (eds.), *BAIL 2010 – Boundary and Interior Layers, Computational and Asymptotic Methods.*

82. B. Engquist, O. Runborg, Y.R. Tsai (eds.), *Numerical Analysis and Multiscale Computations.*

83. I.G. Graham, T.Y. Hou, O. Lakkis, R. Scheichl (eds.), *Numerical Analysis of Multiscale Problems.*

84. A. Logg, K.-A. Mardal, G. Wells (eds.), *Automated Solution of Differential Equations by the Finite Element Method.*

85. J. Blowey, M. Jensen (eds.), *Frontiers in Numerical Analysis - Durham 2010.*

86. O. Kolditz, U.-J. Gorke, H. Shao, W. Wang (eds.), *Thermo-Hydro-Mechanical-Chemical Processes in Fractured Porous Media - Benchmarks and Examples.*

87. S. Forth, P. Hovland, E. Phipps, J. Utke, A. Walther (eds.), *Recent Advances in Algorithmic Differentiation.*

88. J. Garcke, M. Griebel (eds.), *Sparse Grids and Applications.*

89. M. Griebel, M.A. Schweitzer (eds.), *Meshfree Methods for Partial Differential Equations VI.*

90. C. Pechstein, *Finite and Boundary Element Tearing and Interconnecting Solvers for Multiscale Problems.*

91. R. Bank, M. Holst, O. Widlund, J. Xu (eds.), *Domain Decomposition Methods in Science and Engineering XX.*

92. H. Bijl, D. Lucor, S. Mishra, C. Schwab (eds.), *Uncertainty Quantification in Computational Fluid Dynamics.*

*For further information on these books please have a look at our mathematics catalogue at the following URL:* www.springer.com/series/3527

# Monographs in Computational Science and Engineering

1. J. Sundnes, G.T. Lines, X. Cai, B.F. Nielsen, K.-A. Mardal, A. Tveito, *Computing the Electrical Activity in the Heart.*

*For further information on this book, please have a look at our mathematics catalogue at the following URL:* www.springer.com/series/7417

# Texts in Computational Science and Engineering

1. H. P. Langtangen, *Computational Partial Differential Equations.* Numerical Methods and Diffpack Programming. 2nd Edition
2. A. Quarteroni, F. Saleri, P. Gervasio, *Scientific Computing with MATLAB and Octave.* 3rd Edition
3. H. P. Langtangen, *Python Scripting for Computational Science.* 3rd Edition
4. H. Gardner, G. Manduchi, *Design Patterns for e-Science.*
5. M. Griebel, S. Knapek, G. Zumbusch, *Numerical Simulation in Molecular Dynamics.*
6. H. P. Langtangen, *A Primer on Scientific Programming with Python.* 3rd Edition
7. A. Tveito, H. P. Langtangen, B. F. Nielsen, X. Cai, *Elements of Scientific Computing.*
8. B. Gustafsson, *Fundamentals of Scientific Computing.*
9. M. Bader, *Space-Filling Curves.*
10. M. Larson, F. Bengzon, *The Finite Element Method: Theory, Implementation and Applications.*

*For further information on these books please have a look at our mathematics catalogue at the following URL:* www.springer.com/series/5151

Printed by Printforce, the Netherlands